Practical Mystic

Practical Mystic

Religion, Science, and A. S. Eddington

⪻ MATTHEW STANLEY ⪼

The University of Chicago Press
Chicago and London

Matthew Stanley is assistant professor of history at Michigan State University. He received his PhD in the history of science and his MA in astronomy from Harvard University. He has been a fellow at the Institute for Advanced Study and the Max Planck Institute for the History of Science.

The University of Chicago Press, Chicago 60637
The University of Chicago Press, Ltd., London
© 2007 by The University of Chicago
All rights reserved. Published 2007
Printed in the United States of America

16 15 14 13 12 11 10 09 08 07   1 2 3 4 5

ISBN-13: 978-0-226-77097-0 (cloth)
ISBN-10: 0-226-77097-4 (cloth)

Library of Congress Cataloging-in-Publication Data

Stanley, Matthew, 1975-
    Practical mystic : religion, science, and A. S. Eddington/Matthew Stanley.
       p.   cm.
    Includes bibliographical references and index.
    ISBN-13: 978-0-226-77097-0 (cloth : alk. paper)
    ISBN-10: 0-226-77097-4 (cloth : alk. paper)   1. Religion and science.   2. Eddington,
Arthur Stanley, Sir, 1882–1944.   I. Title.
    BL240.3.S725 2007
    261.5′5092–dc22

                                                            2007005482

⊗ The paper used in this publication meets the minimum requirements of the
American National Standard for Information Sciences—Permanence of Paper for
Printed Library Materials, ansi z39.48-1992.

for Janelle

Contents

Acknowledgments

As with any piece of scholarship, the single name of the author conceals the amazing network of people, places, and organizations that made it possible. This project originated under the guidance of three formal mentors—Peter Galison, Everett Mendelsohn, and Andrew Warwick—who taught me what it meant to be a historian by both instruction and example. My informal mentor, Michael Gordin, was selfless in helping at each stage of this book's development, despite his own considerable responsibilities. It would be impossible to fully acknowledge everyone whose support was critical to the well-being of this project (and of myself), so I can only humbly list some of them: Geoffrey Cantor, Deborah Coen, Lorraine Daston, David DeVorkin, Owen Gingerich, Karl Hufbauer, David Kaiser, Jude LaJoie, Elizabeth Lee, Ole Molvig, Sharrona Pearl, Denise Phillips, Kate Price, Suman Seth, Grace Shen, Tom Stapleford, and Deborah Valdovinos. Special thanks go to Catherine Rice, Christie Henry, Pete Beatty, and Tisse Takagi at the University of Chicago Press.

Institutional support while writing this was provided by Iowa State University's Department of History, Harvard's Department of the History of Science, Harvard's Department of Astronomy, the Harvard-Smithsonian Center for Astrophysics, the Max Planck Institute for the History of Science, and the Center for the History of Science, Technology, and Medicine at Imperial College in London. Research grants were made available to me by the Minda de Ginzburg Center for European Studies, the Center for Excellence in the Arts and Humanities at Iowa State University, and the Harvard Graduate Student Council.

Historians can make little progress without diligent librarians and archivists, and I would like to thank those who unfailingly helped with my endless inquiries: the staff at Widener Library (particularly Fred Burchsted), Cabot

Science Library, and Wolbach Library; Parks Library (particularly Edward Goedeken); Jonathan Smith at the Wren Library of Trinity College; Peter Hingley at the Royal Astronomical Society Library; Adam Perkins at the Cambridge University Library; Josef Keith at Friends House Library, London; David Dewhirst at the Cambridge Observatory; the staff of the Cambridge County Record Office; and Chris Jakes at the Cambridgeshire Public Library. Materials, permissions, and access were also granted by the Einstein Papers at Hebrew University, the Royal Society Library, the British Library, the Princeton University Library, University College London Library, the London School of Economics Library, the Astronomical Society of the Pacific, the Huntington Library, the Rylands Library at the University of Manchester, the Harvard University Archives, the Cambridge County Records Office, the Public Records Office, the Science Museum Library in London, and the Department of Rare and Brittle Books at Indiana University, Bloomington. Permission was granted by Dr. Michael Weyl to quote his father's letters. Some of the material here appeared in papers in *Isis* in 2003 and *British Journal for the History of Science* in 2007, and I am grateful for permission to use it. Dr. Volker Heine and the Jesus Lane Friends Meeting in Cambridge were warmly hospitable during my work there, and I am grateful to Dr. Heine as representative of the Eddington Trustees for granting permission to use the Eddington papers. Some material from Quaker records is quoted with permission of the Cambridge and Peterborough Monthly Meeting of the Religious Society of Friends in Britain.

    Writing a book is as much a personal odyssey as an intellectual one, and the support given by my friends and family was essential to making this a wonderful experience instead of a harrowing one. My parents, Anita and Bruce Stanley, invested in me a love of the intellectual life from an early age, and my siblings, Jennifer and Chris, have showed me how to combine my career with my family life. Matt Gregory, my best friend, has provided a stable shore and unending counseling over the years. Of course, the most credit for getting me through this process has to go to the most important person in my life, my wife, Janelle. She rode this project's waves of elation and suffering right along with me and kept me from capsizing on a daily basis. Without her constant encouragement and inspiration, I would be both a lesser scholar and a lesser person.

FIGURE I.1
Arthur S. Eddington, circa 1940. Courtesy of the Cambridgeshire Public Library.

Introduction

In 1988, the *Scientist* published an inflammatory editorial by the Cornell historian of biology William Provine. He declared, in no uncertain terms, that science and religion are "incompatible." Those scientists that hoped to reconcile their faith and their work were told simply to "Face It!": "A thoughtful attorney from San Antonio, Tex., wrote recently to ask, 'Is there an intellectually honest Christian evolutionist position? Or do we simply have to check our brains at the church house door?' The answer is, you indeed have to check your brains. Why do scientists publicly deny the implications of modern science, and promulgate the compatibility of religion and science? Wishful thinking, religious training, and intellectual dishonesty are all important factors."[1] His column generated more mail than the magazine had ever received on one topic. Scientists with positions on both sides wrote vehement attacks on or defenses of Provine's argument. Those defending declared that there was no way for an honest scientist to also believe in God. Those attacking protested that they themselves found no conflict between their science and their religion.

Provine had clearly touched a raw nerve. The readers of the *Scientist* were not opposed to discussion of religion as a whole; indeed, a column attacking religious fundamentalism in the same issue received almost no response. Provine generated such emotive discussion because his claims were based on the impossibility of the idea of the religious scientist. The secular scientists engaged in the debate could not conceive of a religious scientist; conversely, their religious colleagues could see no barrier between their faith and their work. During the debate, the first group was startled to see a significant strength of religion within their ranks when religion had been so successfully driven from the content of science. How had this fifth column penetrated the fortress of science?

The answer is that it had never left. Even as the metaphysics and ontology of religion had become less important in the public face of twentieth-century science, religious scientists maintained their identity. This book is a focused study of one religious scientist in particular: the Quaker astronomer A. S. Eddington (1882–1944), head of the Cambridge Observatory and famous for his work in relativity and stellar physics. Eddington's career was of exactly the sort that so puzzled Provine's supporters: a world-class scientist who not just maintained his religious beliefs, but who also brought together the religious and scientific aspects of his life in powerful, meaningful, and productive ways. Further, he defended this interrelation even under tremendous pressure from the astronomical community and the wartime British state. The links between his religion and his science were subtle and pervasive and could not simply be discarded when convenient.

The presence of religious scientists like Eddington (both in the past and the present) raises questions about many of the common assumptions about the relationship of religion and science today. There is no lack of discourse on religion and science, but it often takes a sweeping, normative approach that assumes science and religion are each homogeneous, unchanging, abstract bodies of thought. Concerns about a firm philosophical basis for conflict or harmony dominate. The religious scientists themselves—the most obvious, active, and essential interaction of religion and science—vanish in the chaff of overgeneralization.

There are better ways to discuss these issues, ones in which historians and historical methods have an important contribution to make. In contrast to grand philosophical projects that lay down rationalized boundaries, this book uses a careful, historically specific methodology that looks at both science and religion as synchronically diverse and diachronically evolving. This requires investigating the particulars of spatial and temporal location and, specifically, the way both categories come from individuals—science is done by scientists, religion does not exist apart from religious believers. Providing these particulars is exactly what historical investigation does well, and this book is a study in how to bring to bear the techniques of history onto the problems of science and religion.[2]

By "synchronically diverse," I mean to emphasize the staggering number of "species" of both religion and science that are present in the world at any given time. Both categories suffer from certain broad labels that are useful in everyday conversation but create serious difficulties when used in serious scholarship. Religion is typically broken down into categories such as monotheistic, Judeo-Christian, pantheistic, or Protestant. These obviously obscure the dramatic differences between, for example, Islam and Christianity, or even Baptists versus Pentecostals. Any exploration of religion requires recognition that beliefs and practices are associated with particular and specific groups of believers. Groups do tend to try to enforce homogeneity within themselves, so it often does make sense to talk of a coherent group (e.g., "Zen Buddhists"). But it is necessary to find the correct level of analysis for a given investigation; there are few questions about religion that can be answered on the level of "theism." This is not to say that a focused study cannot provide lessons applicable to other religious groups. *The Varieties of Religious Experience* is largely a study of white Christian men, but William James's insights are surely useful for thinking about religious experience in other contexts. Before expanding conclusions to religion as a whole, though, specificity is needed to ensure proper historical placement.

The same historical specificity applies, of course, to science. Although science tends to be referred to in broader terms than religion, its diversity must be taken into account as well. Émile Durkheim's "beliefs and practices" are useful for categorizing scientific communities as well as religious ones. Certainly biologists undergo different training, hold different fundamentals in their inquiry, and wield different skills from those of physicists. Even within a subdiscipline such as physics it is important to find the correct level of analysis: a solid-state experimentalist is not likely to be mistaken for a superstring theorist, although they are both "physicists."[3] There are historical questions in which they might blend together, but more often the subtleties are necessary elements.

Anchoring this study of science and religion to an individual scientist and religious believer like Eddington immediately situates us in a precise time and place, namely, Great Britain in the first half of the twentieth century. Further, it gives us disciplinary and sectarian specificity: this study is not so much investigating "science and religion" as it is "astronomy and Quakerism." I do not mean to restrict all the conclusions and implications of this book to such a narrow focus, but it is important to understand that interactions between science and religion occur on a profoundly *local*, and often individual, scale. The structure of the microrelationships must be properly understood before we can speak of the macrorelationships. We must always be sure we are talking about the right thing. This is a particular danger when speaking of "science and religion," which for too many people simply means "evolution and creationism." Appreciating the diversity of science and the diversity of religion allows us to break out of the stale discourses in which being a good Christian means rejecting evolution and being a good scientist means criticizing

Intelligent Design, and instead to look for other interactions lost beneath the clouds of rhetoric. Reducing the intersection of science and religion to a binary question discards the mass of what it means to be religious and what it means to do science. By focusing on astronomy and Quakerism, I do not mean to avoid those issues, but rather to provide an opportunity for a fresh approach that will let us see aspects we have missed—the history of science and religion is much deeper and more interesting than just debates over design, and this is a useful opportunity to widen our understanding.

Using an individual as a focus provides us with a natural way to approach synchronic issues, and it similarly frames the diachronic evolution of both religion and science. An individual lets us carefully locate our investigation: just as we need to be clear on our specific *place*, we must be clear on our specific *time*. British science in the first half of the twentieth century saw several important transitions that dramatically altered the way scientists worked and thought about their discipline. The realities of modern warfare forced widespread government involvement in science for the first time, and the centuries-old dominance of Newton finally crumbled under the weight of relativity and quantum mechanics. While British science was, in a sense, being forced to recognize that it had been changed by historical events, British religion had always been keenly aware of its own history. The nation remembered well the effects of the religious schisms of the English Civil War in the seventeenth century, and its leaders worked hard to maintain the delicate tension between a Christian nation and a sectarian one. Ever since the Glorious Revolution of 1688, the British polity had had to grapple with religious diversity in a way few other Western states would do for centuries. This resulted in constant negotiations about the nature and role of religion, and further, every generation reexplored the justification and logistics of this compromise of diversity. The relationship between, for example, Anglican and Irish Catholic was not static for any significant length of time. One of the most important evolving relationships was that between religion and science. *Natural theology*, the use of observations and reasoning about the natural world to deduce the existence and attributes of the deity, was the umbrella term for many of these projects, and the importance ascribed to it was distinctly British. Partly this was because natural theology provided a way for different religious groups to talk to each other: everyone could agree on the divine creation of the world and the physical manifestation of God's benevolent omnipotence. However, the *kinds* of natural theology people found persuasive and useful were temporally, spatially, and theologically contingent.[4] Therefore, this is a British story, and a story about Britain in the particular moment of the Great War and its aftermath.

Values

The historical specificity I am demanding rejects large-scale analyses that seek to characterize "science" and "religion" as abstract, homogeneous entities. Popular books such as those by Stephen J. Gould or Paul Davies represent these categories as pure domains of thought or bodies of facts, disconnected from the realities of scientific practice or religious communities.[5] This is largely the result of authors seeking high-level conclusions of harmony, exclusion, or conflict. Their drive is normative, and they are forced away from the details of descriptive approaches into grand overgeneralizations. To achieve such sweeping results, they cannot afford the complex entanglements that come with tangible scientists and active theologians; their analyses are transcendentally philosophical, not historical. Instead of using broad epistemological or metaphysical considerations, a more fruitful way to approach these issues is through the question of *values*.

Both religious and scientific life can be productively analyzed through the values held by practitioners. Value as a category has been studied by a wide range of fields, from anthropology to clinical psychology, and I will make no rigorous analysis of it here. Generally, we can think of values as defining principles, desired objectives, and preferred methods. They help shape decisions, particularly difficult ones. When a choice must be made, how does one decide? When a project is to be undertaken, what distinctive ideas and goals shape it? Values are that which people are willing to hold to even in difficult circumstances. Clifford Geertz described values as that which instills in a person "a certain distinctive set of dispositions (tendencies, capacities, propensities, skills, habits, liabilities, pronenesses) which lend a chronic character to the flow of his activity and the quality of his experience."[6] Geertz's analysis brings out the crucial element that values serve as moderators and stimuli of action and decision, without being reductively deterministic. A further element is that values are often *moral* values, invested with greater significance and import than simple conventions or practices, and values thus function as normative guides for behavior. When we examine a scientific project or a religious task, we should look for the *ought* that drives it: What values make this task worth doing?

I think few people would contest the statement that historically specific values are important in religion, and surely the history and sociology of science have demonstrated the role of values in science.[7] I am particularly interested in tracing how certain values can both transcend and underpin an individual's understanding of both science and religion. This will fundamentally be a study in how a scientist can have values that support and involve both science and religion, and how that affects his work as a scientist. I will examine values as mechanisms for mediating the relationship of science and society and thus help bridge the gap between externalist and internalist approaches to the history of science. Values are themselves the result of diachronic development—there are no eternal Quaker values, for example. They emerge, change, and wither like anything else and thus enforce a certain historicity on their effects. Values, as historically specific sources of actions and belief, provide a concrete way to think about the relationship between culture and science that avoids overbroad, Zeitgeist-like claims.[8]

This bridging function of values is the primary theme of this book. It will be pursued through the category of what I call "valence values." This is in analogy to the valence electrons in chemistry, which are the outermost ring of electrons that allow atoms to interact with each other. Valence electrons facilitate bonding through their ability to be, in a sense, shared between atoms. Similarly, I am interested in those values that facilitate interaction between science and culture. Not all electrons help with bonding, and not all values link science to other categories. In chemical terms, some values have greater valency (that is, tendency and ability to interact). Here I am interested in exactly those valence values that *do* move between the scientific and the cultural. Which values do this well? And more importantly, what effect does that have on the practice and role of science?

This is an especially fruitful approach to thinking about religion and science in the twentieth century. Important current work on the relationship between religion and science in the nineteenth century and earlier, such as that on Newton's religion, often rests on considerations of religious metaphysics and cosmological commitments.[9] These types of investigations are, unquestionably, important and useful. But it becomes difficult to look at modern physics (in a broad understanding of that term) through these same lenses. At a certain point, physical science becomes a sophisticated and complex enough system that it is no longer feasible simply to insert ontological religious commitments.[10] Religion and physical science continue to interact, however, and this is where a values-based analysis is useful. Values permeate the life and actions of a religious believer and find their way into realms such as science indirectly through their valency without requiring a direct importation of religious claims. In particular, this study explores some of the values underpinning the growth, reception, and interpretation of relativity and astrophysics in the twentieth century. An analysis of values in religion also helps ground the problem of "post-Christian" Britain. The historiographic controversy over the fate of religion in the interwar period has often turned on different senses of religion, and looking at particular religious values allows us to make sense of both declining church attendance and the continuing significance of religion in public discourse.[11] By being careful about *which* religious values are being discussed, we can find a coherent view of British religion at the time. Valency is not restricted to scientific values, and we can expect to find valence values linking religion, economics, and politics.[12]

A value-oriented approach surely requires analyzing groups and shared dispositions, but the individual remains a productive point from which to begin an investigation. Values are put into action only by someone who holds them, so a study of how values affect science should begin at the individual level and then explore how those value-driven individuals interact with their social context.[13] Fundamentally, any reconciliation or rejection of science and religion takes place on the level of an individual coming to terms with two systems that hold different values, identities, epistemologies, and social meanings. Certainly, there is no assurance that an individual *can* accomplish such reconciliation; history is littered with scientists losing their religious beliefs, believers turning away from science, and individuals who aggressively isolate their scientific and religious lives. I have no particular agenda in defending the a priori or eternal possibilities of reconciliation or contradiction. What I am interested in here is how an individual *does* construct a life that combines religion and science through a system of values. Without simply reducing science to religion or vice versa, or imposing the categories of one on the other, what strategies make it possible to defend the idea of being a religious scientist? And most importantly, what *difference* do those strategies make in the practice of science?

Eddington

To pursue these questions, this book examines the British astronomer and Quaker Arthur Stanley Eddington. Eddington was raised a traditional Nonconformist (that is, a non-Anglican Protestant, also called a Dissenter) and studied physics at Manchester, but he spent most of his life as Plumian Professor at Cambridge. He became famous for his work to confirm Einstein's general theory of relativity in 1919, his pioneering stellar models, and his influential science popularizations. One of the features of Eddington's life that makes him useful for exploring historical trends is that he occupied a peculiar position as both outsider and insider that helps reveal tacit and otherwise unspoken assumptions about the role and nature of both science and religion. As a Cambridge don, president of the Royal Astronomical Society, and knight of the realm, Eddington was unquestionably a part of the core of traditional

British power. But he was also a product of Manchester, the center of gravity of liberal, Dissenting England. He was the quintessential turn-of-the-century Quaker, trained in religious ideals that were distinctly alien to those who ran Great Britain. His scientific attitudes also marked him as an outsider: his astronomical colleagues were shocked by his claims that *"there are no purely observational facts about heavenly bodies"* and that it is "a good rule not to put overmuch confidence in the observational results that are put forward *until they have been confirmed by theory.*"[14] As a conscientious objector to the Great War, he was nearly imprisoned for rejecting nationalism. And natural theologians, still reliant on centuries of scientific proofs of God, were frustrated that Eddington's Gifford Lectures contained the claim: "I repudiate the idea of proving the distinctive beliefs of religion either from the data of physical science or by the methods of physical science."[15] These tensions between inner and outer make Eddington a useful point with which to track not just the relationship between science and religion but also shifts in astronomical practice, the history of relativity, developments in international science, and the significance of popular science writing between the world wars.

For much of Eddington's career, there was sufficient overlap between his Quaker and Cambridge lives that he fit smoothly into his surroundings. His usefulness as a historical tracer, however, comes from those incidents where his role as an outsider becomes clear. For example, his Quaker values of internationalism, absorbed during his years at the knee of the pioneer Quaker activist J. W. Graham, were difficult to distinguish from his fellow astronomers' transnational views until the beginning of the Great War. When his colleagues waved the banner of patriotic science after the fighting began, Eddington clung stubbornly to his internationalism. As a religious scientist, he often crystallized existing tensions, making them visible to historians simply by his presence.

The base of Eddington's identity as a religious scientist was that he held that many of the same values were important to both his scientific and religious lives. This is not merely saying that his science and religion were identical or that one depended on the other. Rather, Eddington's valence values linked his science and his religion in a deeper and more dynamic fashion. He held a self-identity as a religious scientist, and much of this book will be concerned with the question of how he understood his values to support this. The corollary investigations will be why such an identity was of interest in the first place and how other historical actors in Britain reacted to it. This was not just a matter of Eddington's philosophy; it is also a question of scientific practice. One of the central tasks of this work is to show how Eddington brought his living values into the formation of his astronomical methods.

Eddington's personal reconciliation of the categories of science and religion did not take place solely on an abstract philosophical level, but affected concrete features of his day-to-day life and scientific practice. This specificity means the relevant categories are not so much "science" and "religion" as "astronomy" and "Quakerism," since these were the specific social institutions, beliefs, and practices with which he allied himself. His identity was both as astronomer and as Quaker, with little meaningful separation between the two. To borrow Shapin and Schaffer's use of Wittgenstein, Quakerism and astronomy both occupied chief roles in Eddington's form of life. Shapin and Schaffer argue for an "experimental life" that is crucial for understanding the controversies around instrumentation in Restoration England. Thus, the divisions between Boyle and Hobbes were not merely philosophical, but came out of a dense web of values and practices that integrated religion, politics, and society in addition to intellectual and philosophical standards. A similar web must be developed to show the salient factors in Eddington's life and work.[16]

It is this aspect of unified values that is absent from the vast majority of work on Eddington. As a major scientific figure, he has received the usual hagiographic treatments that describe his scientific contributions without any attention paid to those views and activities that fall outside their strict boundaries of mathematical science.[17] There are biographies that take a wider perspective and include Eddington's writings on philosophy and religion, but they take no critical perspective and use few archival resources.[18] The few attempts to investigate specifically Eddington's work on religion and philosophy have the common flaw of considering it as a separate field of inquiry distinct from his science and life. Even the excellent work of David Wilson and Loren Graham overlooks the crucial reality that Eddington's public writings were only epiphenomenal to valence values that manifested throughout his scientific, political, and intellectual life.[19] Without an understanding of Eddington's valence values, he becomes a caricatured elder scientist whose philosophical ramblings bear little resemblance or relevance to his earlier scientific work, instead of a figure working thoughtfully to synergize his religious and scientific lives.

By examining the points in Eddington's life where his religion intersected with his science, we gain new insight into the way those roles were understood and the ways in which they could be contested. Additionally, we have the valuable opportunity to see the interaction of science and religion in concrete historical examples rather than philosophical generalities. The touchstones of philosophy, socialism, astrophysics, and war provide a unique narrative of the role of science in society and of society in science. A detailed exploration of the context of such interaction allows us to understand the science-religion relationship in a more meaningful and powerful way.

This book is a biographical study, though it does not attempt to be a comprehensive biography. Eddington's personal and family life will receive almost no treatment, and large parts of his career will not be discussed. Instead, this biographical approach will help illustrate the coherence of particular instances of values moving between religion and science. A biographical approach is particularly useful for the issues I am interested in, because it automatically forces the investigation to a local, concrete, temporally specific site, while retaining the ability to use the individual as a test point for observing important inflection points in history.[20] I will develop my argument through six episodes in Eddington's life in which his values generated particular interactions between his scientific career and his religious outlook. I pay special attention to those interactions in which his values were contested and opposed by others—Eddington's valence values were rarely accepted uncritically, and this friction helps illuminate the details and subtleties of the historical issues in play. Loosely, each chapter is based on a particular valence value or set of values that Eddington saw as being important to both religion and science.

THE QUAKER RENAISSANCE

The formation of Eddington's values took place in the shadow of a great shift in British Quaker culture in the 1890s. This shift, later called the Quaker Renaissance, brought the Quakers out of their so-called quietistic period of isolation, revitalizing the community for an active role in the modern world just as British Liberal politics began to recover from its late-century disasters. Under the influence of leaders such as John William Graham, the Quakers shrugged off a century of introspection to bring their testimonies of service, peace, and the continual quest for spiritual truth to a Britain growing increasingly uncertain of its political, military, and social position in the world. Graham was the chief formative influence on the young Eddington, who learned from him what it meant to be both a Quaker and a productive member of British society. This was where Eddington developed the values of pacifism, personal religious experience, and exploration (both scientific and spiritual). Eventually Eddington arrived at Cambridge, and at Trinity College his performance marked him as someone to be watched. He was quickly steered onto the path of the intellectual elite, which had been tied for so long to the Anglican elite. Eddington, however, never left his Quaker roots and remained active in the Society of Friends despite his new environment and expectations. His time in Cambridge showed the complex realities of being a Dissenter even in a nominally Liberal England.

MYSTICISM

The work that brought Eddington to a prominent position in the British astronomical community was his development of theoretical models of stellar structure. In 1916 he launched a research project to find a series of equations that could accurately describe the observed characteristics of stars via a pragmatic, exploratory method that valued opening avenues of scientific investigation over any dogmatic reliance on mathematics and certain knowledge. This clashed with the established, deductive method of physical science, which had dominated British astronomy for generations and which held that an investigation should begin from first principles of absolute certainty and mathematically develop consequences and predictions. To bring this distinction into clear relief, I analyze the famous disputes between Eddington and James Jeans. I argue that Eddington's innovative methodology was based on values that he carried with him from his Quaker faith. The Quakers of the early twentieth century rejected any claims for absolute truth in favor of a continual quest for spiritual enlightenment. Dogma was held to be an impediment to the search for religious truth, and insight was valued for its pragmatic benefits and its ability to further the spiritual quest. Eddington argued that physical science should function the same way: scientists should not obsess over the absolute certainty of their physics but instead work with a spirit of exploration that relied on physical intuition and observation. "Seeking" (a concept with specific meaning in Quakerism), not finding, was key to success. He claimed that there was, therefore, a methodological unity between science and religion that, if properly observed, would prove fruitful for both.

INTERNATIONALISM

The early months of World War I were as traumatic for the British scientific community as they were for British society at large. Triggered by Germany flouting treaties and by alleged atrocities, anti-German sentiment surged through the astronomical community. Senior astronomers began to voice the opinion that the Central Powers had forfeited their right to participate in international affairs, including science. Eddington was virtually the only voice of dissent in the scientific community, and his arguments echoed those being made by Quakers across Britain. He argued that those Germans who were accused of barbarism by patriotic scientists were the same colleagues British astronomers had been working closely with just a few years before. Further, Eddington described science as an enterprise that was inherently international and interpersonal. Any attempt to restrict science to national or racial boundaries would necessarily corrupt any search for physical truth. It was in this context that Einstein's general theory of relativity arrived at the Royal Astronomical Society (RAS). At the time, Eddington was the only British astronomer still in contact with German science, and even his correspondence came through Wilhelm de Sitter in the neutral Netherlands. I argue that a significant facet of Eddington's support for Einstein's theory was its importance as an opportunity to reconcile German and British science. Among other reasons, Eddington brought relativity into the public forum as a way to heal the intellectual wounds of years of bitter anti-German sentiment, just as Quaker relief organizations moved into Europe seeking to heal the physical and spiritual wounds of four years of fighting. For him, the famous 1919 eclipse expedition to test general relativity was not just an epochal scientific task; it was also a religious task. Eddington's Quaker peace testimony placed a tremendous significance upon the expedition, and his scientific and religious contexts are both crucial to understanding the event.

PACIFISM

The inception of the Great War brought one of the first serious challenges to Eddington's "way of life" as a Quaker scientist. The war's introduction of conscription placed his identity as both scientist and Quaker under government scrutiny. The University of Cambridge, where Eddington was a Fellow, went to great lengths to secure Eddington an exemption based on the national importance of his scientific work. But the government and military had little interest in science and its possible benefits, and this apathy toward the scientific community was manifested when exemptions based on scientific and technical work began to be revoked in order to increase manpower. Eddington was called before the Cambridge Tribunal, a hybrid civilian-military court that decided that his scientific duties were no longer of sufficient importance to merit his protection from conscription. Much to the anxiety of the university, he declared his Quaker conscientious objection to the war and refused to serve even if called. He stood in a public forum and declared that his religious beliefs could not support the call to slaughter other human beings. The tribunal refused to accept his status as a conscientious objector, however. His prior protected status due to his scientific work could not be reconciled with his claims of a religious objection to the war. Effectively, his occupation as a scientist made him formally unable to register a religious identity as well. Eddington was claiming membership in two groups that the wartime government had declared to be necessarily independent. He refused to separate his Quaker identity from his scientific one, and this chapter investigates the difficult terrain created by the intersection of religion, science, and war.

EXPERIENCE

As quantum physics and relativity advanced in the 1920s, Eddington saw in their success a marked change in the nature and scope of the physical sciences. To him the great advances in modern physics were dependent on a highly positivist outlook that restricted the domain of physics to that which could be reduced to "pointer readings" on an instrument and elements of the abstract analysis of tensors and difference formulas. Many of the followers of the Vienna Circle celebrated these advances for finally eliminating metaphysics from physics and for setting science on the path to positive knowledge. Eddington, however, read this philosophical development from a different perspective. He felt that instead of eliminating considerations of human values, positivism asserted their importance and relevance. In such high profile forums as the Gifford and Messenger lectures, he argued that classical physics, with its assumption of a mechanistic and deterministic universe, was inherently flawed and could not be reconciled with our personal experience as human beings, which clearly integrated free will and idealism. Instrumentalism and the increasing abstraction of mathematical physics, then, restricted physics to the realm of the measurable and allowed the realm of personal experience to be asserted as a valid source of knowledge and values. In particular, the restriction of science to the quantifiable meant that a nonquantifiable realm of experience was *needed* to accurately reproduce our experience of the world. Eddington said the counterintuitive notions of relativity and quantum mechanics were actually more believable than classical mechanics because modern physics was explicitly divorced from our everyday experience, which clearly did not obey deterministic physics. Religion, then, no longer needed to defend itself from scientific attacks so long as it was firmly based in religious experience. Eddington argued that a Quaker approach to religion would relieve all the tensions that had grown between science and religion under the dominance of materialism.

RELIGION IN MODERN LIFE

Eddington's values-oriented reading of positivism became embroiled in the post–Great War crisis of science, religion, and society found throughout the

English-speaking world.[21] Many religious leaders spoke out against science and modernism as having brought the horrors of the Great War, and Marxists accused religious values and institutions of prolonging the economic suffering of the interwar Slump. Eddington took up a position against the Marxist community, who claimed that scientific thought required a dismissal of religion in favor of materialism. Against the deterministic materialism of the Marxists, he presented the idealistic defense of human will and experience that he saw at the foundation of modern physics.

Eddington's popular writings came out of this debate, and one of the reasons he became a cultural icon in Britain was that he defended the position that science was not opposed to traditional values, but instead affirmed religious thought and personal experience. Eddington's popular science writing demonstrated a changing relationship between religion and the physical sciences in the Anglophone world in the twentieth century. The natural theology of Victorian physics gave way to arguments like Eddington's, in which religion and science were instead considered as they relate to human experience. This was a fundamental premise of the liberal theology that dominated interwar British religion and came directly from the Quaker Renaissance understanding of religion's role in the modern world. His work created a space for public consensus around the divisive questions of the nature and future of Britain as a modern, but moral, state. Not everyone was pleased with this, however, and Eddington was attacked by philosophers, atheists, and his fellow scientists as having stepped outside the boundaries of his profession. But Eddington was motivated by his values, which demanded that he demonstrate the feasibility and attraction of a life that could integrate modern ideals of progress and rationality with traditional religion, aesthetics, and morals. He saw the growing influence of socialism and materialism as a force that needed forceful refutation from a proper interpretation of modern physics.

Through the lens of valence values, we can use Eddington as a test point for exploring the relationship and interaction between science and religion in Britain in the first half of the twentieth century. His personal efforts to link his lives as astronomer and Quaker will illustrate important interactions between religion and science. Further, the social response to and impact of his efforts in this direction reveal wider social assumptions and investments about the function and nature of those categories. Eddington lived at a historical junction in Britain when it seemed every intellectual needed to take sides either with the forces of secularization and science or with traditional religious values. Religious groups, including many nineteenth-century Quakers, were increasingly hostile toward science and industry largely because they thought they had no other choice: scientists appeared to be relentlessly attacking everything in which they believed. This was not inevitable, however. Some groups wanted to embrace the modern world and integrate its benefits and promise into their religious worldviews. The Quaker Renaissance began with exactly this goal in 1895 Manchester. The task before them was monumental: How could a sincere Quaker also be part of the rational, scientific, modern world? The answer to this was to make a new kind of Quaker, one who could live productively in the Meeting House and in the laboratory. This was Eddington, and his story must begin not with his childhood but in a crowded Manchester conference that set out to redefine what it meant to be religious in the modern age.

The Quaker Renaissance
Making a Religious Scientist

Instead of selecting *either-or*, I prefer to take *both*. There is no line that splits the outer life and inner life into two compartments. . . . Personality is the most complete unity in the universe and it binds forever into an indissoluble and integral whole the outer and the inner, the spirit and the deed.

RUFUS JONES, *The Inner Life*

Manchester, 1895

In November 1895, the Cambridge paleographer J. Rendel Harris took the podium at a massive Quaker conference. He was speaking at a session titled "The Attitude of the Society of Friends toward Modern Thought," which in many ways was the focal point of the conference. The critical question was whether a modern, rational person could also be genuinely religious. He posed it thus:

> Let us imagine someone trying to be at once a man of accurate scientific thought and a sincerely religious person. If he succeeds in being both, how admirable is the fusion. But if, on the other hand, he is either afraid to think or afraid to believe, he can no longer be happy with himself, not truly useful to his neighbor. He is trying to be two people, and he will not succeed in being either. . . . What I mean is that unless we are prepared to regard our spiritual and mental faculties as a part of the same Divine life within us, and entitled to an equally free expansion, we shall presently find one of them becoming the victim of the other. This theory of the detachment of science and religion from one another never has been a working theory of the universe; the two areas must overlap and blend, or we are lost.[1]

But *how* do the two areas blend? This was the issue on which the history of the Quakers was to turn. In 1895 the Society of Friends was deeply divided on how to best deal with the perils of modern scientific, historical, and social thought. The resolution of this division began what is often called the Quaker Renaissance, a period in which values of mysticism, pacifism, and social action came to characterize the Society and greatly increased their public profile and influence. The emphasis on these new values was especially efficacious with regard to younger Quakers just taking their positions as adult members of society; these were expected to be valence values that would support both spiritual and practical lives. One of these young Quakers was Arthur Stanley Eddington, the future astronomer, author, and British icon of science. Eddington was an exemplar of the Quaker Renaissance and internalized and exemplified its values. The valence values that shaped Eddington's career were a direct consequence of his education during the most active years of the Renaissance and the personal influence of its most important leaders.

The schism within the Society of Friends in 1895 resulted from disagreements regarding both belief and practice. The Society was formed originally from the followers of the itinerant preacher and mystic George Fox in the 1650s in northern England. Fox preached that Christians could know God directly through personal experience of the "Inward Light" (sometimes "Inner Light"), which was a spark of the divine in all humans. Seeking a purer Christianity, he rejected the authority of any hierarchical church in favor of direct mystical experience. The Society of Friends was explicitly egalitarian, with no priesthood, no laity, and no set service of worship. Instead of a mass, the Friends had "meetings," which were held in silence until someone was "moved" by God to speak and offer their mystical testimony. Decisions were made by consensus, since no one could have a privileged position; everyone had the same access to God. Quakers were expected to have an unswerving commitment to truth in words and deeds, and they refused to swear oaths (because that implied that one might otherwise not be truthful). The name *Quaker* was originally a pejorative term that mocked the shaking motions Friends were said to have as they touched the divine. They were persecuted heavily during the Civil War and the Restoration period as disrupters of society and religion, and imprisonment of Friends was common. Fox himself was jailed repeatedly, and this victimization at the hands of the state became an important part of Quaker identity. Despite these obstacles, the Quakers collected large numbers of followers with their enthused appeal to each person's individual religious authority and their egalitarian outlook. They called for social change and sought to build a new Britain free from the errors of ossified tradition.

Despite their turbulent origins, the Quakers became a dynamic force in Britain for several decades. However, once their existence as a religious body was assured (by the Toleration Act of 1689 and other developments) they retreated into more reclusive, inward-looking practices. The eighteenth century is generally referred to as their "quietistic" phase by historians and was marked by an extreme emphasis on the personal experience of the Inward Light and lack of interest in the outside world. The Quakers erected numerous barriers between themselves and others, insisting on a strictly policed austere, puritanical lifestyle that was discouraging to potential converts.[2]

The Society became decreasingly active and influential throughout this period, until it was swept up along with most other British religious groups into the Victorian evangelical movement.[3] This wave of evangelicalism dominated Christian thinking in the first half of the nineteenth century and emphasized the fallen state of humankind, a literal reading of Scripture, and the insufficiency of any human action to earn God's grace: salvation was given by divine dispensation only.

Mid-Victorian controversies among the Quakers became attributed to erroneous interpretation of Scripture, giving the evangelicals their first point of entry.[4] The evangelicals, led by figures such as J. B. Braithwaite, saw Quakers in America with sacraments, hymns, and revivals: the full spectrum of Christian worship. They sought to bring those practices back to Britain to help the Quakers shake off their meeting house stupor. Further, they initiated characteristically evangelical missionary work. That took the form of Adult Schools and the Home Mission work, both intended to mimic similar contemporary projects being carried out by other, more active, Protestant sects. Evangelical attitudes also helped reinvigorate Quaker concerns for social justice, and the Quakers helped lead the antislavery movement. By the late 1880s, Braithwaite and his ideas wielded great influence at London Yearly Meeting, which was the closest thing the Quakers had to a central decision-making body. Braithwaite decided that the time had come for the Society to adopt the Richmond Declaration, a formal statement of faith and doctrine in a fundamentalist evangelical vein.[5]

Ironically, it was this action that mobilized a "third way" among the Quakers, opposing both reactionary evangelicalism and paralyzed quietism. Many Friends were strongly opposed to Braithwaite's proposal, not necessarily because of the content of the declaration but rather because of the tradition (going back to George Fox) that Quakers had no creed or dogma of any kind. Further, the evangelical mission work endangered another Quaker value, the lack of a professional ministry. The Home Missions used trained and paid men, which was worryingly like having formal clergy, and many Friends saw this as a dangerous step toward the elimination of the Society of Friends as a distinctive group. It seemed that evangelical values were going to reduce Quakerism to "warmed-over" Protestantism.[6]

This initiated a period of reflection: What *was* it that made Quakers distinctive? It had been clear for some years that younger Friends did not know how to think of themselves with respect to the changing, modern world. The evangelical emphasis had been more of a handicap than a help.[7] The issue was how to frame the Society's beliefs and practices in a world where Darwinism and biblical criticism were powerful forces. In the early 1890s, the liberal

approach to this question was defined by young, energetic Quakers such as
W. C. Braithwaite (the son of the evangelical Braithwaite), John Wilhelm Rown-
tree, Janet Morland, and John William Graham. Far from asserting that the
Society needed a new declaration of faith, these liberals asserted that the
Society could adapt to modern science and history precisely because they did
not have, and had never had, a doctrinal creed.

In 1895 W. C. Braithwaite wrote a striking essay titled "Some Present-Day
Aims of the Society of Friends." This was effectively the liberal Quaker manifesto. He argued that three things were necessary for the Society to enter
the twentieth century: "the growth of the scientific spirit," a perspective on
social questions, and "craving after reality in religion and life." He argued
that Quakerism was the perfect place for those individuals with deep religious
faith who also wanted to acknowledge and benefit from modern thought.[8]
This also helped define the tradition as unique; its message was that one
"should continue to seek spiritual enlightenment through the Inward Light
while labouring to reveal to the wider world that this Divine gift was available
to Seekers of every stripe."[9]

Late Victorian liberal thought resonated with these young Friends and
their understanding of Quaker history. They argued that the trajectory of modern history and society showed the powerful truth of certain elements of early
Quaker life: religion's base in personal experience, the limited utility (and
sometime danger) of formal creeds, and the religious life necessarily resulting
in a life of service and public activity.[10] The leaders of the new liberal Quaker
movement navigated the schism-weary waters of the 1895 Yearly Meeting and
organized a separate meeting in Manchester to discuss the relevance of faith
to modern-day issues and also to "dispel ignorance" of the general public with
regard to the Society of Friends (the press was specially invited).[11] The choice
of Manchester was not accidental; that industrial city had been the center of
attempts to shrug off evangelical constraints for a generation.

The 1895 Manchester Conference drew huge numbers of Quakers from all
across the country. It was later seen as the beginning of the "Quaker Renaissance," and it seems even at the time Friends expected it to be a definitive
moment for the Society to address its difficulties. Thousands attended, with
thirteen hundred coming to the first session alone.[12] The opening presentations, under the aegis "Has Quakerism a Message to the World To-day?"
began with Theodore Neild clearly announcing that all in attendance could
no longer indulge their fear of addressing modern issues. Matilda Sturge put
forward the summary of the liberal case, which was that the evangelical obsession with dogma and doctrine had led to ossification. The world had changed
since George Fox, and both theology and social thought needed to change

with it. Thomas Hodgkin emphasized that Quakerism as a faith had a unique opportunity to grapple with the modern world because it allowed engagement without the retreat into pessimism, socialism, and other perils that had marked so much of the response to modernity. It was "divinely guided" that Friends refused to treat Scripture literally, because it was now that very attitude that would allow the Society to be strengthened, rather than crippled, by the embrace of the secular world.[13] The sense of the opening speeches was that religion should be a leading force, rather than a retarding one, in the development of modern social change, and Friends needed to bring their unique values to that process.

The first part of the main meeting was titled "The Attitude of the Society of Friends toward Modern Thought" and was the centerpiece of the conference. The list of speakers was dominated by liberals (no surprise, given the identities of the organizers), with a sole evangelical voice provided by the aging and stiff J. B. Braithwaite. J. Rendel Harris, the most respected New Testament scholar of the time, spoke eloquently on the relationship between science and religion. He argued that it could no longer be maintained that science and religion were exclusive parts of humanity: "I express myself very strongly because people are everywhere soothing themselves with the threadbare fallacy that science and religion (as represented by the Bible) have nothing to do with one another." The theory of compartments (the idea that the human mind could maintain hermetic separations between incompatible ideas) was dismissed as "hopeless." This route would lead to nothing but a painful arrest of progress, both individual and collective. Such an outcome was anathema to the historical roots of Quakerism, and needed to be avoided in favor of an attitude toward the modern world that would allow both religion and science to progress together. Personal religious experience needed to take its place next to trained intellect as complementary sides of humanity: "For we have already made it clear that the ideal of human development is not an atrophy of the intellect in favor of the emotions nor a fossilizing of the heat in order to allow a more close study of the universe, but it is a development by the law of equable growth."[14] Quakers needed to ally with those who were working in the spirit of truth and seeking and to take their "right place amongst the intellectual forces of the world." Science and history came from within humans, therefore it had been touched by the Inward Light as much as any mystical experience. "Modern thought is not altogether external to ourselves."[15]

Silvanus P. Thompson, a successful physicist and engineer, educationalist, and one of the most prominent Quaker scientists of his generation, spoke under the title "Can a Scientific Man Be a Sincere Friend?" Thompson was well aware of the significant historical presence of Quakers in sciences such as botany, but he set out to present a distinctly modern view of the relationship between religious belief and scientific practice. In particular, he rejected the natural theological framework that had guided many Quaker naturalists.[16] Thompson argued that science could never prove religion to be true: the truth or falsity of religion came through personal, intimate religious experience and conviction, not a logical or experimental demonstration. Thus, religion was simply exempt from the domain of the scientific method.[17] The apparent conflicts came from people speaking of things they did not understand (e.g., a biologist speaking of religious scripture). But this was not to say that there was no link between science and religion: "Human nature is not built in such compartments that a man's religious convictions can be kept from influencing his whole nature, from directing the whole tenor of his life and thought. On the other hand the habits of accurate thought and careful expression acquired in the scientific training, cannot but follow a man into all his dealings with religious questions."[18]

In this way science and religion were *complementary*. Experience in one fundamentally aided the pursuit of the other. Thompson taught that they were both aspects of our "sacred duty" to explore the world around us, a perspective that was an important link with the attitudes of earlier Quaker scientists.[19] The Creator gave us the faculties to investigate our experiences, and we must use them to the utmost of our ability: "He who neglects his intellectual powers or refuses to be guided by the mind to the discovery of truth, is not only an intellectual coward, he is defying the purposes of the Almighty, just as truly as if he were to deliberately starve himself or to put out his own eyes."[20] God created humans as beings designed to seek out truth in the world, be it material or spiritual. The quest for truth was a difficult and painful road, however. Working in both religion and science required an acceptance that our understanding of the world was provisional and changing: "He who is thus a whole man in Christ, who can thankfully rejoice in an abiding consciousness of light within his soul, may fearlessly investigate the problems of thought and life that crowd upon him. . . . He may have unexpected lessons to learn. He may have to learn that not all of that which was for centuries received as truth will pass the test; but he will not learn in vain if amid all he preserves the unsullied heart of the little child. That was the spirit that animated Fox and Penn."[21]

Just as Maxwell revised Ampere's understanding of electricity, so would the Quaker revise the spiritual teachings of his forebears. Fallibility was an inescapable element of human activity and experience, and it needed to be accepted: "Man of science he may be, if such be his bent of mind and his training; and man of science none the less sincerely because he is a true

Friend. For what is a Friend but one who, illuminated by the quickening spirit, has learned to cast off the incrustations which ignorance or intellectual pride or intellectual folly have built up around the simple core of Christ's teaching?"[22] The result of all of this was that the "scientific man" and "sincere Friend" would be recognized not by his achievements and perfection but rather by the way he lives his life: "We have learned that there is no infallible man, no infallible church, no infallible book. We have learned that creed is not separable from conduct; that a man's religion is not what he professes, but what he lives."[23]

John William Graham rounded out the session on modern thought with a call to reestablish the Inward Light and personal religious experience as the foundation of Quakerism. A return to the Light would bypass the issues of organization and interpretation that had bedeviled Christianity in the nineteenth century. He emphasized that only in the individual could God be found: "We have given up the idea that any organization can be the sole vehicle of the Infinite Spirit."[24] Quakerism was remarkable in that it, like modern society, relied on everyday, human experience for guidance. This meant that mundane life could be a source of religious inspiration. "This reverence for the ordinary world order, not craving for irregular marvels, is characteristic of our time. Thereby is fulfilled the essential purpose in all religion of binding man and God closer together."[25] Graham vehemently rejected the evangelical claim of an inerrant Scripture as little better than wishful thinking. Modern thought had made great contributions by drawing attention to the stifled, weak ideas of the past, thereby reassuring truly religious people that religion was not in a book or building, but in the heart. People had no need of revelation to understand science and history; as Graham wrote to his sister in 1895, "That sort of thing we have the brains to study for ourselves."[26]

The liberal Quakers saw the conference as a tremendous success. The excited perception of the meeting as a genuinely new event in the history of the Society of Friends (and a landmark event in the history of Christianity) was captured in a letter from H. S. Newman to the American Quaker Rufus M. Jones: "Every Christian Church *must* face modern criticisms & modern scientific thought.... This Conference is the effort *for the first time* in our Society to *face* this emergency."[27] The values of the "Quaker Renaissance," now having been promulgated publicly, could begin to transform the Society. These values were the revival of the Inward Light, rejection of absolute and unchanging truth, rejection of a professional ministry and biblical literalism, and an embrace of an outlook that combined a dynamic grasp of modern thought with an appreciation of religious traditions. Above all, these values were supposed to *mobilize* the Quakers: "Christ's spiritual presence becomes a great moral force not only in the sphere of *being* but also in the sphere of *action*."[28]

The liberals sought to capitalize on the conference's momentum to enact change throughout the Society of Friends. The Quaker decision-making process of patiently working for consensus does not lend itself to quick change, however, and their influence was felt first in the education and mentorship of the next generation of young Friends. It was these Friends that would truly bring the values of the Manchester Conference to life. Exemplary of this generation was the young Arthur Stanley Eddington, who arrived in Manchester as a student in 1898, just as the conference's goals were being formed into concrete matters of action in education.

Young Life

Arthur Stanley Eddington ("Stanley" to family and friends) was born in Kendal, in England's Lake District, on December 28, 1882. A "birthright" Quaker, both his parents came from Quaker families stretching back to the time of George Fox. His father studied philosophy and taught at a Quaker school until his death from typhoid in 1884. Young Stanley's mother moved with him and his infant sister to Weston-Super-Mare in southwest England, where her family lived. There, Stanley was educated at home in a rather conservative environment—his mother was of the generation of Quakers that forbade alcohol, tobacco, and the theater.[29] He showed both interest in and talent with numbers from an early age. He learned the multiplication tables before he could read and attempted to count both the stars in the sky and the words in the Bible (reportedly making it to the end of Genesis). Later he was sent to a small preparatory school run by former colleagues of his father's. He later credited exceptional teachers there with instilling in him a love for literature, physics, and mathematics.[30]

In the spring of 1896, at the tender age of fourteen, Eddington received a first class in the Cambridge Junior Local Examinations, with a distinction in mathematics. This was an event celebrated by the local Quaker community with an "entertainment" exhibiting scientific instruments such as a Crookes tube.[31] Soon after, he discovered what would be a lifelong passion for bicycling, a pastime to which he brought his characteristic thoroughness: he kept a cycling diary documenting every trip's route and length for the rest of his life. Unsatisfied with straightforward distance totals (which peaked at 2,669 miles for the year of 1905), he developed an "n-number" to judge his progress, where n was the number of times he had ridden n or more miles.[32]

His family was fairly poor and would have had difficulty paying for Stanley to go to university, a problem solved by his winning a Somerset County Scholarship worth £60 a year for three years. He also made a habit of sitting for any available scholarship competitions, including Victoria University and the University of London, which brought in another £80 or so a year.[33] This was enough to pay for a modest lifestyle at a not-too-expensive school. Social pressure limited his options as well. It had only been a generation since Quakers and other Nonconformists were even allowed to attend Oxford and Cambridge, and few students had taken advantage of the opportunity to go to universities that were none too happy to have them. This combination of financial and social concerns meant going to Owens College (recently part of the University of Manchester) would be the best choice. Manchester was in the heart of Nonconformist England and was a safe place to send both an impressionable young Friend and a budding mathematician. Best of all, there were safe family connections there (a critical consideration for a religious community that had only recently allowed marriage to outsiders): the principal of Dalton Hall, the Quaker residence at the university, had been a coworker of Stanley's father. This principal was John William Graham (see fig. 1.1), one of the leaders of the Manchester Conference, who had been working hard to make Dalton Hall a root of the Quaker Renaissance.[34]

Once the administrative difficulties of his being not quite sixteen years of age were resolved, Eddington quickly became a stellar student. At the end of his first year, Graham wrote to Eddington's mother to report on the boy's progress. His preliminary exam results were full marks in physics and top of his class in Latin, English history, and mathematics. He was excelled by other students only in chemistry and English language. Graham expressed concern only about overloading him and "keeping up his exercise," neither of which seemed to be a real danger.[35] Graham was quite struck by Eddington's manifest talents and took him under his wing.

Education in Manchester

John William Graham was probably the single most important influence on Eddington as a student. Graham had just taken over as principal of Dalton Hall (he had been mathematics tutor there previously) and was using it to fulfill the vision of the 1895 Manchester Conference. He had been educated at University College London and King's College in Cambridge and afterward devoted his life to teaching and reinvigorating the Society of Friends.[36] Graham was an outspoken leader of the liberal Quaker movement and was especially vigorous with respect to the need to show young Friends that their religion was a natural match for the needs of the modern world.

FIGURE 1.1

J. W. Graham, on the far right. Courtesy of the Cambridge County Record Office and Cambridge and
Peterborough Monthly Meeting of the Religious Society of Friends in Britain.

As principal, he had the opportunity to shape the experiences of many
of the country's brightest young Quakers as they were first learning how to
engage with the world. Many students, including Eddington, had never left
their home counties before coming to Manchester. Graham saw his task as
demonstrating how Quaker values were something that would aid them in

their interests in science, history, and scholarship. In a pamphlet entitled "The Meaning of Quakerism," which he wrote while Eddington was in residence at Dalton, he laid out the framework of his understanding of liberal Quaker thought. Quakerism was presented as a religion based on experiential and experimental knowledge of God: "The appeal which will be made in this essay will be made to religious experience, far more than to religious theory. It is a contribution toward a handbook for the Christian's laboratory.... It is an invitation to join in a spiritual meal, rather than an address on the mode of its acquisition, or the physiology of its digestion." Contemporary science, such as F. W. H. Meyers's psychological work, was showing that experience was something *sui generis* and not dependent on any particular conception of the world. It was thus much more reliable than the evangelical emphasis on inerrant scripture, which would always be vulnerable to new understandings of science and history.

> The difficulties brought against religion by the attack of science and history upon statements in the Bible, are therefore no attack at all to us. We are all for Truth, and born critics by our very first principles. No theological dogma has any binding force, unless it carries conviction. The revelation of God in science and history is our delight. And it is clean contrary to Friends' principles to let authority stand between ourselves and ascertained truth. We are not, however, without due Quaker caution as to what is ascertained truth.... Our principle of unity has never been theological or speculative, it has been a participation in united reliance upon individual guidance.

The Inward Light was the key to understanding religion and religious experience. Dogma, creed, and tradition were of no significance. Only an experiential willingness to embrace new knowledge could lead to a measure of truth.[37]

Religion, science, and history were all knowledge that came from human sources and were therefore fallible. In a lecture on this topic, Graham argued that it was this fallibility that made all these things *useful*. Truth that could not be questioned or changed simply became doctrine to be worshipped from afar, with no impact on the life of the religious person. Mysticism and science were just the opposite: because there was no unquestioned truth to be ossified, the religious person must always be active in seeking truth, and this seeking would inevitably lead to aiding society. "Mysticism not irrational, not unpractical; active, not passive. Freed from ritual, dogma, and church, the mystic is left with the service of humanity." The Society of Friends, in its new, invigorated, liberal incarnation was the best possible instance of this: "Quakerism is mysticism in a pure form."[38]

This emphasis on personal experience was, of course, one of the fundamental positions of the modernist Quakers. Graham was a highly public figure
in the community, and his writings helped articulate many of these new positions. The key to convincing the entire Quaker community that these should
be the values to shape their future was arguing that these values were rooted
in the history of the society. He and his colleagues were generally successful;
a review of his pamphlet "The Meaning of Quakerism" assured readers that
Graham "represents that vigorous Quakerism of to-day which is keenly intellectual and abreast of the latest thought while it holds fast the central
principles and the ideals of the Society founded by George Fox."[39]

One of the stated principles of the Manchester Conference was to educate
society at large about the Quakers, and Graham was also an active force in
that task. The general perception of the Quakers was that they were a dying,
anachronistic group that refused to go to the theater or dance. In addition
to making the next generation of Friends a distinctly modern community,
he had to convince non-Quakers that this modernism was representative of
the religion. In a column in the *Methodist Register*, he denied that Quakers
were rule-bound or intolerant of outsiders. The question of why the Society
of Friends was so small was actually answered by the rigor of their faith. Their
austere mysticism was difficult:

> But the great reason why Quakerism remains small is that it is not an easy
> thing to be a Friend. . . . Self-reliance is, as we say, attractive, but it is taxing
> also, and most people do not care to pay the tax. Think what it means to
> have no one to sing to you, no music in your worship, no aesthetic help in
> carved column or stained glass, no one to tell you what to believe, no creed
> to cling to, no Sacrament to solemnize you, no clergyman to look after you in
> pastoral fashion. Every Friend has to take a share in all these things for himself.
> You must be ready to preach if inwardly called, to teach if you can, to visit
> the sick and the poor, to attend to all the extensive business of the Society.
> And a Friend's inward exercise makes no less demand. Silence must not be
> to him a time of idle vacancy, but a communion, with no outward aid. The
> majority of men cannot enjoy this. . . . Let it be remembered that no excellence
> is cheap, and not many kinds are common. The Quaker temperament is not
> the creation of a day in anybody.[40]

The road of the Friend was not an easy one, but this was equally true for
any form of modern thought. Learning from experience was critical to moving
forward but could be painful.

One aspect of George Fox's Quakerism that Graham felt needed to be
resurrected, but was not addressed in 1895, was the peace testimony. Fox had
been famous for refusing to fight in any of seventeenth-century England's

bloody conflicts, citing the pacifism of Jesus in the New Testament. Quakers followed his example by forgoing service in or support of the military. The peace testimony had received little active attention in the nineteenth century, but Graham wanted to revive what he saw as an essential element of early Quakerism. He called for Quakers to ally with anyone pursuing peace, whether Christian, socialist, or otherwise. The key was to make the fight against war an active one, seeking out and eliminating the sources of conflict.[41]

The anemic Quaker response to the Boer War at the turn of the century was, for Graham, a disturbing example of how passive their understanding of peace had become. Under the leadership of Graham and W. C. Braithwaite, Quakers after the war decided that instead of addressing each wartime crisis as it came, they needed to think about the imperialism and inequalities that led to war in the first place. A special deputation was sent out from Yearly Meeting to all the Monthly Meetings "with a view to arousing our members to their responsibilities . . . of maintaining our 'testimony of peace.'"[42] Graham proposed that Quakers embrace arguments against war from whatever source (economic, Marxist, etc.), not just Christian principles. He later wrote a widely read book, *Evolution and Empire*, that was essentially a compendium of contemporary Quaker thinking on peace.[43] A critical aspect of reviving the peace testimony was linking it with the core concept of modern Quakerism: the Inward Light. After a decade of work by Graham and others, the London Yearly Meeting produced a document doing so explicitly. Quakers were no longer expected to reject war because of scripture or even because of the example of Jesus; rather, they rejected violence because it was a violation of the divine spark that was found in all people. One historian of religion marks this as a definitive point in the creation of the modern Quaker identity: "This official recognition of an explicit connection between the peace testimony and Inward Light theology represented not only the triumph for the liberal tenets of the Quaker Renaissance but also the creation of a rock upon which future Quaker war resistance would be anchored."[44]

Eddington came under the mentorship of J. W. Graham at the peak of the elder Quaker's fervor and energy. From Graham he absorbed the essentials of the identity of the renaissance Quaker just when Eddington was first venturing into the world. Thus Eddington took with him a belief in the fundamental importance of the Inward Light and its consequences. These consequences included the importance of mysticism over scripture, pacifism based on respect and tolerance, and the value of pursuing human knowledge. The search for experience of God was driven by the same source as the search for historical or scientific knowledge, and therefore they all demanded equal reverence. All came from human investigation and demanded constant revision as only approximations of truth. But this was no weakness, since it drove one to continual, unending service in the quest for human betterment. Eddington was literally immersed in this philosophy; Graham actively constructed life in Dalton Hall to be a crucible for making a new kind of Friend.

While Dalton Hall and J. W. Graham were no doubt critical elements in Eddington's decision to go to Manchester, he must have been in awe of the opportunities to pursue his love of and talent in physics and mathematics. Manchester University was built around Owens College, an institution that had developed to fill a perceived gap in the science education provided by the traditional universities.[45] Physics education at Owens College became defined by the dynamic personality of Arthur Schuster, originally appointed to the applied mathematics chair. He was from a Jewish German family (but was prudently baptized by his parents) and became an expert in the burgeoning field of spectrum analysis, taking his PhD at Heidelberg under Gustav Kirchhoff. He had been working at the Cavendish Laboratory with James Clerk Maxwell and Lord Rayleigh for five years when he accepted the invitation to come to Owens.

Schuster became chair of physics in 1887 with the death of Balfour Stewart. Horace Lamb took over the applied mathematics position, giving Schuster free reign to reshape the university's approach to science education. He was a firm believer in the importance of both applied mathematics and practical applications to modern science but felt that the purpose of the educational institution was the training of scientific *judgment*. Physics was to be taught as a "unity of thought and practice." Lab work and original research were made integral to undergraduate education. He saw the separation of science and life as artificial and dangerous; science was an organic part of society and held responsibilities toward it. Science students needed to be taught to be aware of the world around them and how their work might affect it.[46] This was the ideal of the civic scientist. Robert Kargon argues that in nineteenth century Britain there was a transition from seeing science as a collection of God's laws that were discovered by gifted individuals, to a method of expertise that should be applied to all areas of human knowledge.[47] The expertise of the scientist was a resource for making society a better place.

Eddington worked closely with Schuster during his four years at Manchester. Indeed, he could not have avoided it, since Schuster personally taught all but the most elementary physics classes. In addition to his view of the importance of the civic scientist, there were two elements of Schuster's view of science that particularly influenced Eddington. The first was his conception of the relationship between mathematics and physics, and the second was the importance of an international outlook for the scientist.

Schuster's inaugural lecture at Manchester was "The Influence of Mathematics on the Progress of Physics."[48] According to him, mathematics could act
on its own only after the physical investigator had done his work. The physicist
must first develop conceptions that can be expressed as mathematical symbols
and then conduct quantitative experiments or observations that show the use
of these conceptions. However, neither the experimentalist nor the mathematician could work independently of the other; it was the way they supported
and inspired each other that gave their specialties value. "For no subject can
stand by itself, and the utility of each must be measured by the part it takes in
the play of the acting and reacting forces which weave all sciences into a common
web." The mathematician was weak because he needed the world translated
into quantitative symbols, the experimentalist was weak because an unexpected result tempted him to invent a wholly new theory for something that
might only be an instance of a long-understood truth. It was in this tension that
progress was made. The scientist needed to understand the delicate nature of
his task: "All our theories are necessarily incomplete, for they must be general
to avoid insurmountable difficulties." Approximation and partial understanding
were the tools essential to discovery. Caution was always needed in going from
hard-won physical results to universal theory: "It often happens in mathematical
explanations of physical phenomena that the equations originally deduced
contain a series of constants which are then determined to fit the experiments.
This process, which is perfectly legitimate, does however often prove that the
theory is successful in giving us a useful formula of interpolation, and need
not be conclusive in favour of the ideas which have led to the formula."

Once those approximations had become robust enough to serve as fundamentals, however, theory could be given over to mathematics. Speaking of the
development of the wave theory of light after Thomas Young and Augustin
Fresnel, Schuster said: "The undulatory theory now entered on a stage in
which it could be taken up by the mathematician pure and simple. Its foundations had to be rendered more secure, and its consequences worked out
to a greater extent than even Fresnel had done." Mathematics could then ex-
trapolate conclusions beyond the original experimental base. The duty of the
physicist was to "investigate how far we can safely push certain assumptions
and where a new hypothesis must be brought into play. . . . The undulatory
theory . . . shares the common fate of all theories, and leaves a vast quantity of
facts unexplained and waiting for more complete investigations. . . . History
then does not teach us of any royal road to success." Final truth was not to
be the goal, but rather a progressive and aggressive exploration of the boundaries of experimental and mathematical truth. There was no analytical, clinical
method by which physics could be guided. Rather, a *fingerspitzengefühl* was

needed to intuitively grasp and sense what methods would be fruitful. The university's job was to teach this skill and, in the process, teach the students what do with their roles as citizen scientists: "But more important for the ultimate progress of truth than a solitary success is the training of the faculty which enables the scientific man to judge correctly, and to appreciate the results of those who strike out new roads and extend the boundaries of knowledge. It seems to me to be one of the chief objects of an institution like this is to bring up men who, by conscientious consideration of scientific speculations, may help to give that solidity and elasticity to public opinion which is necessary for the rapid advance of science."

Schuster emphasized that his students learn the experiential lessons of *doing* science. This lesson could only be taken away from first-hand contact with the operation of research. Eddington would carry this with him for his entire career; science was a kind of doing, not a body of knowledge. This meshed smoothly with J. W. Graham's insistence that modern knowledge such as science came from the same root as religious knowledge—experience. Experience was Eddington's first valence value, and it was reinforced from both the scientific and religious sides.

Another element of Schuster's view of physics that would have resonated with Eddington's life in Dalton Hall was his fervent support of internationalism in science. An article Schuster wrote for the *University Review* spelled out his thoughts. The basic argument was that there were certain problems in science that could be investigated only with international cooperation, such as astronomical observations and geodesy. Fortunately, special organizations had been brought together to coordinate science across national borders. But there was an added, less tangible, benefit as well: "The co-operation of different nations in the joint investigation of the constitution of the terrestrial globe, of the phenomena which take place at its surface, and of the celestial bodies which shine equally upon all, directs attention to our common interests and exposes the artificial nature of political boundaries. The meetings in common discussion of earnest workers in the fields of knowledge tend to obliterate the superficial distinctions of manner and outward bearing which so often get exaggerated until they are mistaken for deep seated national characteristics." Thus, Schuster believed that the very act of working with scientists from other nations showed that political and racial distinctions were artificial and pointless. One can sense Schuster's own multinational history coming into play here. As a German Jew in England, it is unlikely he ever felt at home; perhaps his work with physicists across the world helped him forge an identity independent of his own nebulous national origin. But the bonds forged by international science could have ramifications far beyond the individual, or even the scientific community: "I do not wish to exaggerate the civilising value of scientific investigation, but the great problems of creation link all humanity together, and it may yet come to pass that when diplomacy fails—and it often comes perilously near failure—it will fall to the men of science and learning to preserve the peace of the world."[49] The future of the world could well lie in the hands of scientists who correctly understood the international nature of their work. Certainly, this kind of thinking would have resonated with Eddington. He was being encouraged toward internationalism from both Graham and Schuster, in both the physics building and the Meeting House. The values he was absorbing were explicitly positioned as being critical to both his scientific and religious education.

A further overlap for Eddington would even have been found in his textbooks. The texts used in the physics tutorials were written by Schuster and the Quaker physicist Silvanus P. Thompson. Thompson was one of the key speakers at the 1895 Manchester Conference and had written extensively on the methodology of physics. For him, physical science was a kind of "mental and moral training."[50] His Quaker views of physics as a kind of spiritual exploration of the world were paramount: "Now the test of the truth of a complete theory, as of a temporary hypothesis, is to be found in its deductive applicability to the further discovery of fact. If the theory be true, then certain consequences must necessarily follow. We forthwith proceed to investigate the facts and see whether they bear out our theory. If they do, we step forward to grasp further truths."[51] Physics shared certain elements with religion, poetry, and all other aspects of human experience: "And when his further studies lead him on to higher and more occult problems; when the beauty, the inevitableness of natural law dawns upon him as he traces, from point to point, from system to system, the workings of the mystery that surrounds him on every hand . . . then the emotions which these things arouse within his breast, fill his mind and possess his spirit; emotions, I venture to say, no less pure, or true, or real, than those excited by the most thrilling piece of history or romance."[52] Schuster no doubt picked Thompson's textbooks because they agreed with his views on physics. That Eddington would have seen the same alignments from a religious perspective should come as no surprise: he was being taught by Graham that all the elements of modern life should come together in a unified life. Eddington fused his lessons, whether he learned them in the classroom, laboratory, or Meeting House. To him, they were but different facets of the same jewel of modern life, and it was his duty as a Quaker to bring them together.

Eddington finished his time at Manchester in 1902, receiving a first-class degree in physics. He had focused his classwork on physics and applied mathematics, but also excelled in English and Greek.[53] Most of his time was spent in honors classes working closely with Schuster and with Horace Lamb. He continued to pay for his education with scholarship competitions, and in 1901 he sat for an entrance scholarship to Cambridge. He was awarded a £75 scholarship in Natural Science at Trinity College.[54] Eddington was overjoyed, and many of his teachers felt that his ability in physics had simply outgrown Owens; if he was to fulfill his potential, he needed the resources and community of Trinity. Graham wrote to Eddington's mother (who was perhaps worried about Stanley venturing into the heart of Anglican England) that "he will find a larger sphere and more competition at Cambridge."[55]

From Cambridge to Greenwich

At Cambridge Eddington immediately threw himself into preparing for the Mathematics Tripos Exam. His background with Lamb (himself a Wrangler) and Schuster proved extremely useful, and he was able to contract R. A. Herman as a coach.[56] It was still traditional for mathematics candidates to learn much more from their coach than from their lectures, though that was slowly changing. He took lectures from such impressive minds as E. T. Whittaker, A. N. Whitehead, and E. W. Barnes.[57] Neither Whitehead nor Barnes had begun thinking seriously about religion yet, and it is unlikely their work on this subject had much influence on Eddington.

Eddington's ability was quickly recognized at Trinity, and before the end of his second term the college had changed his entry award to a major scholarship of £100 yearly. This was still not quite enough to support a student, and Eddington continued to sit for scholarship and degree competitions.[58] He sat for the Tripos exam in his second year—unusual but not unheard of. What was extraordinary was that he placed as Senior Wrangler; that is, he scored the highest grade of all taking the exam. This was the first time a student in his second year had done so, and it caused something of a sensation. This was also the first time a Friend had achieved Senior Wrangler status (see fig. 1.2), and Eddington was invited to a party in his honor in Manchester to celebrate. The party opened with a game of cricket at Dalton Hall, and speeches were made by the Vice Chancellor of Manchester University, Horace Lamb, and J. W. Graham.[59]

It was at Trinity that Eddington met his close friend C. J. A. Trimble, with whom he would spend a lifetime of nature hikes and holidays.[60] As with his cycling diary, Eddington recorded their travels with great precision ("We were out for 126 hours in 14 days").[61] His trips with Trimble were just one part of Eddington's passion for sports of all kinds (he made time to see the Red Sox

MR. A. S. EDDINGTON (TRINITY)
Senior Wrangler.

Mr. Arthur Stanley Eddington, Senior Wrangler, is a son of the late Mr. A. H. Eddington, of
Kendal. He was born at Kendal in 1882, and
was educated at Brynmellyn School, Weston-
Super-Mare, and Owen's College, Manchester.
His private tutor was Mr. R. A. Herman. Our
portrait is by A. H. Legg, Weston-Super-Mare.

FIGURE 1.2
Arthur S. Eddington, 1904. Courtesy of the British Library.

play during a trip to Harvard in 1936), though his enthusiasm often overran
his ability (he was a notoriously bad golfer).[62] At Cambridge Eddington also
began to drift from the conservative social mores of his family by enjoying the
theater, becoming inseparable from his pipe, and on a handful of occasions
even sampling champagne and beer.

Eddington received his degree in 1905, and afterward he was somewhat
adrift. He had not yet decided what sort of career to pursue. There was no
immediate financial pressure: coaching work at the university was plentiful
(thanks to James Jeans's departure), and he even taught briefly in the engineering faculty.[63] He kept active in the Nonconformist Union and in politics (his
journal mentions a lively party in the wake of the Liberal election victory).[64]

Although undecided on his path in science, teaching apparently caught Eddington's interest. In January 1906 he joined the Friends' Guild of Teachers, an organization originally founded to help train teachers in Quaker schools, but which welcomed Friends who taught in any context.[65]

He may have been spurred to join the Guild when his mentor, J. W. Graham, was appointed to the presidency in that year.[66] Graham exerted a great deal of influence on the formative years of the Guild, and his attitudes were manifest in its philosophy. Both teachers and students were expected to integrate their religion with an active, informed, rich life: "The aim of education is the formation of a strong, sensitive and balanced character—the attainment of a developed and complete life.... [One must] regard the acquisition of learning as an instrument, and not an end; again, he will hold throughout a strong belief in the unity of the human consciousness, and, as a consequence, the unity of education."[67]

The Guild was very much influenced by the goals and attitudes of the Manchester Conference, and it actively sought out new and innovative methods in education. The 1902 presidential address, given by S. P. Thompson, proudly proclaimed that they were trying "experiments in education" and thus adopting the processes of "modern science."[68] Guild members hoped students would be active in making the world a better place by applying Quaker values to the problems of modern society. Most important was the development of an attitude of constant improvement and seeking after new and better methods: "It is one of the first principles of education that it is our duty to arouse the spirit of investigation."[69] Eddington went to Dublin for the annual meeting of the Guild and thrived among like-minded teachers and intellectuals. He would remain a member for life and served as president for some years as well.[70]

Eddington's career trajectory was given a sudden impulse by events on the outskirts of London: the Chief Assistant at the Royal Greenwich Observatory resigned in order to take a prestigious post in Scotland. The Astronomer Royal, W. M. H. Christie, sought a replacement in the traditional way: "I have followed the precedents of previous appointments to the office of Chief Assistant and have been making inquiries respecting gentlemen who would be qualified for the post, and in particular High Cambridge Wranglers."[71] The position was seen as something of a training ground to help promising mathematicians start their careers. Christie came across Eddington immediately in his search and quickly offered him the post. He cited Eddington's experience with optical and electrical equipment, as well as glowing recommendations from Schuster and others. He closed his letter to the Secretary of the Admiralty: "Mr. Eddington thus possesses in a high degree both theoretical and

practical qualifications for the post, and I find there is a general consensus of
opinion among those whom I have consulted that, amongst the candidates
available, he is the one best suited to perform the duties of Chief Assistant."[72]
Eddington's experience with instruments was perhaps not as great as Christie
had hoped: upon receiving the offer of the position, Eddington spent day and
night at the Cambridge observatory practicing with the transit circle.[73]

The post was an excellent one. The opportunity to work in astronomy full
time, with access to the newest and best observations and equipment, was
simply too good to pass up. The Chief Assistant position had historically been
filled by some of the best Cambridge mathematics students, and Eddington
was seen by many as an extremely promising young man.[74]

While his scientific career was getting well under way, he was having
some difficulty finding a Quaker community in London in which he felt
comfortable. He switched from Peckham to Woolwich Meeting, but was still
not entirely happy.[75] The ethos of the 1895 Manchester Conference had not
yet taken firm hold in London, and it seems he was uncomfortable with the
conservative and evangelical Friends that still dominated the meetings there.
He began to look outward for those who shared his perspective on Quakerism,
going to Rotterdam for the 1907 Peace Conference, participating actively in
the Society of Friends at the University and Friends' Guild of Teachers, and
attending the Kendal Summer School in 1908.[76]

The Summer School

The Summer School movement began in 1897 when John Wilhelm Rowntree
wrote to George Cadbury for help in improving the intellectual and spiritual
background of the ministry offered at Quaker meetings.[77] Since there were no
professional ministers in the Society, this was essentially a call to improve the
religious and intellectual education of the entire body of Friends. Cadbury
contributed the asked for money and the first Summer School went ahead,
advertised as an opportunity to "stimulate thought, promote helpful reading
and study, and . . . to awaken in the Society a fuller conception of the place in
the service of Christ of the trained and consecrated intellect."[78]

The format of the Summer School was an unusual one. A small number of
"campers" would retreat to the country and spend ten days hearing speakers,
holding silent worship, and walking in the hills. The Summer School was
explicitly designed to help birthright Quakers learn how to bring their beliefs
into the modern world: "This movement is one of the results of a desire on the
part of the more advanced and cultivated Quaker circles to leave the theories

of quietism and of individual perfectibility in which, since the middle of the 18th century, Quakerism has mainly developed.... They are now preferring to try and share their ideals with their neighbors.... Their descendants shall find the communion of their birth sufficiently elastic and capable of evolution to suit the changing needs of ever moving modernism."[79]

In 1908, some three hundred Friends participated, two-thirds of whom traveled a significant distance. The gathering was held in the Quaker heartland, and the community and press were extremely interested in what was seen as a sign of the increasing momentum of the call to engage with the modern world. This task was not expected to be easy. The responsibility was nothing less than to use one's entire life as a visible demonstration of spirituality: "To be a Christian to-day was to meet the challenge of the century by finding out how far our lives can be used to carry on the priceless flame of truth."[80]

The Summer School was especially important for Eddington, as it updated and crystallized the values of the 1895 Manchester Conference. The speakers in 1908 included J. W. Graham and William Franks, as well as the noted Friend Seebohm Rowntree, who spoke on social questions. Eddington found Rowntree's lectures to be "splendid" and was also markedly impressed by the American Quaker Rufus M. Jones.[81] Jones spoke on the nature and value of personal religious experience, as well as specific aspects of Quaker mysticism. The local press reported his lectures as "brilliant attempts to bring into line traditional accounts of religious ecstasy and modern scientific analysis of mental phenomena."[82] His lectures on mysticism reminded the campers that "the fact that God and man do meet is proved by men and women in a way it never can be by logic."[83] Eddington's (and, indeed, much of twentieth-century Quakerism's) thinking on mysticism appears to owe a great debt to Jones.

Rufus M. Jones was a Maine Quaker who showed extraordinary charisma and leadership from an early age. As a historian of religion and a teacher, he devoted his life to overcoming the sectarian schisms that he saw tear apart American Friends in the nineteenth century. From the Hicksite split through the Richmond Declaration of evangelical faith, Jones saw all of these as errors based on an undue reverence for theology and scripture. He argued that theology was completely foreign to the primitive Church. The life of Jesus and the Apostles was based, rather, on religious experience.[84] Like the early Church, the first Quakers relied on an experiential, experimental approach to God: the Inward Light. The effect of such mystical experience was to release a certain kind of spiritual energy that made an individual extraordinary in their actions.[85] Mysticism was "the type of religion which puts the emphasis on immediate awareness of relation with God, on direct and immediate consciousness of the Divine Presence. It is religion in its most acute, intense, and living stage."[86]

Modern thought posed no danger to this experiential approach to religion. Unlike a scriptural approach, which needed to be wary of the discoveries of biology and physics, an experiential foundation needed only a conscious self. "Science has furnished no evidence which compels us to give up believing in the reality of a personal conscious self which has a certain area of power over its own acts and its own destiny, and which is capable of intercourse, fellowship, friendship, and love with other personal selves."[87] He did not doubt that there was a realm of causality and unvarying law, but it could not be disputed that the experience of our own personality was *real*.[88]

Jones argued that there was no need for this view of religion to be proven. It was enough that religious experiences suggested the existence of a spiritual reality. One needed just enough truth to work with: "We cannot *prove* by these somewhat rare and unusual mystical openings that there is an actual spiritual environment surrounding our souls, but there are certainly experiences which are best explained on that hypothesis, and there is no good reason for drawing any impervious boundary around the margins of the spiritual self within us."[89] This sort of inference to the best explanation, in the absence of deductive proof, was valuable because it was productive. Further, Jones claimed the turn of the century was being marked by a shift from authority in all fields of human thought. This was a "shift from 'what is the unbroken tradition' to 'what are the facts? What data does experience furnish?' . . . We take slender interest in dogmatic constructions; we turn from those in impatience." This had brought science to positivism, history to the archives, and religion to mysticism.[90]

But this mysticism was not of the self-denying, cutting-oneself-off-from-the-world variety; this was a pulsing, driving mysticism that thrived in the corporeal world. As William Penn said, "True godliness does not turn men out of the world, but enables them to live better in it and excites their endeavors to mend it."[91] Jones called for "affirmative," "practical," or "conative" mysticism, in which God could be intimately experienced in the course of everyday living. "Religion does not consist of inward thrills and private enjoyment of God; it does not terminate in beatific vision. It is rather the joyous business of carrying the Life of God into the lives of men—of being to the eternal God what a man's hand is to a man."[92] All activity was toward this mystical goal. The daily life of the mystic must bring together his mundane life with his spiritual: "There is no inner life that is not also an outer life. To withdraw from the stress and strain of practical action and from the complication of problems into the quiet cell of the inner life in order to build its domain undisturbed is the sure way to lose the inner life."[93] There was no binary aspect to religion: faith or works, inner or outer. All were one.

Experiential mysticism as part of daily, worldly life required a pragmatic attitude. One could never know absolutely, one could only explore experience.

> We are pragmatists by the very push of our immemorial instincts. Our first question, consciously or unconsciously, is apt to be, what effects will come, if I act so, or so? Will this course work well? Will it further some issue or some interest? And this deep-lying pragmatic tendency—this aim at results—appears woven into the very fiber even of much of the religion of the world. Sometimes the results sought are near, sometimes they are remote; sometimes they are sought for this world, sometimes they are sought for the next world; sometimes the pragmatic aim at results is crudely and coarsely selfish, sometimes it is refined, or altogether veiled, but religion has no doubt often enough been an impressive kind of double-entry bookkeeping, the piling up of credits or of merits which some day will bring the sure results that is [sic] sought.[94]

Real religion was thus intimately tied up with all the intellectual, moral, and social undertakings of the age. And since true religion was adogmatic by its very nature, so too all elements of the mystic's life were dynamic: "The beatitude lies not in attainment, not in the arrival at a goal, but in the *way*, in the spirit, in the search, in the march." The danger of the Pharisee's way is that his aim can be achieved: "Once the near finite goal is touched there is nothing to pursue."[95] It was not necessary to have full and complete knowledge of God before living a spiritual life: "Fortunately we do not need to understand vital processes and energies of life before we utilize them and start living by them."[96]

Jones argued that Jesus spread his message not through books or an organization, but through direct contact with his inspiring life. Modern people who lived the same way were "practical mystics." Speaking of his "practical mystics," he said, "They are very busy persons, overloaded with their own life work, their vocation, but that in no way prevents them from being transmitters of great moral and spiritual forces; quite the contrary, they are all the better transmitters because they are steadied and stabilized with a weighty occupation."[97] A mystic was always practical, and the more effectively practical, the more truly mystical.

This same active, pragmatic attitude needed to be applied to all the aspects of Quaker life that came from the Inward Light. This included the peace testimony. The traditional Quaker passive-resistance model simply did not do justice to the source of their pacifism, namely, respect for the spark of divinity in all people. "Pacifism means *peace-making*. The pacifist is literally a peace-maker. He is not a passive or negative person who proposes to lie back and do nothing in the face of injustice, unrighteousness and rampant evil. He stands for 'the fiery positive.' Pacifism is not a theory; it is a way of life. It is something you *are and do.*"[98]

Jones practiced what he preached. He traveled the world spreading his message through his own activities and teaching. He was staggeringly active in relief work, especially during the Great War. He founded the American Friends Service Committee, an organization that revolutionized the way Quakers sought to eliminate war through international understanding and relationships.[99]

When he stopped in England in 1908, Jones was undertaking his first great series of tours to spread his message of a renewed Quakerism based on the Inward Light and its attendant mysticism. His ideas meshed powerfully with the values of the Manchester Conference, and liberal Quakers on both sides of the Atlantic built on them to create a modern, liberal Society of Friends.

The relentless theme of the Summer School was the transformation of a religious community to meet the world. The self-isolation of the conservative Friends and the rigid literalism of the evangelical Friends needed to be discarded in favor of an organic link between mysticism and social activism: "Perhaps we shall see in our times the Society of Friends again occupying a high place and standing for what is best in the soul's progress by single-heartedness in religious enthusiasm and efforts toward social reform."[100] Personally, Eddington found the Summer School quite beneficial but complained of a lack of time for silent worship.[101] His thoughtful attendance at and reflection on the Summer School shows us that his exposure to Quaker values regarding religion and the modern world was not just a feature of his education in Manchester (where Graham could exert a direct influence), but that he continued to deliberately pursue such matters even once he was a professional astronomer. The essence of the Quaker Renaissance was that religious values were valence values, and Eddington's career is a study in how those ideals could function in practice.

From Greenwich to Cambridge

Eddington's time at the Royal Observatory was well spent. He learned a host of new skills, including eclipse observations and telegraphic latitude determinations.[102] When he began his own research, he immediately distinguished himself with innovative techniques and mathematical skill. His work on statistical cosmology earned him national and international recognition and membership in the Royal Astronomical Society (RAS) in 1906.[103] His work came to the attention of the entire astronomical community in 1907 when he received the prestigious Smith's Prize (as well as a fellowship at Trinity) for his work on stellar movements.[104] Eddington's journal records that he continued to read voraciously, including Karl Pearson, Henri Poincaré, Silvanus P. Thompson, Bertrand Russell, and Alfred North Whitehead in addition to a dizzying number of novels and literary classics—his love for Lewis Carroll's surreal stories would prove useful in his later efforts to explain the strange world of relativity.[105]

Eddington felt that astronomy was the best opportunity for him to blend his varied skills in mathematics and physics and thereby make a contribution to science. In 1909 his former teacher Arthur Schuster invited him to take a chair of theoretical physics at Manchester, but Eddington turned it down to stay on at Greenwich. He said returning to physics and academic life was an attractive prospect, but "I have concluded that the suggestion and opportunities that I meet with in a large observatory are more likely to lead to good research work on my part than any I could hope for elsewhere."[106] The Royal Observatory gave him nearly unparalleled opportunities to work in the exciting new fields of astrophysics being pursued all over the world. Furthermore, he was building important relationships with European astronomers such as J. C. Kapteyn, Ejnar Hertzsprung, and Karl Schwarzschild. In 1910 Frank W. Dyson was appointed Astronomer Royal, and he and Eddington became fast friends. They shared a forward-looking perspective on astronomy, as well as a commitment to international cooperation and partnership in the field.[107]

Eddington was finally lured away from Dyson and Greenwich by an offer that arrived while he was on holiday in Cornwall in 1913.[108] George Darwin's death had left Cambridge's Plumian Professorship vacant, and Eddington had been selected as the perfect candidate to fill the post. The Plumian was officially a chair of "Astronomy and Experimental Philosophy," and a second astronomy professorship oversaw the Cambridge Observatory. However, Robert Ball, the previous director of the observatory, died at nearly the same time as Darwin. It was decided that Eddington, as an astronomer with solid training in mathematical physics, was a perfect choice to fill both posts. There was some controversy as to whether Eddington would take the directorship, but Dyson and other British astronomers supported him strongly and he soon moved into the observatory with his mother, his sister Winifred, and the first of a lifetime of beloved dogs. Eddington never married, and Winifred filled all the social roles of a professor's wife until his death from stomach cancer in 1944.[109]

His teaching duties were relatively light and were generally restricted to matters of practical astronomy (spherical trigonometry, etc.), with one course his first year on his research on the structure of the stellar system. The environment of Cambridge seems to have been a tremendous stimulus to his research work—presumably no longer having to fulfill menial duties for the Royal Observatory provided him with an increase in free time. He rejoined the Mathematical Club, the $\nabla^2 V$ Club (a group for applied mathematicians), as well as occasionally attending the Moral Sciences Club.[110] It should come as no surprise that he also took an active position in the Society of Friends at the University, and the records of the Cambridge Quakers mark the arrival of the Eddingtons as new and active members.[111] They began attending the Jesus Lane Meeting (see fig. 1.3), which would remain Eddington's spiritual center for the rest of his life.

By this time the Jesus Lane Meeting would have felt very comfortable to Eddington. They were using the 1908 edition of *Faith and Practice*, a book used to help informally guide Quaker decision making and worship. This edition relied heavily on liberal thought and encouraged Quakers to see their faith as having come from "a widely spread yearning for a presentation of Christianity, more spiritual and less theological, more ethical and less dogmatic, more practical and less ceremonial." The Inward Light and experience based on it was to be the core of worship, and discrete formalisms of prayer were to be rejected. Most important, Friends should remember that the Inward Light called for a spiritual life to encompass all activity: "The doctrine of the indwelling of the Spirit has been to the Friends neither a philosophical idea, nor a pious opinion only, but an eminently practical faith, embracing within its scope the whole of human life. The presence of the Spirit gives the power to translate the Apostle's advice into practice, 'Whether, therefore, ye eat, or drink, or whatsoever ye do, do all to the glory of God.' Hence, little account is made of the popular distinction between things secular and things religious: all work, all times, every employment that is not wrong may be accounted holy."[112] Eddington's community embraced the idea that his profession as scientist would harmoniously integrate the values of the Society of Friends.

In early 1914, just as Eddington was settling into the Madingley Road Observatory, Silvanus P. Thompson visited the Jesus Lane Meeting. While in Cambridge, he gave two lectures, including one on twentieth-century Quakerism.[113] Thompson would later give the Swarthmore Lecture, an annual presentation at London Yearly Meeting by a prestigious member of the Quaker community. It typically addressed issues of contemporary interest and concern to the Friends. Thompson's lecture was titled "The Quest for Truth" and dealt with the relationship between science and religion. Eddington would

SOCIETY OF FRIENDS, CAMBRIDGE MEETING

FIGURE 1.3
The Jesus Lane Friends Meeting House, in Eddington's time and today. Courtesy of the Cambridge
County Record Office and Cambridge and Peterborough Monthly Meeting of the Religious Society of
Friends in Britain.

certainly have both met Thompson and read his Swarthmore Lecture, and
"The Quest for Truth" provides a useful overview of many of the ideas with
which Eddington began his career as Plumian Professor.

Thompson was a firm believer in the truth and stability of physical law.
Nonetheless, he rejected the claim that physics dictated a deterministic view of
the universe. The Laplacian idea that all future events were determined solely
by mechanical laws was rejected because it allowed no room for the soul and
its efficacy. It was dangerous to think that "the whole world is ruled by fate,
by fixed and determinate necessity, affording no scope for free-will or for the
operation of moral forces. Such a view would reduce the universe to a mere
mechanism and remove all moral responsibility from man; a view to be sternly
repelled." A strict proof of why determinism was wrong was not needed; it
was sufficient that we as conscious beings have a sense of the concrete reality

of our free will. Indeed, it was a common error to think of physics as relying on strict proof. Instead, physics was a constant juggling of provisional hypotheses that should be tested and discarded without attachment to dogma: "What the scientific enquirer does is to hazard a number of hypotheses, some of them probable, others quite improbable, and to test them one by one to see which is right, or which of them is nearest to the truth. For every hypothesis that turns out to be correct he may have to frame a score that prove invalid. Persons who do not understand this mode of arriving at truth often regard this as a very shaky procedure, and condemn it as trying to arrive at truth by means of error. But it is really an every-day process of thought."[114] What Thompson meant by "an every-day process" was that this pragmatic evaluation of ideas according to their utility was in fact the basis for all productive activity in the world. Science, industry, education, and, most of all, religion demanded this attitude, which was fundamentally based on our inherent natures as mystical creatures. Error in all of these realms was to be expected and even embraced. An inability to deal with a changing, maturing perspective on the world would (and in the history of Christianity, had) lead to a fatal rigidity: "The craving for certitude is not in all respects a sign of spiritual health. The very eagerness to be certain tends to vitiate the search by a temper of impatience."[115] Mysticism was the true root of all "seeking," and it demanded a distinctive outlook in all activities.

Thompson argued for the interdependence of religion and secular life. He presented modern thought as wholly harmonious with Quakerism. This was the essence of the lessons Eddington had been taught since arriving in Manchester, and no doubt Thompson, as a successful physicist and a greatly respected Friend, was a clear living example that these lessons were valid. Eddington's religious values did not simply impress themselves on his education in physics and astronomy; instead, he saw those values as being *already fundamental to* physics and astronomy. The Manchester Conference, channeled through J. W. Graham, Rufus Jones, and Silvanus Thompson, resonated strongly with the physical science that Eddington saw being done by Schuster, Dyson, and (again) Thompson. This relevance of the same values to both physics and Quakerism drew Eddington to make those values the foundational elements of his identity. He was a good Quaker because he saw mystical experience as the root of religion, and he was a good astronomer because he cultivated a mystical pragmatism. Internationalism and pacifism were as critical to scientific well-being as they were to spiritual well-being. To Eddington, what let him progress spiritually would allow him to progress scientifically. He was truly a religious scientist, with valence values forming a deep harmony between those two descriptors. The question then becomes, What difference did this make? How did Eddington, as a professional astronomer, make different decisions and choices than his non-Quaker colleagues? Did it matter that he saw an alliance between physical science and the Religious Society of Friends? The answer is yes, and the following chapters will investigate the concrete effect of this alliance in Eddington's career and work.

Mysticism
Seeking and Stellar Models

How difficult to convey the scientific spirit of seeking which fulfills itself in this torturous course of progress towards truth! You will understand the true spirit neither of science nor of religion unless seeking is placed in the forefront.

A. S. EDDINGTON, *Science and the Unseen World*

The previous chapter established that Eddington's outlook was dramatically shaped by the values and aims of the 1895 Manchester Conference, in which a group of Quakers set out a vision of their religion as being compatible and harmonious with modern life. But what difference did this make in his actual work as an astronomer? Did this affect his scientific practice or the technical content of his science? His distinctly Quaker valence values were essential to Eddington's most enduring work, his development of models of stellar structure.

Understanding Eddington's methodology is the key to understanding how spiritual matters pervaded his scientific practice. This methodology was based in contemporary Quaker emphasis on mysticism as the root of true religion. A consequence of depending wholly on mysticism and religious experience was the rejection of a need for dogmatic certainty in favor of a spirit of continual seeking for spiritual truth. This theme is found not just in Eddington's religious and philosophical writings, but also in his astrophysical work, particularly his stellar models. These religious beliefs and values provided conceptual resources for constructing a view of science as an open-ended enterprise in which theories were valued not for their certainty or finality but rather by their ability to allow further scientific investigation.

This pragmatic, exploratory philosophy was fundamental to the creation and use of Eddington's models of stellar structure, which were remarkably successful in their ability to investigate and explain astronomical phenomena where the underlying physics was unknown or inaccessible. His willingness to move forward despite lacking certain foundations was tied closely to his philosophy of science, which in turn presented a unified vision of scientific and spiritual methodology.

Religious Roots

By the time Eddington began his work on stellar structure, he had formed a day-to-day philosophy of science that was based on his valence values. The religious mysticism he absorbed from Rufus Jones fused with the pragmatism he learned at Arthur Schuster's lab bench to form the valence value of *seeking*. This value was the foundation on which Eddington built his astrophysics, particularly his investigations into the then-intractable problems of stellar interiors. His success in addressing those problems was based on his willingness and ability to forgo scientific certainty in favor of opportunities for further progress. Developing his theory of stellar structure was an exercise in contesting the local boundaries of scientific validity. It was a concrete manifestation of the Quaker values he carried with him. Just as the Quakers argued that the import of the spiritual life was not in dogma or final truth, Eddington was comfortable with a scientific method that functioned without certainty.

This seeking was a fundamental aspect of the Quaker faith and involves the idea of an experiential, open-ended quest for knowledge. One of the Queries of the Faith, to be read occasionally in Quaker meetings, reads: "Are you loyal to the truth and do you keep your mind open to new light, from whatever quarter it may arise?" This was understood in Eddington's time to defend knowledge, of God or otherwise, as continually capable of expansion; in theological terms, this is known as continuing revelation.[1] The Friends felt dogma needed to be avoided as the most dangerous impediment to the spiritual quest. One of Eddington's favorite passages in the General Advices to the Society of Friends, first laid down in 1656 and revised annually, says: "These things we do not lay upon you as a rule or form to walk by; but that all with a measure of the light, which is pure and holy, may be guided; and so in the light walking and abiding, these things may be fulfilled in the Spirit, not in the letter; for the letter killeth, but the Spirit giveth life."[2] The "letter" was that which represented unchanging truth, never to be altered or questioned. One of the most important elements of the Quaker identity was that they rejected the comfort of certainty for a life of seeking truth.

Eddington's most important influences in understanding this part of Quakerism were Silvanus P. Thompson and the other leaders of the 1895 Manchester Conference, which dealt with the pressing issues of being a Quaker in the modern world. Thompson argued that scientific investigation was a "sacred duty" that came from the same spirit that motivated George Fox. Rigid dogma was poison to both science and religion, as it crushed the wonder and mystery that drove all seeking.[3] In his later Swarthmore Lecture, Thompson linked the scientific method to mystical pursuit and even everyday life: "What the scientific enquirer does is to hazard a number of hypotheses, some of them probable, others quite improbable, and to test them one by one to see which is right, or which of them is nearest to the truth. For every hypothesis that turns out to be correct he may have to frame a score that prove invalid. Persons who do not understand this mode of arriving at truth often regard this as a very shaky procedure, and condemn it as trying to arrive at truth by means of error. . . . It is better to make wrong guesses than to be in such a muddled state of mind as to be unable to guess at all."[4] The greatest danger was relying on unchanging truth: "The craving for certitude is not in all respects a sign of spiritual health."[5] Eddington met Thompson several times as a student and young scientist and diligently read all of his writings. Thompson's influence is quite clear in Eddington's efforts to forge a Quaker identity that integrated the modern world into the mystical world.

Drawing on Thompson and other Quaker leaders, Eddington saw seeking as the key to the Manchester Conference's vision of a fusion of the scientific and religious identity. In *Science and the Unseen World*, Eddington wrote, after relating the likely path of stellar and human evolution: "So brief a summary cannot convey the true spirit and intention of this scientific probing of the past, any more than the spirit of history is conveyed by a table of dates. We seek the truth; but if some voice told us that a few years more would see the end of our journey, that the clouds of uncertainty would be dispersed, and that we should perceive the whole truth about the physical universe, the tidings would be by no means joyful. In science as in religion the truth shines ahead as a beacon showing us the path; we do not ask to attain it; it is better far that we be permitted to seek."[6] Speaking from his Quaker perspective, Eddington argued that the key to understanding the relationship between science and religion was their respective dealings with experience.[7] The new activist Quakers insisted especially on the individual religious experience as the basis of their faith; similarly, science based its validity on particular experiences of the physical world. Here, mysticism and experiment met as complementary sides of human experience. "As truly as the mystic, the scientist is following a light; and it is not a false or an inferior light. Moreover the answers given by science have a singular perfection, prized the more because of the long record of toil and achievement behind them."[8] Not all science was equally in balance with this mystical view, however. The same attitude that allowed for religious enlightenment needed to be carried into the scientific process. Physics was held to be an excellent example of such seeking science, but "On the other side of [the physicist] stands an even superior being—the pure mathematician—who has no high opinion of the methods of deduction used in physics, and does not hide his disapproval of the laxity of what is accepted as proof in physical science. And yet somehow knowledge grows in all these branches [of science]. Wherever a way opens we are impelled to seek by the only methods that can be devised for that particular opening, not over-rating the security of our finding, but conscious that in this activity of mind we are obeying the light that is in our nature."[9] He experienced a deep cooperation between his roles as scientist and mystic. The imagery of following a light was not accidental, the key was truly to *look for and pursue truth* continually. This was his deeper level of observation that encompassed mysticism and in a sense included the scientific process in the mystical quest.[10]

Unlike the natural theologians of earlier generations, neither Eddington nor his modernist teachers pointed to the material world to establish harmony between science and religion. Instead, Eddington found a philosophical bridge to span the disciplines, where the religious scientist was reassured not by the fingerprint of God in the world but rather by an outlook that brought all forms of true seeking together.

This combined search between science and religion "points the way by which the conception of man as an element in a moral and spiritual order can be dovetailed into the conception of man as the plaything of the forces of the material world."[11] Both were partners in seeking. Understanding seeking in this way had distinct effects on the way Eddington thought about the meaning, process, and value of science. The justification of science was largely internal to the scientist and came from the "striving in man's nature."[12] This striving linked science, religion, and aesthetics: "The desire for truth so prominent in the quest of science, a reaching out of the spirit from its isolation to something beyond, a response to beauty in nature and art, an Inner Light of conviction and guidance—are these as much a part of our being as our sensitivity to sense-impressions?"[13] This attitude of continual seeking meant that the products of science should not be judged on their merits as final knowledge but rather by how well they could support further seeking. For example, Eddington spoke of Hubble's law not as a truth to be evaluated but a tool that provided future exploration.[14] "The spirit of seeking which animates us refuses to regard any kind of creed as its goal";[15] instead, "it is actually an aid in the search for knowledge to understand the nature of the knowledge which we seek."[16] These values underlay the particular talents that he brought to his astrophysics. He thought of his science as a continuing search for truth and was therefore always willing to take unprecedented and sometimes unwarranted leaps so long as they moved his science forward. Like religious insight, theoretical advances were to function as the beginning of a process, not a conclusion.[17] The spirit of seeking was the foundation on which both his scientific and religious identity rested. Eddington's sentiment was laid out simply and elegantly: "We are explorers."[18]

Stars and Spirit

The best illustration of this philosophy is found in Eddington's seminal work in astrophysics. Despite the high profile of his work in relativity, his reputation as a scientist was achieved through his work on stellar structure.[19] His pioneering efforts in creating meaningful models of the interior of stars were not only crucial to the growth of twentieth-century astrophysics, but they provide an excellent example of the way his values had specific consequences in his science. Additionally, the famous controversies between Eddington and James Jeans during the interwar period regarding the subject provide a valuable opportunity to see Eddington's methods thrown into sharp relief as they were defended against opposing views. In order to understand the significance and uniqueness of Eddington's approach, we will need to briefly examine the contemporary state of stellar physics.

It was in the nineteenth century that astronomers and physicists first began to mark successes in investigating the physical nature of stars, an aspect of the physical world notoriously declared unknowable by Auguste Comte.[20] Spectroscopy was extended first to the sun and then to more distant stars, providing the first evidence that they were composed of materials identical to those found on earth. The presence of iron, sodium, and carbon was quickly confirmed by observers such as William Huggins and Norman Lockyer, and the results were incorporated into solar theories developed by Hermann von Helmholtz and William Thomson, among others. They were primarily concerned with solar energy sources, and Helmholtz's contraction theory inspired the American J. Homer Lane to investigate the equilibrium conditions at which the sun would no longer contract. He developed a highly abstract model assuming a liquid or dense gas sphere, where energy was transported by convection. His work was developed into a more significant form by the Swiss physicist Robert Emden, whose investigation focused on polytropic gas spheres; that is, gas spheres held up by internal pressure that varies throughout the interior of the sphere. He investigated a wide range of stable polytropes and looked for solutions that resembled the sun in terms of mass and radius. The method was accepted fairly quickly in the astronomical community but proved to be of limited practical use—the applicability of contemporary thermodynamics to solar interiors was unclear at best. Exactly what stars were and how they behaved remained poorly understood; Eddington, writing the *Encyclopaedia Britannica* entry for "Star" around 1910, admitted that stars "might be solid, liquid, or a not too rare gas."[21] The physics of the turn of the century was simply inadequate to create a meaningful model of stellar structure.[22]

FIGURE 2.1
James Jeans. Courtesy of Mount Wilson Observatory and the Huntington Library. COPC 2957.

This belief in the inadequacy of physics to explain stars was defended by James Jeans (see fig. 2.1), a dominant figure in mathematical physics of the time (and, like Eddington, a Wrangler and winner of the Smith's Prize). Jeans had been educated as a youth at the Merchant Taylors' School, which, as historian Andrew Warwick points out, had an excellent reputation for training students in mathematics.[23] His work there earned him a major entrance scholarship to Trinity College, Cambridge, in 1895. He was coached in mathematics by R. Herman and R. R. Webb, who encouraged him to sit for the Tripos after just two years.[24] He won the Smith's Prize in 1901, and his impressive skill in mathematics led to teaching positions at both Cambridge and Princeton. In 1912 he retired from teaching to live on his wife's considerable fortune, and he was able to devote his complete energies to mathematics and physics. Jeans's ability was remarkable by any accounting: by age thirty he had published three comprehensive books on mathematical physics (*Dynamical Theory of Gases*, 1904; *Theoretical Mechanics*, 1906; and *Mathematical Theory of Electricity and Magnetism*, 1908), each of which was in widespread use for many years.

Jeans had applied his formidable analytic skills to problems in radiation, mechanics, and astronomy, and although he was interested in astrophysics, he thought it was a severely limited field. In a 1909 review of Emden's textbook *Gaskugeln* for the *Astrophysical Journal*, he laid out the difficulties any investigation into stellar structure would face:

> The mathematician of today who wishes to devote his skill to the service of astronomy finds himself in a particularly difficult position. Problems in which mathematical analysis can start from the basis of assured physical facts . . . form a class of strictly limited extent. The class comprises, roughly speaking, those problems whose physical basis is the law of gravitation alone. . . . Dr. Emden's book must not be judged by its agreement or disagreement with the known facts of astronomy; his main service to us, I think . . . is that he has given us a most valuable summary of all that is known about the abstract theory of the configurations of a sphere of gas acted on by gravity and gas pressure alone. The gas has to be supposed to be acted on only by agencies which are fully understood, for exact analysis can obtain no foothold where unknown or partially unknown agencies are postulated. The author reminds us of Green's words: "I have no faith in speculations of these kinds unless they can be reduced to exact analysis."[25]

Jeans's perspective was from the British tradition of celestial mechanics, which held to a mathematical model of truth in which valid knowledge came from rigorous deductions based on certain (that is, not in doubt) premises. The recent heroes of this tradition were Henri Poincaré and George Darwin (to whom a young Eddington had often been compared), who were justifiably honored for advancing the theory of bodies moving under dynamical laws. Jeans's book *Problems of Cosmogony and Stellar Dynamics* laid out the essentials of the method: "The main object of the essay is to build a framework of absolute mathematical truth; the backbone of the structure is the theoretical investigation into the behavior of rotating masses. . . . When a firm theoretical framework had been constructed, it seemed permissible and proper to try to fit the facts of observational astronomy into their places."[26] Jeans's own work in astrophysics, mostly dealing with the stability and behavior of rotating fluid masses, was solidly within this group, which continued to wield great influence in British astronomy.

Eddington came to polytropic theory by a winding path through cosmology, star streaming, and stellar evolution. The details are interesting but not crucial for this discussion.[27] Essentially, during a discussion with Henry Norris

Russell regarding stellar evolution, he realized that a theoretical model of pulsating stars, known as Cepheids, would be useful. Astronomers were ignorant of how such variable stars might function or how they might be related to ordinary stable stars. Attempts to model variable behavior with binary stars, the only known phenomenon that seemed likely, were shown to be inadequate. The only existing stellar model (Lane-Emden) could barely describe normal stars. It was not an obvious candidate for further expansion: there were neither new physics nor new observations that could be used to develop it further.

When Eddington began his work on stellar structure, he rarely followed the accepted mathematical practice of defending his assumptions and only then proceeding. Instead, he skillfully and rapidly moved beyond what he could prove and simply attempted to advance the theory. The uncertainty of his foundations was justified at the end of his work when he would demonstrate that his theory was insensitive to variations in the basic parameters.[28] In a dramatic departure from the tradition of celestial mechanics, the value of the theory was to be found in its ability to provide understanding and enable further investigation, not in its deductive relationship to established facts.

Eddington's first presentation of his work on stellar structure was at the December 8, 1916, meeting of the Royal Astronomical Society, where he read a paper entitled "The Radiative Equilibrium of the Sun and Stars."[29] In this paper he extended Karl Schwarzschild's earlier work on radiation pressure in the solar atmosphere and extended it to interior conditions, specifically inside giant stars. Eddington said he was inspired in this line of work by the discovery that some variable stars (the only stars for which density could be calculated) had low enough densities to make the use of the perfect gas law reasonable.[30] He warned: "It is scarcely necessary to say that the conclusions here given are tentative, being based on analysis which is only concerned with obtaining a probable approximation; but there seems to be a satisfactory accordance with observation, so far as is known."[31] In what would become a familiar pattern, Eddington argued for the validity of such tentative results based on accordance, or even a possible accordance, with observation. The key was to find foundations of general applicability and then invoke special assumptions about unknown or unknowable quantities. These special assumptions were justified technically by the insensitivity of the results to variations in the assumptions and were justified methodologically by their usefulness in opening a problem to investigation:

> It is clear that we cannot arrive at much certainty with regard to the conditions in a star's interior, except so far as the treatment can be based on the most general laws of nature. There are some physical laws so fundamental that

we need not hesitate to apply them even to the most extreme conditions; for instance, the density of radiation varies as the fourth power of the temperature, the emissive and absorbing powers of a substance are equal, the pressure of a given gas varies with its temperature, the radiation-pressure is determined by the conservation of momentum—these provide a solid foundation for discussion. The weak link in the present investigation is that I have assumed without much justification that a certain product $\kappa\epsilon$ is constant throughout a star. I have given some evidence that if it is variable the general character of the results would not be greatly altered; and, as a step toward the elucidation of the problem of stellar temperatures, I plead to be allowed provisionally one rather artificial assumption.[32]

When calculating likely values of κ, he came up with a "quite absurd" value and rejected it. He then selected a lower value, which gave him too high a temperature. After bracketing the likely values, he discussed the assumptions of the calculations, amongst which were the stellar boundary conditions (justified by showing variation in central temperature would not largely affect κ), low variation of ϵ (only small changes from different cases), neglect of scattering (it could be accounted for in the absorption calculations), and finally the radiative hypothesis for energy transfer, which was implicitly justified by the success of the calculations.[33]

When discussing the opacity, he admitted the lack of either firm experimental or theoretical foundation for the analysis: "We have found that κ has a value of about 30.... Unfortunately, there seems no other evidence to indicate whether this value is at all near the truth.... It happens that the experimental values of κ for hard X-rays absorbed by solid material are usually about equal to the value here found. In the absence of a theory to guide us, we cannot infer that the same values would hold at a temperature of 10^6 degrees and for considerably longer waves. The agreement is probably accidental."[34]

The great benefit of this paper was the new accuracy and confidence with which conditions inside a star could be modeled, particularly temperature. In the audience, H. H. Turner, Frank Dyson, and H. F. Newall were all excited by the new investigation. Newall said that "the opportunity which the giant stars gave of treating certain assumptions has led Prof. Eddington to a daring conception of the inside of a star that gives one far more faith than any arrived at before." However, many present at the meeting advocated caution, including H. H. Turner, who questioned Eddington's assumptions of certain uniformities among stars (temperature and mass, respectively). Ludwik Silberstein doubted whether it was safe to apply Maxwell's laws for radiation pressure to an extended mass of gas, which was a situation not conceived of in the traditional derivation.[35]

Jeans, in the first of many commentaries on Eddington's work, suggested that atoms in the stellar interior would likely be dissociated, bringing the mean molecular mass to much lower levels. This suggestion, which Eddington later attributed to F. A Lindemann as well, was taken into the investigation in a note in the *Monthly Notices of the Royal Astronomical Society*. Eddington commented that the resulting changes were good, "but since the theory is necessarily only a rough approximation, it is not desirable to lay too much stress on the removal of small discrepancies. A more important confirmation is that the theory, when thus modified by introducing ionisation, provides an explanation of why stellar masses are of the order of magnitude actually found."[36] Thus, the benefit of the addition was not the increasing precision of calculation, but its ability to link the calculation with observation and physical meaning. Eddington was always concerned to move a problem into this domain, where it could be extended into further investigation.

The importance of dissociation drew Eddington's attention to the problem of energy production, as significant energy needed to be present throughout the star to ensure dissociation. He introduced a dimensionless coefficient to represent differing, and unknown, rates of energy production throughout the star. Characteristically, he set it to unity and argued that it would not deviate significantly or, even if it did, that only small changes would result.[37] It was in this same note that Eddington first proposed the possibility that this stellar energy was produced by the annihilation of protons and electrons.[38]

A similar calculation was applied to κ. He assumed it was constant throughout the star, which let him derive a formula that showed the proportion of radiation pressure to gravity for any giant star. "In default of information as to how κ varies with the temperature, I make this assumption of the constancy of κ in what follows." He then looked at how much variation in κ one could expect before it showed up observationally (the limit to which was the observation that giant stars do not differ by more than 2 magnitudes): "The observations, therefore, seem to show that κ cannot vary very much."[39]

As a capstone, Eddington calculated the temperatures of giant stars as a group. He warned his readers that his goal was to be more utilitarian and exploratory than rigorous: "It is true that in this (as in some other places) I am pressing the theory further than it could be expected to go. But it may be interesting and suggestive to follow out fully the behavior of a theoretical model; and since a comparison with observation is often possible, we may seek to find out whether the model is defective.[40] Eddington's models were never supposed to be complete projects. In an important sense, they were intended to be found wanting: they were to be tested against observation until they failed, and that failure provided the key for improving the models.

His stellar models were originally developed to shed light on the puzzle of variable stars, and specifically the suggestion that they were pulsating stars. A few months after his first paper on radiative equilibrium, Eddington presented "Pulsation Theory of Cepheid Variables," a piece specifically intended to apply his new equilibrium conditions to the possibility of a pulsing star. Eddington's new work on stellar structure made the existence of such stellar pulsations physically plausible; he constructed a model in which the pulsing star behaved like an adiabatic engine, thus opening it to attack via the well-developed tools of thermodynamics. The mysterious opacity variable was postulated to function like a governor, increasing and decreasing the radiation pressure as necessary to allow steady oscillations. Many critics of the pulsation hypothesis, Harlow Shapley among them, argued that regardless of whether the vibrations could be *maintained*, they could never have been *initiated*. Eddington accepted this as true: there was no way to know how the pulsations began. But this was not important to his task, which was to explore the observed behavior of an apparently pulsating star, not demonstrate with certainty that such entities exist: "Though we cannot offer any adequate theory as to *how* the star manages to behave as an engine, we can point out some evidence that it *does* so behave. I am not sure whether the following mode of regarding the question is strictly allowable; but I venture to put forward the suggestion tentatively."[41] The suggestion was to investigate the decay or reinforcement of stellar vibration with temperature and density changes via analogy with sound waves. Clearly, he had no way to know whether waves of this sort existed in stellar matter or the strict conditions of their behavior if they were present. He had instead seized upon an idea that provided an avenue by which to proceed with the investigation, even if it had no foundation beyond its simple utility and physical sensibility.

The first formal reply to Eddington's papers on stellar interiors did not warm to such pragmatic thoughts. At the November 1917 meeting of the Royal Astronomical Society, James Jeans launched a multifront attack on Eddington's work of the past several months. The particulars of the attack varied widely but were always based on accusations of a lack of mathematical rigor and physical precision and an unacceptable tendency toward unfounded generalization and speculation. Jeans's presentation explicitly replied to Eddington: "He obtained some results which seemed to me to be rather astounding, and I thought I would examine whether the mathematics is altogether above suspicion."[42] It is telling that the first in this series of papers was an attempt to establish the fundamentals of the subject, radiative transfer of energy. His goal was to rigorously rederive the basic physics that underpinned the entire discussion. "An accurate determination of the laws of radiative transfer is a necessary preliminary to many problems of stellar physics."[43] His implicit rebuke of Eddington was that his investigations made no attempt at determining exactly what could be known about the basic physics. Eddington's willingness to use any tool that would advance the problem was attacked as unforgivable sloppiness.

According to Jeans, this sloppiness appeared not only in mathematical laxity, but also in baseless physical speculation. In his second paper, Jeans demonstrated his view of the correct way to build stellar models, using his new and accurate equations of radiative transfer. He criticized both Eddington and H. N. Russell for not being properly careful in declaring universal, as opposed to local, physical laws. Russell's evolutionary model was flawed because it could not be known what characteristics of stars were particular to our neighborhood of space and which universal; Eddington's stellar model was flawed because it postulated laws that were general throughout the interior of stars, of which we had literally no knowledge: "It is surely obvious without discussion that there can be no perfectly general laws of the type enunciated by Professor Eddington. . . . The rate of emission of energy, being also the rate of generation of energy in the star's interior, must depend on the ultimate source of this energy. . . . Hence a preliminary to any attack on the general problem must be a decision as to the source of the energy."[44] Jeans, quite prudently, argued that known energy sources (chemical reactions, radioactivity) apparently were not significant, so the best, and truly the only, approximation one could make is that no energy other than gravitational contraction was available. Eddington's unfounded suggestion of an unknown process (annihilation) as an energy source was improper, but his use of it in his theoretical exploration was a violation of the deductive method, whereby conclusions were guaranteed by the soundness of the premises.

Jeans's rebuttal began a series of debates that have become legendary in the astronomical community. The halls of the RAS were well known for a being a site of rough-and-tumble intellectual discourse, but Eddington and Jeans elevated that to an art form. Both men possessed great rhetorical skills and deployed them to the fullest to promote their ideas and crush those of their opponent. Verbal barbs accompanied mathematical rejoinders, and both men excelled at one-upsmanship. Eddington proved especially prickly and competitive: once, upon hearing that Jeans would be speaking on the topic of radiative viscosity, he quickly calculated what Jeans's result must be and gleefully sent it to his rival on a postcard before the meeting.[45] Their debates were so lively that G. H. Hardy and a number of others joined the RAS for the sole purpose of watching Eddington and Jeans have at each other.[46]

Eddington was irritated by Jeans's initial attack and felt that Jeans was putting unfair emphasis on the "universal" qualities of his theory. He thought he had been quite careful to make the limits of his model clear and complained to H. N. Russell that Jeans was misrepresenting Eddington's work in public forums such as *Nature*.[47] Jeans claimed that because Eddington's work was based on unfounded assumptions, it was nothing but a tautological derivation of his original premises.[48] Eddington replied that he made no special claims to universality; he was making a pioneering investigation into a little-known area. His theory was designed to be robust regardless of the particulars: "In the one place where it was necessary to consider the source of stellar energy, I attempted to show that my formula fairly represented both the radio-activity and the contraction hypotheses—having regard to the necessarily approximate character of the investigation."[49] Despite Jeans's accusations, Eddington's work was not merely qualitative speculation about the nature of the stars. His approximations dealt with unknown quantities, but he was finding consistent, mathematical results.

Jeans continued his attack into later RAS meetings, where he argued that Eddington's sound-wave analogy in variable stars was useless, because it could not possibly be compatible with conservation of energy. Eddington replied, defending his techniques: "I am not surprised that Mr. Jeans was not convinced. I was not myself. But it seemed of interest to find some sort of analogy between what happens in a sound wave and in a Cepheid variable. . . . Whether it is possible for this to occur by dynamical means I do not know. I have made no suggestion as to how it can be done, although I have considered it a lot, but the mere fact that the vibrations are in the right phase to go on without dissipation seemed worth while putting on record."[50]

Many of the exchanges between Eddington and Jeans were based on misunderstandings of the other's methodology. Jeans's reaction to Eddington's papers was irritation at an apparently unfinished model, but he was unable to see that this was Eddington's *intent*. Eddington was in no way claiming that his theoretical developments were certain fact; rather, they were investigations into promising routes, first attempts to build a beachhead on an unknown shore. His attitude made it perfectly reasonable to take shortcuts and bypass unknowns with intuitive speculation. Eddington's first presentation of an apparently simple relationship between the mass and luminosity of a star ($L \propto KM$, where K was a constant) was an excellent example of the misunderstandings between the two scientists. Jeans complained that the relation gave us no new understanding because the mass-luminosity proportionality constant K was not given exactly, and Eddington replied that he had simply been trying to show the *existence* of the proportionality.[51] His argument was not one of mathematical certainty in the ability to calculate luminosity but was a demonstration that the theory could be connected to the observed facts of stellar populations, even if that connection was tenuous.

At the next RAS meeting, Jeans further critiqued the assumptions that allowed the postulate of a mass-luminosity relation. Luminosity, Jeans said, was really total energy, so Eddington's claim was that the total energy output of a star was proportional to its mass. Jeans insisted that any energy considerations begin with the basic laws of conservation of energy, and therefore the total outflow of energy could not be proportional to anything but the total energy developed in the star. Relentlessly, he accused Eddington of taking mathematical shortcuts to bring his calculations in line with observation quickly.[52]

"Shortcuts" was perhaps too harsh, but Eddington freely admitted that he guided his investigations so that they would meet with observation as quickly as possible, even if this meant bypassing exact formulations. In 1919 the *Monthly Notices of the Royal Astronomical Society* published a pair of his papers tackling the problem of stellar pulsations. He set out his methodology on the first page: "A complete solution of this problem would be very difficult; but it seems possible to determine the general character of the oscillation, and to obtain results which may be compared with observation."[53] The mathematics of his investigation quickly ran aground when he arrived at an equation for the oscillations that could not be successfully treated mathematically. He got around the difficulty by assuming the vibrations were adiabatic, which simplified the equation enough to be solved. Instead of arguing for the applicability of this approximation, he used it freely until it was clearly invalid (i.e., it returned physically inadmissible results). "Our adiabatic approximation will be found to break down near the boundary, so that we cannot formulate the exact conditions. But this scarcely matters."[54] The inexactitude of the approach was not a problem, because its purpose was only to hold long enough to use trial and error to get a result that could be compared with observations.

An important factor in Eddington's derivation of the period of stellar oscillation was the ratio of specific heats, which could not be known without more information on the specific composition of the stellar interior. Eddington escaped this trap by assuming his theory was, for the moment, correct. He then used the known period of Delta Cephei (the second Cepheid variable to be discovered) and calculated the hypothetical value of its ratio of specific heats.[55] This value, of course, was perched on a rather precarious structure of assumptions and guesses. But *it worked*, allowing Eddington to show that the observed periods of most Cepheid variables could all be explained by the pulsatory theory. Further, the theory allowed better calculations for interior characteristics such as temperature. Eddington made no claims for final truth, however: "In this calculation we have pressed the theory to an extreme degree. Our object is not so much to assert the truth of the conclusions, as to use every opportunity of discovering by comparison with observation the directions in which our approximate treatment may be improved."[56] This "pressing" of the theory was essential to Eddington's technique. He described the first two iterations of his model as pushing against different limitations on astronomical knowledge: "In the new theory I purposely went to the other extreme, so it is likely that the truth is between the new results and the old."[57] Going to the extremes let him find the points of overlap between an embryonic theory and established observations.

This methodology allowed Eddington to take a crucial first step toward verifying the theory that variables were oscillating stars: showing that predictions of the theory could be consistent with observation. The second part of his paper dealt with an important objection to the theory. Many opponents of the hypothesis argued that any such oscillations would be damped fairly quickly and that we should therefore observe a significant slowing of the star's vibration. To defend against this, Eddington again used his classical heat engine model of the star to begin investigating the nature of such damping.[58] His calculations were successful in showing that any dissipation of energy in the vibration would take place over thousands of years, far too slowly to be seen. He warned, however, that this investigation was designed only to show the possibility of further investigation, not establish certainty: "It should be understood that we have calculated the dissipation due to one particular cause, and we have not considered possible counterbalancing causes, such as work done by radiation pressure. The calculation is not intended as a determination of the actual life of the variation; it deals merely with one formidable objection to the pulsatory theory and shows that the objection is less serious than it appeared at first sight."[59]

When Eddington presented these papers at a meeting of the RAS, Jeans was quick to attack. He said that the situation simply could not be resolved with any dynamical theory; the necessary facts were not available and the problem was thus insoluble. Eddington agreed that relying on dynamical methods led to obviously incorrect conclusions (such as stellar lifetimes that were much too short), but argued that this showed the necessity of considering phenomena outside dynamics rather than making the problem inaccessible.[60] He said that the effects of new energy sources were clear, so we should be open to considering the implications of such new and surprising phenomena, even if their characteristics are hidden.

Jeans held to the position that dynamics was the only reliable knowledge that applied to stars and that astronomers should not try to push beyond it.

His alternative theory of Cepheids reflected this. His hypothesis was that the variations in brightness came from explosions on the surface of a rotating body. He arrived at this theory by fitting a function to the observed Cepheid curve and then matching that function to a known dynamical process: a rotation of a single elongated body with an embedded periodic phenomena at a certain orientation.[61] Here Jeans too moved quickly to bring his results in accord with observation, but it was telling that he chose an explanation (perhaps the only explanation) that could be described exactly with the accepted laws of celestial dynamics and required none of the intuitive leaps made by Eddington.

In the spring of 1919, Jeans presented work further attacking Eddington's models. This was a full-scale operation that sought both to demolish the assumptions on which Eddington built his work and to show the correct deductive method of analysis. A central point was that the deviations of the ratio of specific heats in varying environments made it impossible to establish any predictions regarding the behavior of radiation: "For these reasons Professor Eddington's analysis cannot be said to prove anything definite as to the emission of radiation. This agrees with the obvious common-sense view of the question: something more than the mass must be known about a star before its emission of radiation can be determined." The something more was the pressure, density, and temperature at all points in the star's interior. However, "Before these three quantities can be determined, a third equation is needed, and this equation is necessarily very complicated, depending upon the whole past history of the star as well as upon its mechanism of generation of energy. It is to avoid this intricate equation that Professor Eddington introduces his assumed relation $H/g =$ constant."[62] Jeans was correct: a mathematically rigorous treatment would require such an equation. Such an equation was also completely impossible to solve, since virtually none of the information needed (the energy mechanism and the "whole past history of the star") was available. What was remarkable in Eddington's work was that he was able to produce so many results with such simple, if formally unjustified, assumptions. Jeans derived a new equation for radiation pressure that was to fulfill his criteria: "This equation, it will be noticed, is perfectly general, and involves no simplifying assumptions or approximations at all."[63]

The simplifying approximation with which Jeans was chiefly upset was Eddington's claim that the stellar energy source must be effectively constant with respect to time, an idea that Jeans called a "somewhat enigmatic assumption." He found it "very difficult to imagine any source of energy that would not vary with the time, so that the analysis points to the conclusion that the energy must be purely gravitational." Jeans again refused to admit the evidence for new and unknown energy sources and championed gravitational contraction as the only mathematically sound source of energy. Lindemann concurred: "I agree that there does not seem to be any alternative source." Eddington, absent at Principe Island to test Einstein's theory of gravity, could not respond.[64]

The next few years saw a lull in the dispute between Eddington and Jeans, as each was occupied by other projects, Eddington chiefly with relativity. In this period both parties put forward some minor papers on stellar physics, but the next illuminating controversy came in 1924, when Eddington presented his more developed theoretical work on the mass-luminosity relation. Eddington came across the relationship while working on giant stars, where his assumption of perfect gas law behavior seemed to have more merit. He derived a formula expressing the luminosity of the star dependent only on its mass and one independent constant, which he determined by calibration with Paul Merrill's recent excellent observations of the star Capella.[65] This formula yielded a smooth curve of luminosity versus mass (see fig. 2.2), on which Eddington plotted the observed values of the giant stars available. To his surprise, the resulting curve fit not just the giant stars, but *all* the stars, including dwarf stars, for which both mass and absolute luminosity were known. Excited by the implications of this unexpected accord between his theory and observed stellar characteristics, he presented his paper at the RAS in March.

In his paper Eddington warned that he was unable to find an absolute form of the relationship between luminosity and mass, but he was able to conclude that "the form of the law seems to be fixed within narrow limits." The theory appeared to predict correctly the magnitudes of all thirty-six stars available, and he noted that "it would be surprising if the accordance . . . arose from mere accident."[66] His results, while largely derived from theory, integrated observed data, and Eddington was careful to distinguish between "first-rate" and less reliable observations (and how that affected his argument).

The paper was explicit in discussing the approximations on which the results rested, particularly the use of the perfect gas law (which, surprisingly, seemed to apply to both giant and dwarf stars). Eddington closed the paper by assessing the "exactness" of the fundamental equation involved. It was built on two assumptions: the first was a constant value for the opacity in the star, which he called "rather crude" but which had support from either Kramer's opacity theory or Eddington's own nuclear capture theory.[67] The second was an assumption that μ, the mean molecular weight in the star, was constant. Unlike the opacity assumption, this had no robust theoretical justification, so Eddington performed a series of calculations in which he varied μ systematically and assessed the effect of that change on the luminosity. Satisfied that any changes would be slight, he let the assumptions stand.[68]

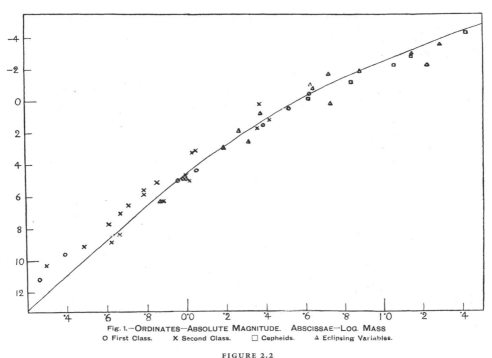

Fig. 1.—Ordinates—Absolute Magnitude. Abscissae—Log. Mass
O First Class.   X Second Class.   □ Cepheids.   ʌ Eclipsing Variables.

FIGURE 2.2

Eddington's 1924 Mass-Luminosity Relation. Courtesy of the Royal Astronomical Society.

Jeans's rebuttal was brutal: "All Eddington's theoretical investigations have been based on assumptions which are outside the laws of physics."[69] He wanted to show how Eddington's reliance on special assumptions led him to see relationships that were, simply, not there. The mass-luminosity relation was an illusion: "When the problem is treated in a sufficiently general way, the rather surprising properties which Professor Eddington attributes to the stars are found to disappear. In particular, it is found that there is no general relation between the masses and luminosities of the stars—the supposed 'mass-luminosity' law disappears entirely as a theoretical law."[70] Again, Jeans demanded that any expression predicting the energy output (luminosity) of a star include a detailed description of the actual rate and manner of the generation of energy in the star.

Jeans said Eddington had neglected to mention a critical assumption that underlay the entire project. He chastised Eddington for assuming that λ (here the ratio of gas pressure to radiation pressure) was constant throughout the star. He argued that basic physics required that it actually vary with the fourth power of the temperature, and thus could vary by as much as a factor of 1,000 inside a star.[71] This error, Jeans said, was the source of Eddington's dramatic graph: "There is no limitation to a single curve or narrow band across this diagram such as is exhibited in Eddington's figure. If the stars are found in actual fact to lie on a narrow band, there must be extraneous reasons for the absence of stars outside this band; they are in no way driven from these exterior regions by the laws of physics. . . . Although our theory adds very little in the way of numerical results to what has already been given by Eddington, it leads to a very different interpretation of these results."[72] His claim was that the only way to do a physical investigation was from the fundamentals upward; Eddington's beautiful curve was only an artifact of his sloppy physics.[73]

Eddington's reply clearly staked out what he saw as the middle ground between stifling mathematical certainty and irresponsible speculation:

> If we had no knowledge of λ [the rate of energy generation], then it is true that the stars might wander all over the diagram. As in most problems in mathematical physics, we have some known data to work with and some that we can only make a fairly likely estimate of, and the only question is whether the uncertainty of the estimate is likely to be such as to make progress impossible. I came to the conclusion that unless the conditions in the star were much more extreme than there is reason to expect, λ would be reasonably constant in different parts of a star. The fact that the actual stars appear to follow the curve closely seems to show that I was right in regarding this approximation as satisfactory, and I do not see why Dr. Jeans thinks that this is accidental and that the clustering of the stars to the curve has some entirely different cause.[74]

Thus, the assumption was acceptable because it allowed progress on the problem, and even better, it revealed a regularity in the stellar population that had previously been unknown. For Eddington, this was defense enough.

Jeans refused to agree, citing Eddington's apparently irresponsible speculations. Further, he claimed that the regularity of the curve was simply Eddington making up his own physics: "I cannot agree with Professor Eddington that λ = constant is a permissible approximation in the present state of our knowledge. . . . But my dissent from Professor Eddington's work is not one of numbers, it is one of mathematical principle. I do not say that his curve is numerically inaccurate, but that it should not be in his diagram at all. One unjustified physical assumption has reduced him from an area to a curved line; two such assumptions would have reduced him to a point; and three such assumptions would have proved that stars cannot exist at all."[75] Unsurprisingly, no accord was reached on the subject.

Soon after, Eddington published his groundbreaking book *The Internal Constitution of the Stars*, in which he systematized and elaborated on the stellar models he had been developing since the end of 1916. In it he sought not only to provide astronomical insight but also to show the benefits of his open-ended, pragmatic approach to astronomical problems. The introduction to the book addressed the criticism of Walther Nernst and Jeans, who maintained that a firm understanding of subatomic energy was necessary for any meaningful investigation; Eddington simply said that we can make do with what we have.[76] At an earlier RAS meeting he had argued that the difficulties of stellar energy did not mean it was unapproachable. Indeed, there were perhaps too many opportunities: "My own experience has been that the astronomical facts are so perverse that, however many constants one introduces, it is difficult to find any reasonable theory of sub-atomic energy that will fit them all. I think it is sometimes thought that in speculating on sub-atomic energy we have very little to go on in the way of fact. But really the facts (or, at any rate, supposed facts) are abundant."[77]

In *Internal Constitution*, Eddington repeatedly set his method apart from that of the mathematician or, as he sometimes said, the pure physicist. The early chapters of the book were devoted to radiation pressure and its function in the stars and discussed opacity at length. He admitted that a physicist, before beginning the problem of opacity, would want to know the simple fact of the material involved. But, said Eddington, "If he asks this, we are done; because we have little, if any, knowledge of the proportionate composition of the material in the star."[78] He bypassed the deductive approach to the problem, which would be to establish the opacity and thermal gradient and therefore determine the radiation outflow, and instead used the radiation outflow (which was observable) and apparent thermal gradient to determine the opacity. He invoked the constancy of $n\kappa$ as "an especially good approximation. . . . [This] is the law which appears most likely according to our limited knowledge and it is appropriate to develop its consequences fully for comparison with astronomical observation. The function ν depends on the relative distribution in different parts of the star of the source of stellar energy—a distribution at present unknown. What makes progress possible is that ν is comparatively insensitive to very great changes in the assumed distribution of the source."[79] Lacking the basic physics to calculate the behavior of the material, he started from observation and determined the likely limits of that behavior. When considering how much quantum mechanics and thermodynamics were necessary to achieve a meaningful answer, he asserted that "although a reasonable degree of rigour is required, the laborious exploration and closing of every loophole is of secondary importance."[80]

The equations governing this radiation flow were the ones to which Jeans attempted to give a mathematically exact treatment, and which he argued required knowledge of all parameters before use. Eddington's defense of his own treatment is lengthy but revealing:

> A certain amount of controversy has occurred with regard to the derivation of this equation which reflects the time-long difference of view between the physicists and the mathematician. Perhaps a short digression on this antagonism may be permitted, for it is likely to give rise to many misunderstandings in problems of the kind we have to consider. I conceive that the chief aim of the physicist in discussing a theoretical problem is to obtain "insight"—to see which of the numerous factors are particularly concerned in any effect and how they work together to give it. For this purpose a legitimate approximation is not just an unavoidable evil; it is a discernment that certain factors—certain complications of the problem—do not contribute appreciably to the result. We satisfy ourselves that they may be left aside; and the mechanism stands out more clearly freed from these irrelevancies. This discernment is only a continuation of a task begun by the physicist before the mathematical premises of the problem could even be stated; for in any natural problem the actual conditions are of extreme complexity and the first step is to select those which have an essential influence on the result—in short, to get hold of the right end of the stick. The correct use of this insight, whether before or after the mathematical problem has been formulated, is a faculty to be cultivated, not a vicious propensity to be hidden from the public eye. Needless to say the physicist must if challenged be prepared to defend the use of his discernment; but unless the defence involves some subtle point of difficulty it may well be left until after the challenge is made.

Eddington here justified his use of "insight" as a tool for overcoming difficulties. Note that his justification was based on the efficacy of such techniques for allowing investigation into the problem, not giving its final conclusion. He went on to describe the difference between his critics and himself as a disciplinary one:

> I suppose the same kind of insight is useful to the mathematician as a tool; but he is careful to efface the tool marks from his finished products—his proofs. He is content with a rigorous but unilluminating demonstration that certain results follow from his premises, and he does not realise that the physicist demands something more than this. For the physicist has always to bear in mind a thousand and one other factors in the natural problem not formulated in the mathematical problem, and it is only by a demonstration which keeps in view the relative importance of the contributing causes that he can see whether he has been justified in neglecting these. As regards rigour, the physicist may well take risks in a mathematical deduction if these are no greater than the risks incurred in the mathematical formulation. As regards accuracy, the retention of absurdly minute terms in a physical equation is as clumsy in his eyes as the use

of an extravagant number of decimal places in arithmetical computation. . . .
We may turn to appreciate the luxury of a rigorous mathematical proof. If
the results obtained do not agree with observation the fault must assuredly
lie with the premises assumed. . . . If space were unlimited we might try to
duplicate investigations where necessary so as to satisfy both parties. But
if one investigation must suffice I do not think we should give way to the
mathematician. Cases could be cited where physicists have been led astray
through inattention to mathematical rigour; but these are rare compared with
the mathematicians' misadventures through lack of physical insight.

The mathematicians' obsession with certain knowledge was a luxury that
was actually harmful in the quest for physical understanding. Their self-
imposed restriction to the goal of a "proof" limits the applicability of their
results. Physicists, in comparison, sought understanding more in accord with
physical reality:

> The point to remember is that when we *prove* a result without understanding
> it—when it drops unforeseen out of a maze of mathematical formulae—we
> have no ground for hoping that it will apply except when the mathematical
> principles are rigorously fulfilled—that is to say, never, unless we happen to
> be dealing with something like aether to which "perfection" can reasonably be
> attributed. But when we obtain by mathematical analysis an *understanding* of
> a result—when we discern which of the conditions are essentially contributing
> to it and which are relatively unimportant—we have obtained knowledge
> adapted to the fluid premises of a natural physical problem.
>
> I think the idea that the purpose of study is to arrive at a string of proofs
> of propositions is a little overdone even in pure mathematics. Our purpose in
> studying the physical world includes much that is not comprised in so narrow
> an ideal. We might indeed say that, whereas for the mathematician insight is
> one of the tools and proof the finished product, for the physicist proof is one
> of the tools and insight the finished product. The tool must not usurp the
> place of the product, even though we fully recognize that disastrous results
> may occur when the tool is badly handled.[81]

After this methodological defense, Eddington discussed the mass-luminosity relation and, in particular, its unexpected result of the applicability of
perfect gas law to dwarf stars. He presented this as an interesting situation in
which his assumption of the perfect gas law turned out to be right, but he
only discovered its wider applicability by looking to progress further on the
problem. No earthly laboratory work or contemporary physical principles
would have suggested that such outrageous stellar densities and levels of
ionization would exist; no one could have or would have given a good reason
why the gas law would apply to any but giant stars. Indeed, H. N. Russell

expressed surprise, even years later, that the equation worked so well: "I wonder more than ever *what* it is that makes the mass-luminosity relation as close as it is."[82]

The assumption of perfect gas conditions applied in a way Eddington had never expected, which gave new insight into unusual stellar conditions. This sort of success could not have been achieved with the mathematical conservatism of Jeans. Rather, it grew out of Eddington's willingness to work on the problem without "proof," with a spirit of exploration instead of reification.[83]

At the end of the book, Eddington included an entire chapter on the subject of stellar energy and, specifically, on the possibilities of subatomic annihilation and the construction of elements as sources of that energy. He speculated widely on the characteristics of such energy sources and the effects they would have on stellar bodies. Eddington justified this, as he called it, astronomer's approach: "I think the pure physicist may be inclined to regard our discussions of the details of subatomic energy as airy speculation; if so, he greatly misunderstands the position of the astronomer. It is not a question of unrestrained conjecture remote from observational facts.... Surely it is permissible to sort [facts] into order and consider what laws and theories they may suggest without being held guilty of vain speculation."[84]

The pure physicists (e.g., Jeans) had been quite vocal in their opinion of Eddington's work on stellar energy. Interestingly, Jeans had himself written on the possibility of alternative energy sources as early as 1904.[85] However, it was merely a suggestion and, unlike Eddington, he made no attempt to draw consequences or new avenues of investigation from that suggestion. Similarly, Henry Norris Russell published an essay on the subject in 1919. He made the modest assumptions of a finite energy supply, a production rate depending on the remaining amount, and an increase in production with temperature. The discussion was very general and could have fit almost any source (e.g., radioactivity). He explicitly demurred on making any claims of an actual mechanism: "The present argument has, however, been purposely kept clear from discussion of the details of any such hypothesis." The only conclusions drawn were that a star has a nucleus, where energy is liberated, and an outer shell. His next suggested step was to find the distribution of density.[86]

Russell probably represented the middle ground of the astronomical community of the time, and Jeans's reluctance to treat the subject as a valid avenue for investigation was indicative of the more conservative, mathematical scientists. Eddington was hardly dissuaded, and *Internal Constitution of the Stars* wanders far and wide among the possibilities of stellar energy. He frequently paused in his mathematical and physical analysis to head off anticipated criticism: "If the astronomical evidence afforded more definite guidance for a formulation of the laws of liberation of subatomic energy, we should still, I suppose, have to submit the resulting theories to the censorship of the mathematical physicist. . . . But [the mathematical physicist's] own position contains difficulties and contradictions and it is doubtful if he is justified in exercising any rigid censorship."[87] Interestingly, Eddington here turned the mathematicians' accusation of lack of evidence back against them, arguing that their "censorship" could only apply to those theories with firm foundations.

Further, the correctness of the theory could be considered to be ancillary. Writing in *Nature*, Eddington argued that, properly understood, subatomic energy was as open to observation as any laboratory experiment:

> I suppose that most physicists will regard the subject of subatomic energy as a field of airy speculation. That is not the way in which it presents itself to an astronomer. . . . The measurement of the output of subatomic energy is one of the commonest astronomical measurements—the measurement of the heat and light of a star. Naturally the astronomer is not content to go on with these measurements indefinitely without an attempt to arrange them into some sort of coherence. . . . If the physicist had in his laboratory unknown sources of energy, the output of which he could measure and the physical conditions of which he could determine, he would not be so backward in speculating on the causes and laws of the phenomena. The astronomical study of subatomic energy is no whit less direct than this . . . the problem is no more speculative than any other induction from experiment.[88]

In *Internal Constitution of the Stars*, Eddington moved on to the possibility of energy production by transmutation of the elements. Recent experimental work on the mass defect, combined with the abundance of helium in stars and its virtual absence anywhere else, had led to the suggestion that hydrogen was somehow being formed into helium, with a consequent release of energy. The problem with the hypothesis was the large Coulomb forces that would need to be overcome before nuclei could meet. Eddington anticipated the objection that stars could not possibly be hot enough to allow transmutation: "But the helium which we handle must have been put together at some time and some place. We do not argue with the critic who urges that the stars are not hot enough for this process; we tell him to go and find a *hotter place*."[89] Eddington's pragmatism argued that the existence of helium was a virtual proof of the possibility of its creation. In the end, however, the transmutation theory received somewhat limited treatment because, among other reasons, it "leads to no interesting astronomical consequences."[90] In contrast, the hypothesis that energy was released by subatomic annihilation suggested important new investigations: "The theory of annihilation of matter is more fertile in astronomical consequences than the other forms of the subatomic theory, and for this reason alone it seems worth while to follow it up in detail. We shall not be greatly concerned with *how* the annihilation is accomplished; but it may perhaps be well to have a scheme in mind."[91]

The book ended with a final reflection on astronomy as a growing, changing science, and its special role in allowing investigations impossible in traditional physics:

> There are two clouds obscuring the theory of the structure and mechanism of the stars. One is the persistent discrepancy in absolute amount between the astronomical opacity and the results of calculations based on either theoretical or experimental physics. The other is the failure of our efforts to reduce the behavior of subatomic energy to anything approaching a consistent scheme. Whether these clouds will be dissipated without a fundamental revision of some of the beliefs and conclusions which we have here regarded as securely established, cannot be foreseen. The history of scientific progress teaches us to keep an open mind. I do not think we need feel greatly concerned as to whether these rude attempts to explore the interior of a star have brought us anything like the final truth. We have learned something of the varied interests involved. We have seen how closely the manifestations of the greatest bodies in the universe are linked to those of the smallest. The partial results already obtained encourage us to think that we are not far from the right track. Especially do we realise that the transcendently high temperature in the interior of a star is not an obstacle to investigation but rather tends more to smooth away difficulties. At terrestrial temperatures matter has complex properties which are likely to prove most difficult to unravel; but it is reasonable to hope that in a not too distant future we shall be competent to understand so simple a thing as a star.[92]

Boosted by the success of *Internal Constitution of the Stars*, Eddington's methods were swiftly adopted by astronomers around the world. And when his pragmatic phenomenology met its American cousin in H. N. Russell's work, theoretical astrophysics began to grow as a robust and promising research program.[93] This is not to say that all of Eddington's ideas were uncritically assimilated. Through the 1930s he became embroiled in astrophysical debates with E. A. Milne every bit as fierce as his battles with Jeans. Nor do I mean to suggest that Eddington and Russell were the only important contributors to theoretical astrophysics; rather, I am arguing that the techniques they pioneered provided the theoretical technologies that were then used by other investigators.

In many ways Eddington remained at the top of the new field of theoretical astrophysics, but as the decade wore on it seemed that he was being increasingly left behind by developments. In an ironic turn, the rigorous deductive methods he had fought against so successfully became integral to the next generation of stellar physics. His infamous controversies with Subrahmanyan Chandrasekhar are an excellent example of this shift. Their disagreements on the behavior of degenerate stars took place while Chandrasekhar was a fellow at Cambridge during 1934–35. I will make no detailed study of this controversy here. As the story is commonly told, Chandrasekhar presented detailed calculations on relativistic degeneracy, which Eddington dogmatically refused to accept because they disagreed with his own theories.[94] Eddington is usually portrayed as unfairly bringing his great authority to bear against an upstart with new ideas.

Eddington was certainly vigorous in intellectual debate, but his treatment of Chandrasekhar appears no different than his battles with Jeans or Milne. Rather than ascribe a vague sense of hostility to Eddington, I believe we can better understand what happened by looking at his astrophysical methodology, as described in this chapter. Eddington saw in Chandrasekhar's work the same sort of rigid faith in deduction that he saw in Jeans's arguments nearly twenty years before. He maintained that the study of stellar physics required an investigator to look for physical understanding, not just slavish mathematical proofs. As with Jeans, he did not take issue with the mathematical correctness of the degeneracy formulas, but rather with whether they were physically meaningful. He wrote that Chandrasekhar's "result seems to me almost a *reductio ad absurdum* of the relativistic formula. . . . I do not think any flaw can be found in the usual mathematical derivation of the formula. But its physical foundation does not inspire confidence, since it is a combination of relativistic mechanics with non-relativistic quantum theory."[95] Eddington simply saw Chandrasekhar making the sort of mistake (thinking that straightforward deduction without concern for physical *meaning* was the best way to approach stars) that he had seen Jeans use to little profit.

Unfortunately for Eddington, the situation in astrophysics had changed. When he was first building his models, the physics needed for a complete description was not yet available, and the only path forward was the sort of phenomenological pragmatism that underlay his mass-luminosity relation. But by 1935 the quantum theory had advanced tremendously and was robust enough to provide useful results through conceptually straightforward (if mathematically difficult) derivations. In his astrophysics, Eddington remained wedded to the methodology that had been productive for him, even as the disciplinary variables (e.g., incomplete knowledge of the physics of opacity) that made it so important changed underneath him. This is not to say that theoretical astrophysics suddenly supplanted Eddington's pragmatic approach with Jeans-style deductivism, but rather that developments in physics allowed for rigorous calculation of certain factors that previously could only be accessed by approximation and inference. Eddington simply did not accept, unlike a growing number of his colleagues persuaded by Chandrasekhar, that theoretical astrophysics could begin to comfortably embrace elements of what he would call the "mathematician's" approach.[96]

Practice Becomes Philosophy

By the time *Internal Constitution of the Stars* was published, Eddington had already become a public icon of science, in large part because of his leadership in the 1919 eclipse expedition (see chapter 3), and he was in increasing demand for interviews, lectures, and essays throughout the 1920s. He responded with expositions and popularizations on a wide variety of science-related topics, emphasizing the philosophical basis and implications of modern physics. Although the highly abstract and symbolic elements of this work (his "selective subjectivism") have attracted the most investigation, the process-oriented pragmatism seen in his astrophysics was strongly present in his popular and semipopular writings.[97] There was a significant divergence between these two branches of his philosophy, but both relied on valence values. The value of seeking is the focus of this section, and the value of experience will be addressed in chapters 5 and 6.

The appearance of Eddington's astrophysical methodology in his philosophical treatments of science illustrates his unusual approach to philosophy. He thought the philosophy of science could come only from the practice of science. There could be no philosophy apart from what scientists actually *did*. He admitted that there was no contemporary philosophical consensus in the physical sciences but claimed that there was a de facto philosophy of science embedded in the practice of physical science. It was "the philosophy to which those who follow the accepted practice of science stand committed by their *practice*. It is implicit in the methods by which they advance science . . . and in the procedures which they accept as giving assurance of truth."[98] Given this sense of philosophy of science, I will move beyond the philosophical work that has been written on Eddington in the past, which has almost always analyzed his philosophy as an entity separate from his science. A meaningful understanding of Eddington's perspective on science must be found in his day-to-day scientific work as well as his explicitly philosophical writings.

One of the most important issues for Eddington in these writings was whether science was a dynamic or static enterprise, a concern that came directly from his seeking values. He argued that science was inherently progressive, in the sense of being both constantly changing and constantly improving.[99] In his lectures he often reminded his listeners that he did not want to sanctify scientific results. Even speaking to the British Association for the Advancement of Science (BAAS), he reminded them,

> There is no finality about the results that we have been considering tonight. If the astronomer a hundred years hence ever turns to read the crude ideas prevailing in the early 1900s, doubtless he will in his fuller knowledge find them almost ludicrous. But I believe he will recognize that we had got hold of some underlying kernel of truth however imperfectly we understood it. . . . Whatever fate may befall the hypotheses we now rear, we are impelled to make some attempt to gather together the scattered results; and perhaps in our faulty expression there will be some truth that will survive. A knowledge of nature is the great end of our work; but, if we cannot attain that, there is at least the *struggle* after knowledge, which is perhaps no less a thing.[100]

Eddington felt that one of the benefits of this attitude was "a tidiness of mind."[101] This "tidiness" resulted from understanding that the current results were only current and might need to be replaced at any time. A scientist must not let his mind be cluttered by outmoded ideas. This limitation applied to the most basic facts. In *The Expanding Universe* he warned that even astronomical observations should not be considered eternal or uninterpreted.

The challenge in this sort of popularization was to convey the current state of physical knowledge without reifying that knowledge. He wanted to be sure to express that "in science we do not expect finality."[102] Even his "selective subjectivism" warned that a theory was only correct until a new observational tool or method appears, and then its meaning must be expected to change. Eddington argued that the twentieth century's revolutions in physics and astronomy showed that it was simply the nature of physical theories that they were continually unfinished.[103] He illustrated this with a quote from *A Midsummer's Night Dream*: "It shall be called Bottom's dream, because it hath no bottom."[104]

Even if physical science fell short of reaching a final, permanent truth, it could still be a tremendous tool for generating knowledge. What was required was a change in perspective on the goal of science, which was commonly understood to be truth about the physical world. But here, instead of truth being an ultimate *thing*, it became a process.[105] This also changed the understanding of what qualified as "good science." If one was not looking for a final answer, then what were the standards? In *New Pathways in Science*, Eddington stated: "We strive to reduce what we have ascertained to an exact formulation, but we do not leave it buried in its formal expression. We are continually drawing it out from its retreat to turn it over in our minds and make use of it for further progress; and it is in this handling of the truth that the rigor of scientific thought especially displays itself."[106] Rigorous science was the constant pushing on the boundaries of knowledge and, in a sense, using that knowledge to push beyond itself. This progressive aspect of science had a definite direction, as well; blind groping was not seen as meaningful. Eddington wanted to see the plurality of truths in science begin to converge.[107] As I will discuss shortly, that final point of convergence would be not a single descriptive fact, but a truth about the scientific process itself.

Eddington wanted his audience to take away several lessons about the nature of scientific discovery. He warned against trusting implicitly in scientific data, since the progressive nature of science constantly overturns its own results.[108] The standard image of science needed to be revised, as well. Our "model of nature should not be like a building," but like an engine.[109] That is, it should not be an edifice to be admired, but a device to help one move forward. "Deeper than any 'form of thought' is a faith that creative activity signifies more than the thing it creates. In this faith, the crumbling of hard-won knowledge in the successive revolutions of science is not the continual tragedy it seems."[110] The treasure of science was its ability to explore and always move forward.

The crux of the matter was that Eddington was arguing for an open-ended scientific process. Seeking, not finding, was the essence of science. He lamented the difficulty of conveying the depths of this sort of scientific practice in short lectures and articles.[111] *Stars and Atoms*, the popular version of *Internal Constitution of the Stars*, took pains to explain exactly what it meant to say that observation and theory matched and how this should relate to future investigation:

> It would be an exaggeration to claim that this limited success [the mass-luminosity relation] is a proof that we have reached the truth about the stellar interior. It is not a proof, but it is an encouragement to work farther along the line of thought we have been pursuing. The tangle is beginning to loosen. The more optimistic may assume it is now straightened out; the more cautious will make ready for the next knot.... We have taken present day theories of physics and pressed them to their remotest conclusions. There is no dogmatic intention in this; it is the best means we have of testing them and revealing their weaknesses if any.

In his inimitable style, Eddington illustrated the value of seeking over dogmatism with a reevaluation of one of the Greek classics. The story of Daedalus and Icarus was usually told to admonish those who push too far, but Eddington provided a novel perspective:

In weighing their achievements, there is something to be said for Icarus. The classical authorities tell us that he was only 'doing a stunt', but I prefer to think of him as the man who brought to light a serious constructional defect in the flying machines of his day. So, too, in Science. Cautious Daedalus will apply his theories where he feels confident they will safely go; but by his excess of caution their hidden weaknesses remain undiscovered. Icarus will strain his theories to the breaking-point till the weak points gape. For the mere adventure? Perhaps partly; that is human nature. But if he is destined not yet to reach the sun and solve finally the riddle of its constitution, we may at least hope to learn from his journey some hints to build a better machine.[112]

Eddington worked to promulgate this attitude in education as well. While he was president of the Friends' Guild of Teachers, he addressed the group on "The Purpose of Science." Discussing the history of science, he commented that most scientists did not see the eventual outcomes of their work, but instead could only see one step ahead: "But it is just the clear vision of one step which means progress—whither we know not, but nevertheless progress. Whilst many are flinging themselves blindly against the wall of mystery, here and there a leader can discern places where the wall is yielding, where attack can force an aperture; it is sufficient if it illuminates the stone that has next to be loosened. We cannot ask more of a scientific leader than that his vision shall suffice for the next step. In the country of the blind the one-eyed man is king."[113] At the same meeting, the group recommended that the overall concern in science education be for "the awareness of the pupil in the nature of scientific thought and method; the development in the pupil of the desire for scientific thought and method, and the gradual elevation of this desire to the level of an ideal felt to be worthy of attainment."[114] Thus, the goal of science education was to make scientific investigation a value unto itself.

Given what I have shown of this element of Eddington's values, part of the significance for him of relativity theory begins to be clear.[115] Its epistemological emphasis focused on the *tools* of scientific investigation: measurement, time, and space. Relativity, by examining the nature of those tools, allowed entirely new forms of investigation. Einstein's triumph, according to Eddington, was that he unified physical phenomena while bypassing the intractable nature of gravitation: "If the principle of equivalence is accepted, it is possible to stride over the difficulties due to the ignorance of the nature of gravitation and arrive directly at physical results. Einstein's theory has been successful in explaining the celebrated astronomical discordance of the perihelion of Mercury without introducing any arbitrary constant; there is no trace of forced agreement about this prediction. It further leads to interesting conclusions with regard to the deflection of light by a gravitational field, and the displacement of spectral lines by the sun, which may be tested by experiment."[116]

But Eddington continually warned that scientific results were ephemeral and transitory. Why should relativity be marked as an important milestone if it will be washed away by future progress? Eddington's answer was that different theories have different degrees of stability; Joseph Larmor's ether theory was clearly only a stepping-stone, but relativity was of a slightly different character:

> Is the point now reached the ultimate goal? Have the points of view of all conceivable observers now been absorbed? We do not assert that they have. But it seems as though a definite task has been rounded off, and a natural halting place reached. So far as we know, the different possible impersonal points of view have been exhausted—those for which the observer can be regarded as a mechanical automaton, and can be replaced by scientific measuring-appliances. A variety of more personal points of view may indeed be needed for an ultimate reality; but they can scarcely be incorporated in a real world of physics. There is thus justification for stopping at this point but not for stopping earlier.[117]

The theory was a natural stopping place because it consistently integrated all observations of a particular character (i.e., metrical observations). It could therefore be considered a useful base to build upon.

Principally, relativity "invites us to search deeper" and allowed physicists an entirely new angle from which to approach the world.[118] Eddington always stressed this aspect of relativity, often in direct response to attacks by figures such as Oliver Lodge. Critics such as Lodge in the astronomical community argued that relativity allowed no absolute (that is, exact) knowledge of phenomena and was therefore a sort of negative progress in the physical sciences. Eddington was quite capable of replying directly to this argument, but he often chose not to. Instead, he tended to maneuver around the attack by redefining the issue under discussion. In his Gifford Lectures, he stressed that he did not want "to demonstrate or confirm the *truth* of the theory, but to show the *use* of the theory." It can be safely said that he was the only contemporary expositor of relativity with a section of his book entitled "Practical Applications."[119]

By "application" Eddington was not referring to nuclear energy, but rather to the new sort of scientific development possible with relativity. One of his favorite examples was the increased mass of beta particles emitted from radioactive nuclei. Relativity did not "correct" for this mass so much as it allowed us to see the electron from a new perspective—its own. The ability to shift frames at will allowed the physicist to view an event from any vantage point, thus providing infinite venues for new investigation.[120]

Eddington saw relativity's progressive character as the perfect complement to his own understanding of science as an unfinished, open-ended quest. Throughout his career we can see the influence of this philosophy on his reception of theories (particularly his understanding of quantum mechanics), but far more interesting was the way his open-ended philosophy appeared in his own science. The value of seeking encompassed the individual mind, distant stars, and the meaning of life itself: "If the human race ever achieved a static perfection, it would be time for it to come to an end, for there would be no possibility of further achievement, and therefore there would be nothing to live for."[121]

Conclusion

In Eddington's astrophysics we see a concrete example of how his distinctive valence values changed his scientific practice. His liberal Quaker emphasis on a mystical approach to life merged with his Manchester and Cambridge training to form a novel methodology for astrophysics. This pragmatic, exploratory approach was successful in addressing problems that had been insoluble to his colleagues for a generation. These techniques were not unique to Eddington (Russell had great success with similar work on stellar physics), but this methodology that would come to dominate astrophysics was in part formed by Quaker Renaissance values.[122] This approach to physics, later known as phenomenology, would become integral to quantum mechanics and particle physics as well as astrophysics.[123]

In Eddington we have a case where the values of a modern physicist were essential to the formation of a discipline and where the origin of those values was immediately forgotten and made generic. Here, religious valence values provided new avenues for investigation in physical science, but this is not the only bridging function they can perform. The next chapter will illustrate the effect of a different valence value, internationalism, on the boundaries of science and politics and the imbuing of a scientific project with cultural meaning. I will show how, as with stellar physics, the explosion of relativity and Einstein onto the world stage was bound inextricably to questions of values.

Internationalism
The 1919 Eclipse and Eddington as Quaker Adventurer

Science is above all politics.
SIR OLIVER LODGE, 1914

Is it not an actual fact that babies have been killed in ways almost inconceivably brutal, and not as a mere individual excess, but as a part of the deliberate and declared policy of the German army?... Is it not a fact that German men of science have gone out of their way to declare their adhesion to these things?
"From an Oxford Note-Book," *Observatory*, 1916

When once asked why Einstein enjoyed such a tremendous public reputation compared to the founders of atomic and quantum physics, Ernest Rutherford replied that it was due to the timing of the 1919 eclipse expedition that provided a confirmation of general relativity: "The war had just ended; and the complacency of the Victorian and the Edwardian times had been shattered. The people felt that all their values and all their ideals had lost their bearings. Now, suddenly, they learnt that an astronomical prediction by a German scientist had been confirmed by expeditions... by British astronomers.... An astronomical discovery, transcending worldly strife, struck a responsive chord."[1] The initial situation that Rutherford described was exactly right: coming out of the Great War, British society was devastated from years of trench warfare, rationing, and hatred of the enemy. However, the note of international harmony is rather misleading. Rutherford's opinion is widespread, and the expedition is often portrayed as a great victory for scientific internationalism over jingoistic militarism. For example, many years later the astrophysicist William McCrea wrote a patriotic, congratulatory article on how much of his success Einstein owed to the good auspices of the Royal Astronomical Society.[2] Unfortunately, the claim that the RAS's support of Einstein was a straightforward scientific step ignores the tremendous anti-German forces that were present in British science during and after the war, as well as the difficult and unpopular struggle waged in support of Einstein, and German science in general. This chapter argues that the expedition's significance as a dramatic and pivotal event in both science and society was far from inevitable, and it was only because of its wartime context and the value-driven actions of particular individuals, chiefly Eddington and his allies, that it achieved its canonical status in the history of modern science.

The values of the Society of Friends most recognized around the world are their pacifism and internationalism. Members of the Society have been well known for their frequent refusal to fight since the seventeenth century, but the liberal Manchester movement Quakers not only refused to participate in war, they worked actively in the world to prevent it and remedy its effects. Eddington learned such activist pacifism from his mentor J. W. Graham, who considered these values to define the modern Quaker. The corollary of pacifism in the modern world was internationalism—the best way to prevent war was to work against the nationalism and historical divisions that drove conflict.

Despite Eddington's groundbreaking work in astrophysics (chapter 2), his name is most widely associated with relativity, both for his role in the 1919 expedition to measure Einstein's predicted light deflection and his aggressive popularization and promulgation of the theory in the Anglophone world. His passion for Einstein's theory is often pointed to as a weakness, and there is a widespread belief that his sympathy for relativity led to inappropriate, or even fraudulent, interpretation of the 1919 data in favor of Einstein. Because of this, the eclipse expedition is often seen as an intrinsically great event in both scientific and social history (for bringing Einstein to the fore and soothing wartime anger) but one based on contextually contaminated science (Eddington's biased observations). I would like to invert this dichotomy: the evidence shows that Eddington's work at the eclipse was well within the scientific standards of the day and that the larger significance of the expedition, both contemporaneously and in the present, was largely the result of Eddington's values.

Eddington became a pivotal figure in these events because of his religious antipathy to the war. His Quaker values meant he saw international science as a site for a moral contestation of war and prejudice. His actions, particularly in regard to the 1919 expedition, shared the motivation of the Friends who organized the refugee camps and ran the blockade to feed German children. These Quakers who braved the chaos of wartime and postwar Europe became known as "adventurers": men and women led by religious conviction to leave the safety and comfort of home to work for peace. For Eddington, the expedition to "weigh light" was as much a religious calling as the food programs, and he saw himself as one of these Quaker adventurers. The expedition was motivated by the valence value of internationalism and thus becomes an event steeped in political, moral, and religious meaning. Einstein and relativity became a focal point through which Eddington could advance both science and international understanding. In practice, these two goals blurred together, as he truly felt that astronomy could not progress in a world wracked by hatred and war.

The conflicts within the RAS emerged from the wider culture of anti-German sentiment and wartime hatred, and Eddington refused to deal with them as though they were unique to astronomy. Rather, he sought to combat jingoism in science by importing the techniques used by Quaker pacifists in British society as a whole. An understanding of this origin for Eddington's efforts and arguments will show that the expedition and its place in the history of science have a meaning that can be understood only with reference to the Great War and its effects on British astronomy. To investigate the context and significance of the expedition, this chapter will follow the historical actors, and specifically Eddington, through the experience of being a scientist in a country at war. I will first establish the sociocultural context of the actors: What values and beliefs were in play? How did it become possible for the expedition to be relevant to the war? Next, I examine in detail the expedition's execution and interpretation. Questions of technical diligence and scientific responsibility will be paramount here. Finally, I will discuss the expedition's presentation to the world community and the influence it had on the first shaky steps of astronomy after the Great War.

Astronomy in the Grip of War

The crisis of the summer of 1914 happened to coincide with the annual meeting of the British Association for the Advancement of Science (BAAS), held that year in Australia. Many of the leading figures in British science, including Eddington, were thousands of miles away from the stormy fields of Europe, considering issues of experiment and organization rather than sovereignty and strategy. After Archduke Ferdinand's assassination on June 28, events in Europe were followed closely by the members of the association, in attitudes varying from excitement to dread. Eddington wrote to his mother on the third of August: "We heard definitely of the war between Germany and Russia. Everyone here seems to take it for granted that England will join in. It all seems incredible. We are anxiously awaiting news."[3] Confirmation of Britain's entry into the war arrived in Australia shortly before an important dinner, at which several foreign members of the association were in attendance. Seeking to reassure the Germans present, Oliver Lodge rose and gave a short speech, declaring that science was beyond all politics.[4] This mood of camaraderie was maintained by the Australian hosts, who went out of their way to make the Germans feel welcome. The war was taken lightheartedly enough that some clever soul rewrote the patriotic anthem "The British Grenadiers" with lyrics praising the valor of physics and astronomy.[5]

Unfortunately, this isolation from the war faded during the journey home. The ship made a detour to Singapore to pick up 140 British soldiers, soon to be assigned to the British Expeditionary Force. Eddington noted sourly that the ship would now make a fine prize for the *Emden*, the feared German commerce raider.[6] Indeed, at one point a false report circulated in London that the BAAS ship had been sunk, with everyone captured and their papers lost.[7] Patriotism was stirring on board as well: Professor William Herdman conducted mysterious experiments in the ship's refrigerator trying to make an explosive shell filled with knock-out gas.[8]

Upon their return to Britain, the impact of the war on astronomy was immediately evident. The commencement of hostilities had led to the severing of telegraph connections to the Central Powers, a move which had disproportionate effects on astronomers. For years the international astronomical community had been using a scheme of coded telegraphs known as the "Science Observer" to ensure rapid dissemination of observations and discoveries (for both priority claims and efficiency). The central hub of this system was in Kiel, Germany, and astronomers were anxious about the disruption to their well-oiled system: "Owing to the cutting of the cables, it is now impossible to communicate with the Centralstelle at Kiel. There is, therefore, no official means of intercommunication of astronomical discoveries."[9]

Astronomer Royal Frank Dyson (see fig. 3.1) in the United Kingdom and Edward Pickering in the United States came up with a stopgap method whereby American telegrams would be sent to Dyson, who would then transmit them individually to allied and neutral observatories in Europe.[10] The confusion of the situation was made evident by the simultaneous organization of another replacement telegraph center in Copenhagen. S. E. Strömgren, the leading astronomer in neutral Denmark, hoped his institute would be able to function as a truly international replacement for the Centralstelle. He was on good terms with many Entente astronomers, and his years working in Kiel meant he was trustworthy to the Germans. It was agreed by all concerned that a bureau that could communicate with astronomers on both sides of the lines would be best, and Strömgren planned to handle the telegraphs for the length of the war.[11]

It was wartime, however, and the British government was unwilling to let streams of coded telegrams pass back and forth with enemy countries. The Astronomer Royal wrote to the Postmaster General asking for permission to continue using the system and included a copy of the code. The response came

FIGURE 3.1
Frank Dyson and A. S. Eddington. Courtesy AIP Emilio Segrè Visual Archives.

quickly: the code was rejected "on censorship grounds," with no further explanation.[12] Apparently this carried no official sanction, and astronomers continued to send telegrams via Copenhagen. It was not until the Defence of the Realm Act was passed and censorship was taken over by the War Office that astronomers required special authorization to communicate with Strömgren.[13]

Not everyone was comfortable with this state of affairs, however. H. C. Plummer, the director of the Liverpool Observatory, came under suspicion of "being in league with alien enemies." He elaborated: "I have communicated [the telegrams'] purpose to my Committee, who have decided that it is prudent to suspend all transactions during the war, that may have the appearance of assisting the enemy."[14] In addition to those receiving pressure from above to institute or follow anti-German policies, other astronomers began to exhibit jingoistic sentiments as well. Typical was R. T. A. Innes in Johannesburg, who, when apprised of the plan to communicate via neutral countries, confessed that "he would prefer to send such messages to a British Institution."[15]

Hostility against the enemy was rampant across Great Britain. Even before the declaration of war, widespread anti-German rioting broke out across the country.[16] The British public had been anticipating conflict with Germany for years, fanned by invasion scare stories and books such as Erskine Childers's *The Riddle of the Sands*. One historian remarked, "Of the mood of the inarticulate public it is difficult to say more than its most obvious features were an intense hatred of the German Kaiser and people."[17] Concerns grew beyond suspicion of German citizens to encompass anything or anyone of German origin. The Royal Society even received an angry letter asking why German continued to be taught at Gresham College.[18] At the end of October the First Sea Lord, Prince Louis of Battenberg, was forced from office because of the homeland of his remote ancestors. Defending the ejection of a high-ranking official based on nothing more than his name, Lord of the Admiralty Winston Churchill argued that "this is no ordinary war, but a struggle between nations for life and death. It raises passions between nations of the most terrible kind. It effaces the old landmarks and frontiers of our civilization."[19]

Quakers and the War

Eddington's reaction to these and subsequent events was largely shaped by his membership in British society's traditional voice of pacifism, the Quakers. The Quakers had resolutely announced their opposition to violence in what is often called the Testimony against War, originally a protest against the King of England: "It springs from our belief of the potentiality of the divine in all men—the Inner Light, as we call it, which is in every man, no matter how hidden or darkened it may be.... Hatred and violence only feed the flame of evil.... If this be true of personal relations, we believe it to be true equally of civic and international ones."[20] Surrounded by the anti-German violence at home and the organized violence across the Channel, they felt clear calls to duty, and they mobilized to testify for peace. The Quakers in Britain held a conference at the opening of the war to discuss their options, out of which came a message to everyone in the British Empire. It read, in part: "We find ourselves today in the midst of what may prove to be the fiercest conflict in the history of the human race. [We reaffirm that] the method of force is no solution of any question [and] that the fundamental unity of men in the family of God is the one enduring reality. Our duty is clear to be courageous in the cause of love and in the hate of hate."[21]

Their pacifism was based on the idea that the war was epiphenomenal to the dangers of nationalism. A Friend spoke in this spirit: "whatever may be the guilt of the individual countries concerned, it is the system which is much more at fault."[22] Despite its militarism, Germany was not the enemy; the true opponent for the Society of Friends was the human misery that came out of any war. As I will discuss later, it was exactly this attitude that mobilized Eddington in the struggles around international science.

The Battlefield of Science

Hatred toward the Germans was not limited to mobs in the street, however, and it was the anger of the British intelligentsia that would eventually lead Eddington to place the expedition on the world stage.[23] Literati and scientists alike had been alienated by an incident in August as the Germans pushed deeper into Belgium. While trying to take the town of Louvain, the German army destroyed many buildings suspected of harboring snipers, including the magnificent library. The destruction of the fourteenth-century structure and its literary riches was a severe blow to Germany's reputation.[24] Academics, artists, and intellectuals around the world were outraged at what was seen as an assault on culture itself. The German self-identification with the Hun suddenly became shockingly appropriate, and many argued that the war had become one for the preservation of civilization.

The German intellectual community was deeply offended by their country's portrayal in the world press as barbaric. In response, ninety-three leading members of that community drafted and signed "The Manifesto to the Civilized World," in which they indignantly defended their nation and its *Kultur*:

IT IS NOT TRUE THAT OUR TROOPS BEHAVED BRUTALLY IN REGARD TO LOUVAIN. They were forced to exercise reprisals with a heavy heart on the furious population, which treacherously attacked them in their quarters, by firing upon a portion of the town. The greater proportion of Louvain is still standing, and the famous town hall is quite uninjured. It was saved from the flames owing to the self-sacrifice of our soldiers. Every German would regret works of art having been destroyed in this war or their being destroyed in the future. But just as we decline to admit that any one loves art more than we do, even so do

we refuse no less decidedly to pay the price of a German defeat for the preservation of a work of art.

IT IS NOT TRUE THAT FIGHTING OUR SO-CALLED MILITARISM IS NOT FIGHTING
AGAINST OUR CIVILIZATION, AS OUR ENEMIES HYPOCRITICALLY ALLEGE. Without German militarism German civilization would be wiped off the face of
the earth. The former arose out of and for the protection of the latter in a
country which for centuries had suffered from invasion as no other has done.
The German Army and the German people are one.[25]

The document was signed by twenty-two natural scientists and doctors, including (as one observer noted) almost all those "of real celebrity."[26]

Of course, some in Germany disagreed with this sentiment, and G. F.
Nicolai, Albert Einstein, and Wilhelm Förster drafted their own manifesto,
which was sent out privately:

Never has any previous war caused so complete an interruption of that
coöperation which should exist between civilized nations.... Educated men
in all countries not only should, but absolutely must, exert all their influence
to prevent the conditions of peace being the source of future wars.[27]

Unsurprisingly, this statement received virtually no attention in Germany and
was completely unknown outside the country. The Manifesto of Ninety-three
came to be seen as representative of German intellectuals; the provost of Trinity
College, Dublin, wrote that the "declaration from 93 German men of science
and letters ... was issued by way of influencing neutral opinion.... Most of
them required no coercion, for they signed it not as independent men but as
courtiers ... and so they have adopted all the vices of the most servile courtiers,
with one curious exception. They still maintain their boorish manners."[28]
Louvain had exposed German science, art, and literature to the full spectrum
of contempt and hatred and to accusations ranging from barbarism to simple
rudeness.

This hatred was inflamed by the arrival of refugees from the Continent,
particularly Belgium. These refugees, mostly women and children, brought
tales of murder and arson at the hands of invading German troops.[29] Among
the refugees was Robert Jonckheere, of the Lille Observatory. He made an
appearance at the Royal Astronomical Society (for which Eddington was an
officer), where he described his flight from German shelling on October 3.
Concerns for the observatory itself were addressed by Felix le Roy, who reported that he had just come from Antwerp, where the observatory had not
been damaged, and it was hoped that similar restraint was shown in Lille.[30]
On November 3 Jonckheere gave a more detailed speech describing his escape.

Lille had been declared an open city with the approach of the Germans. There were many reports of Germans shooting civilians for attempted sniping, but large-scale fighting began only in October as the French arrived. Jonckheere fled during the bombardment, desperately trying to escape the advancing German lines, eventually walking in stockings as his shoes wore through. He arrived in London after five days of travel.[31] Stories such as his helped paint the Germans as savage occupiers ignoring all conventions of war and encouraged agitation in England.

The reaction of the British astronomical community was recorded in the *Observatory*, a monthly publication that recorded the transcripts of the meetings of several groups, including the Royal Astronomical Society and the British Astronomical Association. It also included official announcements, notes from individual researchers, and an anonymous column titled "From an Oxford Note-Book" that contained quips, stories, and miscellaneous observations about life in science. It was written by H. H. Turner, Savilian Professor of Astronomy at Oxford (see fig. 3.2). However, the column's anonymity presented it as something like an unofficial voice of the astronomical community, without the author's byline that would mark it as individual opinion. It was in the *Observatory* that many of the battles over the course of British astronomy were fought.

At the beginning of the war, the journal was happy to relate Oliver Lodge's optimistic claim of an apolitical science, although it did note later that many of the celebrations in Australia became "a symbol that the Empire was united and determined in the face of the common enemy, and not to be dismayed by his aggression."[32] By mid-1915 the war was a regular feature, especially reports of astronomers killed or taken prisoner at the front. For example, the nephew of a Canadian astronomer was recorded as having died fighting with Entente troops at Ypres on April 23. The same page reported the wounding of two English astronomers.[33]

The June issue included a meditation on the centenary of Waterloo and a sad comment that England's former ally now aimed at "brute control" of Europe.[34] The 1915 BAAS meeting had none of the optimism of the previous year's. Zeppelin warnings broke up the meeting on several occasions.[35] The May 1916 issue brought an unprecedented burst of anger from the "Oxford Note-Book," in which the possibility of ever returning to a normal international world of science was abandoned. In a discussion of the need for new astronomical journals despite existing ones appeared the unexpected comment: "But shall we be able to resume the use of German organizations in the near future? Have the Germans themselves not made it practically impossible?" Following this was a more extended protest:

FIGURE 3.2
H. H. Turner. Courtesy of the Royal Astronomical Society/Photo Researchers, Inc.

At the declaration of War ... the general attitude of scientific men was voiced by the President, Sir Oliver Lodge, who declared that "Science was above all politics," and took the opportunity of proposing the health of the foreign guests of the Association. It would not be possible to adopt this attitude now. We have seen how engagements and relationships, which we all thought were "above all politics," and safe to be respected even in time of war itself, have, nevertheless, been broken and tossed aside in a moment if Germany took the fancy that it could thereby benefit itself. Many of us do not see how, after such an exhibition, we can face the mockery of new understandings and undertakings with such a nation. There is much to be said for awaiting a less heated moment for decision; but meantime there are matters, both great and small, which call for action—and are we not too sanguine in hoping that decision will be easier in the near future? Is not the die really cast already?[36]

The anger toward Germany was as unmistakable as the resignation of any hope of normal relations.

The piece was even more shocking in regard to the author. Before the war, H. H. Turner was one of the astronomical community's most dedicated internationalists. He was the principal coordinator of one of the first organized international astronomical efforts in 1887. In 1914 he was recommended by Eddington for the post of Foreign Secretary of the RAS on the basis of his experience working with foreign scientists.[37] Turner had worked closely with Germans on all of his projects but was now willing to abandon them; the trauma of the war, even on the home front, was such that he could no longer accept them as colleagues. He quoted the official investigation into German atrocities, which read: "A hybrid nation of this type which is intellectual without being refined, which can discipline its mind but cannot control its appetites, which can acquire the idiom of Europe and yet retain the instincts of Asia or rather of some pre-Asiatic horde, presents the greatest problem that has ever perplexed the civilization of man. . . . The dilemma is inexorable: we can readmit Germany to international society and lower our standard of international law to her level, or we can exclude her and raise it. There is no third course."[38] Germany's prosecution of the war had forfeited their position in the civilized world. Science, no matter how ideally apolitical, was not exempt from this. There is no question which choice Turner espoused.

Turner realized the weight of his words and stepped back to justify his opinion on "whether Science can still adopt the attitude of being permanently outside this sphere" of politics. He argued that it was inconceivable that the International Association of Academies would follow through on its plans to meet in Berlin.[39] He did allow the possibility that German men of science could repent and acknowledge their crimes publicly and perhaps in this way gain back some of their status. But fundamentally, there could never be a return to the prewar status quo; the crucible of war had forever welded science to politics.

A single and lonely response to these thoughts appeared a month after the "Oxford Note-Book" called for the exclusion of German science. In a letter to the *Observatory* entitled "The Future of International Science," Eddington sought a future not embittered by the war. He began by pleading for the dependence of astronomy in particular upon international cooperation, which became a call for science as a brotherhood:

> I think that astronomers in this country realize the disaster to progress which would result from dissolution of partnership, and there is no disposition to belittle the contributions of Germany. Some of the problems of our science can only be attacked by world-wide cooperation . . . the lines of latitude and longitude pay no regard to national boundaries. But, above all, there is the conviction that the pursuit of truth, whether in the minute structure of the atom or in the vast system of the stars, is a bond transcending human differences—to use it

as a barrier fortifying national feuds is a degradation of the fair name of science.[40]

Eddington argued that using science to extend the fronts of the war was a
complete misunderstanding of both the practical needs and higher aims of
science. Progress could be made only if astronomers realized that the pursuit
of truth must cut across all divisions: unity was to be found in the scientific quest for knowledge, not a shallow and selfish patriotism. Eddington's
Quaker values were clear; the ideas and vocabulary could have been applied
to mysticism or the interpretation of scripture just as easily.

His letter continued, trying to show the essential humanity of the enemy, quoting German scientists expressing regret over the Manifesto of 93
and praising Professor Wilhelm Förster (an Associate of the RAS), who was
undertaking the dangerous work of assisting citizens of the Entente interned
in Germany. Eddington pointed out that the Berlin Academy of Science had
twice refused to eject its foreign members and the Astronomische Gesellschaft
had honored its commitments to send its publications to English members.[41]

His strategy, as it was for Quakers around the world, was to humanize the
enemy and thus weaken the hatred that fueled the war:

> It is not any personal attitude of the German scientists that presents a difficulty,
> but the feeling that we are involved in a general condemnation of their nation.
> But the indictment of a nation takes an entirely different aspect when applied
> to the individuals composing it. Fortunately, most of us know fairly intimately
> some of the men with whom, it is suggested, we can no longer associate. Think,
> not of a symbolic German, but of your former friend Prof. X, for instance—
> call him Hun, pirate, baby-killer, and try to work up a little fury. The attempt
> breaks down ludicrously.... The worship of force, love of empire, a narrow
> patriotism, and the perversion of science have brought the world to disaster.[42]

He admitted that healing the wounds of the war would take time but reminded
readers that Britain, not yet two decades from the brutal Boer War, held no
high ground for criticism: "Nations repent of their folly, not by a sudden
revulsion, but by the slow growth of a wiser outlook. There have been wars
in our own history which no one would now defend." Eddington concluded
with a quote from Sir Arthur Evans, president-elect of the BAAS. The quote
referred to antiquarian studies, but was chosen as "a call to astronomers also":
"We have not ceased to share a common task with those who are today our
enemies. We cannot shirk the fact that to-morrow we shall be once more
labourers together in the same field. It is incumbent on us to do nothing
which should shut the door to mutual intercourse in subjects like our own,

which lie apart from the domain of human passions in the silent avenues of the past."[43] In this letter, Eddington strove to inject the same humanitarianism into the scientific dialogue that the Society of Friends was pushing into the national dialogue. Both, however, faced tremendous inertia and anger from those convinced of German barbarity in and responsibility for the war.

The first direct response to Eddington's letter was from his Cambridge colleague Joseph Larmor, who also sent an open letter titled "The Future of International Science." Larmor thanked Eddington for bringing the opposing viewpoint to the table but defended the shutdown of relations with enemy scientists as a conservative measure. He saw it as a tolerant and prudent way to keep scientific relations "frozen" at the status quo. He took exception to Eddington's suggestion that a measure of identification with the enemy had been actively discouraged, pointing out that the government had given many groups (including the Quakers) full rights to oppose the war on conscientious grounds. Eddington's own conscientious objector hearing was coming up in just two weeks, and one senses that Larmor was suggesting that Eddington be content with his personal right to object to the war. Eddington was given a short space to respond, in which he clarified his position as one not critical of the government on this issue, and then went on to point out that a warm-hearted partisan could not also be a neutral judge. This was not only a warning of conscience to his British colleagues, but also a reminder that the same difficulty "must confront German scientists also."[44] Eddington expressed his personal difficulties in this situation in a private letter to Larmor: "My object in the letter [to the *Observatory*] was simply to deal with the matter as affecting scientific relations; and I tried to keep other considerations out of it. I believe that very many English astronomers (you among them, I infer) are more sympathetic to the general view expressed in my letter than to Turner's view; and I did not wish to strike a discordant note. I am sorry if I have failed. It is very difficult."[45]

The "Oxford Note-Book" wasted no time in answering Eddington's initial letter and in particular his query "What stands in the way of a continuance of that co-operation which is for the welfare of astronomy?" The brutal response:

> My reply is that the *facts* stand in the way—hard, horrible facts, such as we should not have believed possible before the war. He proposes to shut his eyes to these facts, and to test the situation by the play of our imaginations in connection with some individual.... Surely Prof. Eddington is here using his preconceptions formed before the war, and his own shrinking from horrors, to help him in ignoring actual hard facts? Is it not an actual fact that babies have been killed in ways almost inconceivably brutal, and not as a mere individual

excess, but as a part of the deliberate and declared policy of the German army? Is it not a fact that the *Lusitania* was sunk with a national rejoicing that puts the cold-bloodedness of former pirates to shame? Is it not a fact that German men of science have gone out of their way to declare their adhesion to these things, and that one of them who ventures some excuse still boasts of a "quiet conscience?" If we cast our memories back before the war, it is easy to recall that we should have vowed these things incredible; but that does not alter facts.[46]

The writer's disgust for Eddington's pacifism is obvious, and the remark about him "shrinking from horrors" was likely an attack on his conscientious objection to conscription. Overall, it is interesting to note that Eddington's entire point of trying to understand the opponent's point of view has been missed, and this is illustrative of what happened throughout Great Britain during the war. The magnitude of atrocities, real and imagined, were such that it became impossible to see beyond them to ask who was responsible and to remember old friends before they became villains.

One such old friend was Karl Schwarzschild, the German astronomer who had just died fighting with the Central Powers against the Russians. Memorializing an enemy combatant was, at best, a risky enterprise, but Eddington penned a moving obituary published in the *Observatory*. He sought both to refute common stereotypes of the German race as well as show how even an enemy soldier could be worthy of friendship:

> Schwarzschild's characteristics were not those which are usually associated with the scientists of his nation. Keen and restless, his nature was remote from the slow plodding German character. His writings are a model of clearness and concise expression; he never inflicts on us the unbalanced detail and the mazes of analysis—the penalty of "German thoroughness"—which render so many German scientific writings unattractive. It would be paying an ill-service to the memory of one who gave the final proof of devotion to his country, to seek by these traits to dissociate him from the rest of his nation. We would rather say that through him a new spirit was arising in German astronomy from within, raising, broadening and humanizing its outlook.[47]

Eddington knew his colleagues in the RAS were becoming less and less sympathetic to his internationalism as the war developed, and he looked abroad for allies. He wrote to Annie Jump Cannon at Harvard, searching for a sympathetic ear unclouded by his own government's propaganda:

> It is very sad after the jolly days in Bonn, that this division should come between us and our German colleagues. If only there was mutual respect between the combatants, it would be a less depressing outlook; but I am afraid the contempt and hatred of Germans has increased over here very much in

the last three months, though personally I have not yet got to the length of imagining, say, Max Wolf, as a "pirate and baby-killer"! The knowledge that we have the sympathy of nearly all American astronomers is much appreciated, because you have so much more opportunity than we have of learning what is to be said in favour of the other side.[48]

Eddington desperately hoped that the hatred he saw around him was not indicative of his discipline as a whole, but of course his optimism that "nearly all" Americans agreed with him was to prove unfounded.

While Eddington was navigating the waters of scientific politics in wartime, he was also beginning his investigations into stellar structure. This would cement his scientific reputation at the RAS, but discussion continually returned to the war as the developing slaughter at the Battle of the Somme hardened attitudes. An American correspondent soon entered the debate, Professor Edwin Frost writing from Yerkes Observatory. He quoted a letter he received from an anonymous German colleague: "And yet not one of us scholars was consulted in advance [of the war]. Each one is innocent, and each one should believe the same of his colleagues, in the hostile country, that he, too, is innocent of the frightful calamity.... We must prepare [after the war] to make our relations closer than before. We must attempt to attain the solidarity of the scholars of all nations."[49] The "Oxford Note-Book" was quick to respond. As with his rejoinder to Eddington, Turner refused to acknowledge the possibility of a citizen disavowing the policies and actions of his or her nation. "As for the question of individual responsibility ... it may be worth remarking that though English astronomers were certainly not consulted about the war in advance, from the moment of the brutal attack on Belgium there are few indeed of us who wish to disavow the responsibility for the action of England."[50] This passage crystallizes the issues of disagreement between the anti- (chiefly represented by Turner through the "Oxford Note-Book") and pro-German (chiefly represented by Eddington) factions. Turner, and most of the British population, refused to believe that German individuals, scientists or no, could truthfully claim to have no part in their government's atrocities. This view was strengthened by documents such as the Manifesto of 93. Eddington, however, was driven chiefly by his Quaker perspective on human nature, in which individual understanding and responsibility were the salient factors in relationships. From his point of view, the German scientists were as much victims of the war as were the Belgian scientists.

The Society of Friends organized their war relief efforts around that very assumption, that the victims of war were not limited to those on the defensive side of the fighting. They made it a point not to limit their services because of national boundaries: refugee camps were set up in France and Holland, but also in the Russian district of Buzuluk, where they cared for displaced Prussians and German prisoners of war.[51] Entreating their American cousins for assistance in relief work, the British Friends wrote: "We are sure that the common work of American and British Friends for the assistance of the stricken in a third country will help obliterate the memory of past misunderstandings and so point the way to a real brotherhood of Nations, far transcending mere temporary alliances for the satisfaction of national ambitions."[52]

One of the first projects for war relief was the formation of a group with the unwieldy title "Emergency Committee for the Assistance of Germans, Austrians and Hungarians in Distress" (hereafter, the "Emergency Committee"). It was formally established August 7, 1914, only three days after the war started. Its stated goal was to "aid innocent 'Alien enemies' in Great Britain rendered destitute by the war." The humanitarian goals of the committee were recognized by a small, if influential, group of high-profile men, including the Archbishop of Canterbury, the bishops of Winchester, Lincoln, and Westfield, and Viscount Haldane. The Emergency Committee, in addition to providing immediate material aid, also had long-term goals of increasing understanding among the combatants and preventing future wars: "It was felt, too, that, with a view to the settlement after the war, an organized effort should be made, at once, to bind up some of the wounds which the present disaster is inflicting upon the nations, however unpopular that task might be." And unpopular it was, both with the government and with the general public. The Emergency Committee's offices were set up at St. Stephen's House office buildings in Westminster, just a stone's throw from the House of Commons Recruiting Committee, ensuring a high profile among those prosecuting the war.[53]

Unhappy with the idea of British citizens aiding Germans in the literal center of the Empire, some members of the government grumbled about whether their work violated the Treason Act (laid down by Edward III in 1351), which made it a crime to give "aid or comfort" to "the king's enemies in his realm."[54] While the operation was allowed to continue, it was constantly under government surveillance. No one was arrested under the Treason Act, but many members of the Emergency Committee spent time in prison during the war. Stephen Hobhouse, the first chairman, was arrested as a conscientious objector and sentenced to two and a half years of hard labor. Two other members of the executive committee were imprisoned, including Eddington's friend and fellow Cambridge scientist Ernest B. Ludlam, who was sentenced twice, to six months and then nine months of labor at the infamous Wormwood Scrubs.[55] His wife, Olive, wrote that "Ernest went to prison with a light heart, feeling it a privilege to suffer in such a righteous cause."[56] There was virtually no public support for the Quakers' activities. Worse, the Emergency

Committee received several threats of violence, including a promise to shoot the secretary "at sight."[57]

The initial work of the committee dealt with relieving difficulties faced by Germans who were living in or visiting Great Britain when the war broke out. By government order, all such persons were interned. The Quakers saw to their health and other needs and maintained contact with families separated by the fighting.[58] The committee scheduled regular visits to internment camps to ensure humane conditions and extended their work to include POW camps when those began to appear. British media began to spread rumors about the barbaric conditions British POWs were experiencing in Germany, which led to anger against anyone advocating proper treatment for enemy prisoners. An anonymous Englishwoman, when told of work programs for starving enemy aliens, bitterly remarked: "Our own people find it hard enough in this war time. Whenever you give work to a German you are taking bread out of the mouth of an Englishman. Let the Germans *suffer* I say. *They* got us into the war."[59] Quakers tried to relieve such misconceptions, showing that the prisoners on both sides needed better treatment, but the press was uncooperative. Some actively inflamed the public's anger toward attempts to improve the camps in England. The *Evening News* (London) called the members of the Emergency Committee "Hun-coddlers," and the name stuck.[60]

The root of anger toward the Quakers was the impression that, because they were opposed to the war, they were supporting the aggression and atrocities of the Central Powers. An American observer sought to correct this notion: "The Friends who thus kept alive their human sympathies and humanitarian instincts were not 'pro-German.' They did not approve at all of German military aims, policies or methods."[61] Anna Braithewaite Thomas, one of the chairpersons of the committee, insisted that they wanted to destroy German militarism as much as the front-line soldiers but felt that "Christ's method for doing it would prove the most efficient."[62] The Emergency Committee, and the Society of Friends in general, argued that pacifism, not violence, was they key to ending the war. Compassion for individuals and their suffering would lead to a new and more robust internationalism that would prevent future wars. The American Quaker Rufus Jones described their task in beautiful prose:

> In the clash of arms such waves of hate are generated that everybody who belongs even remotely to the enemy peoples is supposed to be himself an enemy and is therefore treated as an outcast to be shunned by everybody. It seems difficult to remember, under such circumstances, how many innocent sufferers there are and how tragic are the experiences of those who are free from all complicity in wrong-doing but who have been caught in the great net spread for really dangerous enemy aliens. Friends could not forget there are innocent

sufferers and they knew, furthermore, that there were very many persons, even in the enemy countries as well, who were not enemies in act or spirit, who were not responsible for the war, who did not approve of barbarities, and who were eagerly praying for the tragedy to come to an end so that men and women and children might once more *live*.[63]

The Adventurers and the Aftermath of War

The end of the war did not end the calling of these Quakers, but instead brought new opportunities. The Cambridge Friends Meeting, of which Eddington was a prominent member, made this statement of purpose upon hearing of the armistice:

> Suffering and unrest are by no means over, nor is the conflict, tho' it may be removed to another plane of action. And while we could take no part in the late warfare we feel that the inevitable struggle now before the world, social, political or economic, is one to which we may rightly have some thing to give, indeed we feel that it is precisely here that our place may lie. The value and power of this contribution depends on one thing, the ability in which our Church and each one of us as individual members of it can lay hold of the power of the Spirit. Nothing else can avail in this great moment.[64]

British Quakers headed to Germany immediately to help ease suffering in the wake of the fighting. Hopes that the blockade would be lifted and material conditions in Germany improved were dashed as the public began calling for reparations and revenge. During the war, many in Britain had expressed high hopes for a just peace. H. G. Wells wrote that, in victory, Britain would "save the liberated Germans from vindictive treatment," but by the armistice such ideals had been lost in the horrors of the trenches.[65] Public opinion quickly gathered behind leaders such as Sir Eric Geddes, who called for "squeezing Germany until the pips squeak."[66]

Part of this squeezing was the maintenance of the wartime blockade of Germany between the armistice and the signing of the peace treaty. One postwar commentator explained the reasoning behind this: "The Germans are . . . suffering as they deserve to suffer from acute humiliation as well as from all kinds of impoverishment."[67] Starvation was rampant in Germany, which Ferdinand Foch and Winston Churchill hoped to use as a lever to gain German acceptance of the Paris peace terms.[68] Quakers, led by the Emergency Committee, purchased food and humanitarian supplies in England and then shipped them to Germany, in contravention of government orders. Much damage had already been done, however; in midsummer 1919 (as Eddington was returning from the eclipse expedition) American and British Friends estimated the effects of the post-Armistice blockade at 700,000 deaths, in addition to endemic starvation and disease among children. As evidence of the appalling conditions inside Germany mounted, Georges Clemenceau and David Lloyd George were persuaded to loosen the blockade.[69]

Those Friends who ventured to Europe to relieve this suffering, both during and after the war, worked in difficult and sometimes dangerous conditions. These relief workers came to be known as "adventurers" and hold a special place in Quaker history as men and women who journeyed into far and foreign lands as a duty of conscience.[70] The strategies used by these adventurers became the models for Eddington's efforts to use the expedition as a tool in repairing international relationships. Their personal presence in the devastated areas was seen as being as important as the material relief they brought: social and intellectual relationships needed to be stimulated as well. The problem was this: "The entire population was shut up within the frontiers of the country, unable to get any relief from the monotony of suffering, unable, too, to exchange ideas." The Friends' goal was to demonstrate "the brotherhood of man overstepping all artificial barriers of race, politics or creed, which we believe to be the only true foundation upon which the family of nations can rest." Many Quakers were called to these adventures when the plight of the German educational system was discovered. Convinced that resurrecting the German intellectual system was crucial even in times of famine, Quakers undertook to feed hundreds of students in Berlin. Among the German faculty involved was Albert Einstein, who was reported as being "closely connected" with this work. The program grew eventually to feed nearly 16,000 students.[71]

Quakers saw one of the most important benefits of this work to be the intellectual connection with a young generation of students, which would hopefully repair or replace those connections severed by the war. In a sense, this was the Quaker war: "While chemists are testing out the deadliest types of poison gas for future wars . . . it is well that there should also be some notable attempts made to conquer the hearts of men by kindness and to demonstrate that one person who heads an expedition to heal the wounds and desolation of war is stronger than a battalion of men under arms."[72]

Relativity and German Science

Like the Emergency Committee, Eddington had begun his own efforts to repair the intellectual connections of the scientific community. He was one of the few British scientists to maintain contact with scientists working in enemy or neutral countries. Among those scientists was Willem de Sitter of the Netherlands, who alerted Eddington (then the Secretary of the RAS) to a new theory of gravity being developed in Germany.[73]

De Sitter submitted a pair of papers on general relativity for the *Monthly Notices of the Royal Astronomical Society* in 1916, and Eddington was immediately fascinated by Einstein's work. He had been vaguely familiar with the principle of relativity as formulated by Einstein in 1911, and a full year before receiving de Sitter's letters Eddington mentioned the gravitational implications of the principle. A significant point of interest was its ability to link the forces of nature: "a positive result [of relativity's prediction] would mean that gravitation has been pulled down from its pedestal, and ceases to stand aloof from the other forces of nature."[74]

Einstein's full theory of 1915, however, was a very different beast. General relativity's successful explanation of the advance of the perihelion of Mercury was significant, but it was the explanation's origin in a coherent and revolutionary understanding of space and time (not to mention the impressive mathematics) that held Eddington's attention.[75] Beyond the scientific import of the paper, he saw an opportunity to repair some of the damage caused by the war. Writing to de Sitter, he reported that he was "interested to hear that so fine a thinker as Einstein is anti-Prussian."[76] Finding a fellow pacifist (and a brilliant one at that) in the German physics community was just what was needed to restore internationalism to science. Einstein was equally excited at the opportunity, and he praised de Sitter for his work to "throw a bridge over the abyss of misunderstanding."[77]

The debate over the merits of general relativity developed quickly, with Eddington virtually the only defender of the theory against vigorous attacks from many high-profile figures, not the least of whom was Sir Oliver Lodge.[78] The theory was heavily constrained by the lack of experimental evidence and its general technical difficulty. There were few scientists who could manipulate the mathematics, and there was widespread suspicion about a theory that was both highly abstract and made sweeping claims about gravity being merely a property of space. Astronomers were probably better placed than physicists to grapple with a new theory of gravity, though; the motion of Mercury had inspired several attempts to modify Newtonian gravitation. Some members of the RAS felt the theory should be tested solely because of its possible impact.

Much of this moderate opinion was derailed by the general unease at dealing with work from an enemy country. The "Oxford Note-Book," discussing an astrophysical hypothesis of German origin that had recently been shown to be false, speculated on the possibility of any German producing good science: "We have tried to think that the exaggerated and false claims made by Germans today were due to some purely temporary disease of quite recent growth. But an instance like this makes one wonder whether the sad truth may not lie deeper."[79] The subtle claim was that German scientific claims were tainted not only by wartime propaganda but also by a deep flaw of barbarity and corruption hidden in their character. How could they ever be trusted? The apparent success of German science in many fields was attributed by one observer to "plagiarism and piracy" of more civilized nations.[80] And because these flaws were of racial origin, anyone of German birth was now under suspicion: Eddington's old teacher Arthur Schuster, who was born in Germany but had served British physics for nearly his entire adult life, was under extreme pressure to resign his position as Secretary of the Royal Society and his presidency of the BAAS because of his heritage.[81]

The argument that Germans could not be trusted, a priori, formed the cornerstone of the continuing argument for German expulsion from international science. Scientific associations, Turner claimed, rested on certain intangibles: "The basis of such organizations is the good faith of the contracting parties: can we accept in scientific matters assurances which are, by some of the parties, not considered binding in other connections?"[82] Just as the German violation of Belgium had made the claims of their politicians unreliable, their scientists' reports were now worthless. No experiment, no mathematical analysis could be beyond suspicion.[83]

It was no longer possible to imagine a scientific community untouched by the war: "The future of Astronomy, and of Science generally, is so intimately dependent on the issue of this terrible war that no excuse is needed for a single glance, each month, at some significant word or deed." The "Oxford Note-Book" went on to quote Woodrow Wilson: "German power, a thing without conscience or honour or capacity for covenanted peace, must be crushed, and, if it be not utterly brought to an end, at least shut out from the friendly intercourse of nations."[84]

Scientific intercourse became one of the first channels to be closed. In 1917 Strömgren traveled to London to try and secure copies of the Royal Astronomical Society's publications, with the aim of passing them on to RAS Associates in Germany and Austria-Hungary. He tried to spread some goodwill by bringing thirty-two copies of publications of the Astronomische Gesellschaft for distribution in London. The RAS Council decided against making any journals available to him.[85] In part, this was due to concerns about the possible military use of the journals. Recently, the Board of Scientific Societies had offered its services in advising its member societies in cases where information of value could be made available to the enemy, and such worries were salient throughout the British scientific community.[86] But generally, the reluctance to make journals available was a result of exactly the attitude

Turner represented: the belief that the Germans had forfeited their right to participate.

This is particularly evident after the end of the fighting, when the British scientific societies still refused to make their journals accessible. At the RAS, Turner, James Jeans, and Harold Jeffreys pushed through a resolution that "no copies of the Society's publications be sent to former enemy foreign Associates."[87] It was not until a year later that the Royal Astronomical Society and Royal Society finally decided to make some exchange of journals with scientific organizations in "enemy countries."[88] Bitterness remained, however, and the RAS often continued to deny requests from libraries and universities in Central Europe, while providing extra materials for countries such as Serbia.[89] Eddington pushed his colleagues to renew communications with former colleagues and institutions, often pointing to the socioeconomic devastation of the former Central Powers. Scientists in the former German and Austrian empires tried to beg, borrow, or trade for British publications, and they often directed their requests through Eddington, given his known sympathies to internationalism. Writing to A. C. Crommelin in 1919 about a message he had received, Eddington wrote, "Will you bring before the next Council the enclosed application for our publications from the Czecho-Slovak Observatory in Prague. . . . In the dreadful condition of Central Europe I think we ought to accede even if it is stretching a point."[90] This request was finally granted and both back and current issues of the *Monthly Notices of the Royal Astronomical Society* were sent. Some time afterward Eddington was able to take advantage of confusion in the mail to see that both Prague and a previously isolated German observatory would receive the issues dating from the war.[91]

A similar story can be told of the fate of the wartime arrangements for international astronomical telegraphs. It was in the dark days of 1917 when the first cracks appeared in Strömgren's delicately negotiated arrangements. The Bureau de Longitudes in Paris communicated to Dyson that they would "prefer" to send scientific messages to an allied country, rather than Copenhagen.[92] Dyson struggled to find an alternative that would retain the international character of Strömgren's system while placating the French. Some British astronomers were interested in locating a telegraph center in a neutral country, but were concerned about Strömgren's past activities in Kiel. Others were happy to support any scheme that inconvenienced the Germans.[93] The Americans were getting uneasy as well and preferred that the center be located "someplace like Brussels."[94] Dyson finally drafted a letter to Strömgren: "I am very sensible of the good feelings towards Astronomers of the belligerent countries which prompted you to undertake the transmission of Astronomical telegrams during the war, instead of the Kiel central station. Further I appreciate the efficiency of the service as carried out by you. Nevertheless I have reluctantly come to the conclusion that for the present at any rate the Greenwich Observatory should withdraw from this society. I am forwarding payment of the bill owing for the years 1915–6–7." Dyson asked Turner for comments, who replied that he had already stopped using Strömgren's service two years prior.[95]

As the war raged on, scientists, both individually and through professional organizations, began seeking to sever any remaining ties with German colleagues, no matter how indirect or informal. After the collapse of the ersatz astronomical telegraph network, the Royal Society moved to expel its German members: "In view of the war having continued nearly 4 years without any indication that the scientific men of Germany are unsympathetic toward the abominable malpractices of their government and their fellow countrymen, and having regard to the representative character of the Royal Society among British scientific bodies as recognized by the patronage of His Majesty the King, this Council forthwith take the steps necessary for removing all enemy aliens from the foreign membership of the Society."[96] The virulent anti-German sentiment found throughout Britain, France, and America had begun to have concrete consequences for the possibility of resurrecting scientific relations after the war.[97]

The Expedition: Eddington as Adventurer

Help for Eddington's efforts against this mentality came from his friend and former colleague at the Royal Greenwich Observatory, Astronomer Royal Frank Dyson, who was seeking permission and resources to mount an expedition to test Einstein's theory at the 1919 eclipse.[98] Einstein's theory predicted that a ray of light traveling near a massive object, such as the sun, would experience a small but measurable deflection of its path. This was one of the three "classic" relativistic effects predicted by Einstein: the advance of the perihelion of Mercury was already established, and the measurement of the redshift of the solar spectrum was proving difficult. This left the gravitational deflection as the only realistic hope of confirming general relativity. Earlier attempts to observe the deflection had been made by the German astronomer Erwin Freundlich, with no success.[99] According to Eddington, Dyson "was at that time very skeptical about the theory though deeply interested in it."[100] Dyson felt that while the theory was speculative, its implications were so important that it needed to be investigated. He was something of an internationalist but had ambivalent feelings about the war: "I hardly know whether to be glad or sorry that our boys are too young to go [to war]."[101] Personal friendship with

Eddington and technical interest were Dyson's motivations; he had no love for the Germans. He began the project by examining older eclipse photographs in search of this effect but found nothing. At the upcoming eclipse of May 29, 1919, the sun would be located in the center of a bright field of stars, the Hyades, perfect for measuring the deflection. The path of the eclipse was in an inconvenient location, however, and a large expedition would need to be mounted to make the observations. This expedition was just the sort of work that required the international cooperation that Eddington had argued was fundamental to the spirit of science. Further, it was an opportunity for him to bring a peace-loving, insightful German to prominence in both science and society. In this sense, the expedition was a "Quaker adventure" as much as a journey to war-torn Europe would have been. Unlike many adventurer Friends, he was never in physical danger, but Eddington's experiences on the eclipse expedition were fraught with excitement and the exotic flavor of West Africa. Further, his later exposition of the journey strongly reflected the litera-ture reporting the adventures of Quakers working in the aftermath of the war.

Eddington had been on an eclipse expedition once before, while Chief Assistant at the Royal Greenwich Observatory, but he had clearly been thinking about them for a long time. At age thirteen he read an essay in front of the Brynmelyn Literary Society meeting: "Expeditions are sent out from the dif-ferent countries of Europe, frequently accompanied by some of the greatest astronomers in the world in order to make observations. Nothing, except per-haps the much rarer event of a transit of Venus, is so zealously watched as the total eclipse. After almost every expedition of this kind we hear of valuable observations made and photographs taken, but still these eclipses are watched with great care, pounds are spent over the expeditions and still we seem no nearer solving the great problems which the eclipse presents."[102] The "great problems" involved with the 1919 eclipse would be not only the important scientific ones imagined by the young Eddington (though relativity provided for plenty of those) but also the terrible rifts that had been caused by the war.

The eclipse's scientific significance had gradually become clear over the course of the war years. The first mention of relativity's prediction of the bending of light in the *Observatory* was an anonymous note on "Gravitation and Light."[103] This early mention referred to Einstein's 1911 prediction (which was only half the deflection predicted later) made long before he had devel-oped the general theory or even adopted the Minkowskian formulation of space-time.[104]

Einstein's full theory, along with the new value for the deflection, arrived in 1916 via de Sitter, and Eddington wasted no time in arguing for its importance. At the same time that he was embroiled in controversy at Cambridge over the treatment of pacifists such as Bertrand Russell,[105] Eddington set himself up as the chief exponent of relativity. He answered concerns about de Sitter's papers, defended the theory at the RAS, and ameliorated confusion caused by the new mathematics. He had become Einstein's "bulldog." Through the next year he worked on his *Report on the Relativity Theory of Gravitation*, a small volume that was the first complete treatment of general relativity in English.[106] Soon, enough interest in the theory had been generated to begin investigation into logistics for the test. Geographic considerations dominated the discussion, because the eclipse path did not go near any observatories or even any easily accessible areas. It seemed likely that whoever carried out the expedition would have to make their way to Africa or South America.[107]

Unsurprisingly, many astronomers thought the expedition would be a waste of time. The refugee scientist Jonckheere warned that there were several different mechanisms that might duplicate a deflection, making any observation useless. The best-developed objection was that refraction in the solar atmosphere could create an effect identical to gravitational deflection. Other possibilities included optical effects from the particles responsible for Zodiacal light or a condensation of ether around the sun. Either could produce deflection effects.[108] The Oxford physicist F. A. Lindemann replied with pages of detailed calculations showing that the necessary density of stellar atmosphere for appreciable refraction was doubtful: there were several good observations of comets passing close enough to rule out the possibility. The Zodiacal particles were too rarefied, and while an ether condensation would duplicate the qualitative effect, no one had a good enough ether model to make a quantitative prediction. Such a vague objection, Lindemann said, should carry little weight compared to Einstein's detailed and consistent theory.[109]

General relativity's successful explanation for the advance of the perihelion of Mercury was impressive to astronomers, and many thought the eclipse test should be carried out on those grounds alone.[110] The third test, the gravitational redshift, was already being conducted. Charles St. John, an astronomer at Mt. Wilson, had produced preliminary results, and they were not favorable for Einstein. The supporters were persistent, however, and an anonymous note following St. John's report reminded the scientific community that there were "those who have welcomed Einstein's relativity theory of gravitation, both on account of its great beauty of conception and its remarkable success in explaining the discordance of the perihelion of Mercury."[111]

Discussions such as these were important in the scientific debate, but had little effect on the actual planning of the expedition. The planning was chiefly in the hands of two astronomers who were also interested in the expedition's impact beyond the scientific test: Eddington and Dyson. Dyson had been making public statements in favor of the expedition since March 1917, and as Astronomer Royal he was perfectly placed to begin preparations without the full support of the astronomical community. Planning for the expedition was formally the duty of the Joint Permanent Eclipse Committee, a group set up by the Royal Society and the Royal Astronomical Society to pool the intellectual and logistical resources of the two societies.[112]

Preparations for the expedition began in earnest at the November 10, 1917, meeting. Eddington, not formally a member of the committee, was present at the special invitation of Dyson. Relativity was the sole focus of discussion: "Attention was drawn by the Astronomer Royal to the importance of the Eclipse of 1919 May 28–29, as affording a specially favourable opportunity for testing the displacements of stars near the Sun, which are predicted by the Theory of Relativity. It was pointed out that such favourable opportunities are of rare occurrence, and there would certainly be no equally suitable Eclipse for many years."[113] It was decided that there were three possible stations from which to observe: northern Brazil, the island of Principe on the west coast of Africa, and near the western shores of Lake Tanganyika. It was hoped that expeditions could be sent to two of these to guard against poor weather.

Money was an immediate problem, and Dyson planned on applying for a government grant that would include £100 for modifying instruments, and £1,000 contingent on the feasibility of the expeditions. The war in Europe continued to rage (a massive German offensive had just begun), and planning for this sort of scientific travel reflected an optimism that either peace would soon be at hand or that the withering spirit of scientific internationalism could allow its execution. Discussion settled on Brazil and the west coast of Africa as the best sites, and a subcommittee was set up to undertake serious preparations.

The subcommittee worked feverishly to account for all the logistical, scientific and technical issues that would be necessary for an accurate observation. The test would be nearly at the observational limits of the techniques available in the field, and the expeditions needed excellent equipment. It was decided to use the astrographic telescope from Greenwich (which had previously captured excellent starfields during eclipse exposures) and a similar lens from Oxford.[114] As was standard practice in eclipse observations, the astrographs would be "fed" by 16-inch coelostats, which were mirrors rotated by clockwork to keep the image of the moving sun centered on the photographic plate. In this way, the telescopes lay horizontal and unmoving (crucial considerations for mechanical stability), while the comparatively light coelostat mirrors could be pivoted steadily to compensate for the rotation of the earth. The aging mechanisms of these coelostats were a worry, but improvements depended on obtaining a "priority certificate from the Ministry of Munitions" for the use of scarce resources.[115] As a hedge against the unreliable coelostats, one of the planned observers, Father A. L. Cortie of the Stonyhurst College Observatory suggested also bringing a 4-inch telescope fed by a different coelostat, which he had used to good effect on previous expeditions.[116] E. T. Cottingham, from Greenwich, would overhaul all the equipment to the best of his ability in any case.

Actually getting the observers to their sites was proving a challenge. The war had been hard on shipping, and their destinations were far from well-traveled routes. In order to be in the right places on May 29, the teams would have to leave England in February and be gone through June.[117] There was some danger that Eddington himself would not be available: his conscientious objection to conscription seemed likely to land him in prison. He had originally been exempted as an astronomer doing work of "national importance," but the British government desperately needed manpower and revoked his exemption. Eddington refused to fight based on religious grounds, even refusing later opportunities for exemption for which he needed to do nothing beyond simply not stating his religious objection. In an ironic reversal, Dyson (with his connections in the Admiralty) was able to gain Eddington an exemption on the condition that he participate in the eclipse expedition.[118] From Eddington's perspective, the agreement was that he would not be punished for his objection to the war so long as he took part in an activity that he was hoping to use as a tool for peace.

Securing a leave of absence from Trinity was somewhat easier (despite Eddington being the only member of the observatory staff not on war duty), but Father Cortie found he could not be spared. He was replaced at the last minute by C. R. Davidson from Greenwich. The expeditions would leave from Liverpool on the *Anselm*, and go via Madeira. As of their departure, no information on steamers to Principe (Eddington's final destination) was available. There was some discussion of trying to get a warship from the Admiralty, but that came to nothing.[119] They would simply have to hope for the best.

As the departure approached, Eddington became concerned with returning results to Greenwich as quickly as possible. He arranged for each party to take a micrometer (a tool for making very fine measurements), in order that preliminary measures of check plates and eclipse plates might be made at the eclipse stations instead of waiting for the long journey back. He also gathered information on developing photographic plates in tropical conditions, which would likely be quite different than in an enclosed observatory in Europe. Plates were bought from different companies, and different lines within each, in case some particular chemistry proved unsuitable for the tropics. Eddington also arranged a telegraphic code for informing Dyson as to the weather conditions and general character of the eclipse results.[120] There was no particular need for haste in returning the scientific results; certainly there was no danger of them losing priority. It seems, rather, that Eddington's concern was to bring the results of the expedition home in time for them to affect the fluid post-Armistice situation.

Before leaving, Eddington wrote a review article (essentially a distilled version of his *Report on Relativity*) on the expedition, detailing the theoretical and logistical background for the observation. During the five minutes of totality (when the sun is totally obscured by the moon), relativity would be the only topic of investigation. The expected displacement of stars at the sun's limb, as calculated by Einstein, was 1.75 arc-seconds, which would give a displacement of approximately 1/60 mm on the plates. Not an easy matter, but Eddington assured his readers that "this in itself calls for no extravagant precautions of accuracy." Plates taken in the field would be compared with photographs of the same stars already taken at Greenwich and Oxford. Check plates of other parts of the sky could be taken to determine scale, but this was not strictly necessary because the hoped-for effect was distinct from the effect of change of scale.

According to Eddington, there were three possible results from the expedition: no deflection, the predicted relativistic deflection, or a half-deflection calculated from Newtonian mechanics.[121] He pointed out that the results would have important implications for the relationship between matter and energy, which would itself have important consequences for theories of stellar evolution (he made no mention of the important consequences for international science). "The problem of the coming eclipse may, therefore, be described as that of *weighing* light." He addressed Jonckheere's ether condensation hypothesis and assimilated the objection in a rather positivist style by arguing that since this would be deflection of light by the presence of a massive body, it should really not be regarded as a separate hypothesis. Carefully covering all possibilities, Eddington closed with the point that any result, even a null one, would be extremely important, and he held up the Michelson-Morley experiment as an example of the possible importance of a null result.[122] But there was no question that Eddington was hoping for a positive result, both in the physics and in the scientific community.

The expedition itself was far from smooth. The Principe team, Eddington and Cottingham, found themselves waiting for two weeks in Portugal for the next steamer to Principe. Their experiences were dominated by the oddity of traveling while Europe was still technically at war.[123] Eddington remarked on the strangeness of eating without the constraints of rationing and on the war damage in Portugal: "Three ships were torpedoed by submarine in Madeira during the war, and one sees the masts of two of them sticking up out of

the water. The town was also bombarded and there are a few traces visible."
Due to a held-over war-time regulation, passengers on the steamers were not
allowed to know where the boat was at any given time.[124]

In Portugal, Eddington occupied himself with mountain climbing (his
companion was unable to keep up) and a trip to the local casino (he assured
his conservative mother that he went there solely for the good tea). They
arrived on Principe, an island off the west coast of Africa, on April 29, with
a month until the eclipse. The island was described as thickly wooded and
"very charming." Unfortunately, they quickly discovered that Principe had
terrible weather and they would be lucky to get a clear sky. Rain, mosquitoes,
and quinine became the daily regimen. They built waterproof huts for the
equipment, with the help of laborers from the plantation.[125] They were forced
to work under mosquito netting and at least once helped hunt monkeys
interfering with their equipment. Thoughts of the war filled Eddington's
letters home. He wrote of how strange it was to see full sugar bowls and asked
whether his family was still rationed. Above all, he felt frustrated by being cut
off from news of world affairs. "[I wonder] whether peace has been signed."[126]

Eddington and Cottingham began taking check plates on May 16 with only
a little difficulty due to developing photographs in the high temperatures.
Eddington spent his days measuring the plates. The weather worsened as May
progressed, and the morning of the eclipse was not reassuring. It brought a
tremendous rainstorm, which stopped about two hours before totality but left
significant cloud cover. The eclipse, for which they had traveled thousands of
miles, was to reach totality at 2:15 in the afternoon. In Eddington's words:

> About 1:30 when the partial phase was well advanced, we began to get glimpses
> of the sun, at 1:55 we could see the crescent (through cloud) almost contin-
> uously, and there were large patches of clear sky appearing. We had to carry
> out our programme of photographs in faith. I did not see the eclipse, being
> too busy changing plates, except for one glance to make sure it had begun,
> and another half-way through to see how much cloud there was. We took 16
> photographs (of which 4 are not yet developed). They are all good pictures of
> the sun, showing a very remarkable prominence; but the cloud has interfered
> very much with the star-images. The first 10 photographs show practically no
> stars. The last 6 show a few images which I hope will give us what we need;
> but it is very disappointing. Everything shows that our arrangements were
> quite satisfactory, and with a little clearer weather we should have had splen-
> did results. Ten minutes after the eclipse the sky was beautifully clear. . . . We
> developed the photographs 2 each night for 6 nights after the eclipse, and I
> spent the whole day measuring. The cloudy weather upset my plans, and I had
> to treat the measures in a different way from what I intended; consequently I

FIGURE 3.3
The eclipse equipment from Sobral. Courtesy of the Science and Society Picture Library.

have not been able to make any preliminary announcement of the result. But the one good plate that I measured gave a result agreeing with Einstein and I think I have got a little confirmation from a second plate.[127]

He was more succinct in a telegraph to Dyson: "Through cloud. Hopeful."[128]

The expedition to Sobral, in Brazil, was less harried (see fig. 3.3). Davidson and Crommelin were well taken care of by the local authorities and quickly found a good observing station. The team had the service of Brazil's first automobile, as well as the production of ice (useful for developing photographs) from a nearby meatpacker. The coelostat mechanism functioned without difficulties, but the devices displayed a new problem during preparations. The team discovered that the coelostat mirror for the high-quality astrographic telescope had astigmatism, a serious optical defect. All was not lost, however, since the team could remain in Sobral until July to get check plates of the star field, and they would still be able to measure the relative displacement of star images.[129]

In Brazil the eclipse became a public event, and an observatory near the edge of the eclipse path sold tickets to look through the telescope.[130] The weather for the eclipse was beautiful, and the observers took nineteen plates with the astrographic and eight with the small 4-inch lens. They had every expectation of success and cabled home immediately: "Eclipse splendid."[131]

The next day brought an unpleasant surprise. Four of the astrographic plates were developed, and it was discovered that the image quality was far from adequate. A note made at the time reads: "May 30, 3 a.m. . . . It was found that there had been a serious change of focus, so that, while the stars were shown, the definition was spoilt. This change of focus can only be attributed to the unequal expansion of the mirror through the sun's heat. The readings of the focusing scale were checked next day, but were found unaltered at 11.0 mm. It seems doubtful whether much can be got from these plates."[132] The displacements they were looking for were only a quarter of the diameter of the stellar images, so any distortion of the images made the measurement hopelessly unreliable. The poor quality of the coelostat mirror had wrecked the finely tuned optical system of the astrographic telescope, and they would have to rely on whatever results were achieved with the small lens. In any case, Crommelin and Davidson had to stay in Brazil until July to take the check plates, and no measurements could be made before then.

Results, Interpretation, and Acceptance

As negotiators in Paris arranged the final details of Germany's reparations, Dyson waited anxiously in Britain for news from the expeditions. The observers had still not returned by early July, and he had no idea whether the photographs obtained were satisfactory. Eddington and Cottingham had not been able to complete their measurements at Principe because a steamboat strike would have stranded them had they stayed to complete their work. When the two teams arrived home, the tedious task of measurement and analysis began. The photographs and their check plates were clipped into a micrometer, and two sets of screw readings were made by two different people, so as to minimize any human error. The sixteen Principe photographs taken through the cloud yielded only seven plates with star images, but all of these had the crucial stars $\kappa 1$ and $\kappa 2$ Tauri, which had the highest predicted deflection. Only two of the plates had all five of the stars needed for a reliable analysis, however. These gave consistent results, with a calculated mean deflection of 1.61 ± 0.30 arc-seconds. These were good results given the difficulty of the measurement, and Eddington was quite pleased, as this was rather close to the theoretical prediction of 1.75 arc-seconds.[133]

The astrographic photographs from the other expedition were less reassuring. The poor quality of the Sobral astrographic images was confirmed once they were compared with the check plates. They were unambiguous about

showing a deflection, however, and the mean displacement was calculated to
be 0.93 arc-seconds, far from Einstein's predicted value. The results were to be
given little weight because of the clear optical problems of the coelostat, but
they were nonetheless worryingly close to the Newtonian half-deflection.[134]

The auxiliary lens, the 4-inch recommended at the last minute by Father
Cortie, ended up saving the day. Seven of the eight plates taken with it had
excellent images of the seven hoped-for stars. The images were far better than
those from Principe, with deflection results of 1.98 ± 0.12 arc-seconds.[135]
Eddington wrote Dyson in relief: "I am glad the (4 inch) plates give the full
deflection not only because of theory, but because I had been worrying over
the Principe plates and could not see any possible way of reconciling them
with the half-deflection."[136] After a great deal of data reduction, Eddington
presented a calculated mean of 1.64 arc-seconds. This mean included a particular weighting of the results according to how scale determinations made
the Principe and Sobral 4-inch results much more important than the flawed
Sobral astrographic.

Eddington and Dyson began writing the report, which would be given to a
special joint meeting of the Royal Society and Royal Astronomical Society in
November. Eddington wrote the introduction, theoretical background, and
much of the final analysis. He added a graph of the 4-inch Sobral results to
see how well they fit the predicted $1/r$ deflection dependence on distance from
the sun (the Principe results did not have enough stars). He felt that this was
an excellent test to show the internal consistency of the data: "It seemed to
me rather interesting and deals with a point that ought not to be overlooked
and brings out the really remarkable agreement of individual stars at Sobral."
At the last minute, Eddington changed his mind about the presentation of
possibly the single most important element in the paper: the final results. As
it stood, the results from all three telescopes were averaged to give a very good
mean deflection of 1.64 arc-seconds. He felt this was slightly disingenuous
given the obvious problems with one of the three sets: "I do not like the
combination of the astrographic with the other Sobral results—particularly
because it makes the mean come so near the truth. I do not think it can be
justified; the probable errors of both are I think below 0.1″ so they are man-
ifestly discordant. . . . It seems arbitrary to combine a result which definitely
disagrees with a result which agrees and so obtain still better agreement."[137]
It was agreed that the three sets of data would instead be presented separately,
so as to clarify the case for and against each. Both Eddington and Dyson were
quite confident in their results; now they had to convince their colleagues.

On November 6, 1919, J. J. Thomson presided over a special joint meeting of the Royal Astronomical Society and Royal Society assembled for the

sole purpose of presenting the results from the eclipse expeditions, which were unknown to nearly everyone.[138] A. N. Whitehead, one of the many observers packed into the room, described the setting: "The whole atmosphere of tense interest was exactly like that of the Greek drama.... There was a dramatic quality in the very staging:—the traditional ceremonial, and in the background the picture of Newton to remind us that the greatest of scientific generalizations was now, after more than two centuries, to receive its first modification. Nor was the personal interest wanting: a great adventure in thought had at length come safe to shore."[139] Dyson, as Astronomer Royal, began the presentation. He reviewed the background of the expedition and why previous eclipse photographs were inadequate to measure the Einstein effect. He described the techniques used to account for change of scale from the check plates and explained the disappointing problem with the coelostat. He finally came to the results: "After a careful study of the plates I am prepared to say that there can be no doubt that they confirm Einstein's prediction. A very definite result has been obtained that light is deflected in accordance with Einstein's law of gravitation."

Next, A. C. Crommelin, the leader of the Sobral expedition, rose to speak and described his half of the observations. He detailed the difficulties with the astrographic and how they realized there was a problem. The final speaker was Eddington, who not only reported on the Principe expedition but dealt with the larger scientific issues implied by the results and the impact they might have on physics. The observations from Africa were presented, and Eddington argued that the cloud cover was apparently a blessing in disguise, since it may have prevented the coelostat distortion under the sun's heat that ruined the Sobral astrographic. These plates also had a supplementary benefit of having captured a tremendous solar prominence that would keep solar physicists busy for some time. After presenting the data, Eddington moved to the physical consequences of the observation. He described it as a crucial test between Newton's and Einstein's laws, and the results clearly favored the larger deflection. He did qualify his declaration of the confirmation of relativity in light of the so-far unsuccessful attempts to measure the relativistic displacement of solar spectral lines. "This effect (the deflection) may be taken as proving Einstein's *law* rather than his *theory*. It is not affected by the failure to detect the displacement of Fraunhofer lines on the Sun. If this latter failure is confirmed it will not affect Einstein's law of gravitation, but it will affect the views on which the law was arrived at. The law is right, though the fundamental ideas underlying it may yet be questioned." This fine line between theory and law was meant to emphasize that Einstein's numerical prediction had been verified (his law) despite the lack of observational support for the redshift, which was dependent on the metaphysical details of relativity (his theory). That is, the eclipse results were a win for Einstein, irrespective of other investigations.

Despite the lack of spectroscopic evidence, many of those present were convinced that Newton had been overthrown. Thomson spoke for this viewpoint with remarkable vigor and a common caveat:

> This is the most important result obtained in connection with the theory of gravitation since Newton's day, and it is fitting that it should be announced at a meeting of the Society so closely connected with him. . . . If it is sustained that Einstein's reasoning holds good—and it has survived two very severe tests in connection with the perihelion of Mercury and the present eclipse—then it is the result of one of the highest achievements in human thought. The weak point in the theory is the great difficulty in expressing it. It would seem that no one can understand the new law of gravitation with out a thorough knowledge of the theory of invariants and of the calculus of variations.

Thomson then invited Ralph Fowler, the president of the RAS, to speak. Fowler thanked Dyson for his tireless insistence on the importance of the expedition, but concluded with a cautious reminder that further tests of relativity were necessary: "The conclusion is so important that no effort should be spared in seeking confirmation in other ways." Several other speakers voiced similar concerns about the incompleteness of the confirmation, and there were several objections to the attribution of the results to gravitational deflection. Some verbal sparring ensued, and Eddington again insisted that it was Einstein's law and not necessarily his theory that had been confirmed, thus avoiding difficult questions of metaphysics.[140]

Eddington and Dyson spent the subsequent weeks defending the quality and meaning of the expedition results. Dyson was quick to refute "misconceptions which have arisen as to the magnitude of the observed quantity involved in the recent experiment." Many scientists objected that measuring a displacement that was only one-quarter the size of the entire image was dubious, but the Astronomer Royal reassured them that "those who are familiar with the measurement of astronomical photographs will know that it is quite possible to measure quantities of this order of magnitude."[141] The displacements were much larger than those of stellar parallax, which were reliably and consistently measured, and Dyson argued that the results were quite reasonable to those conversant with the methods.[142]

Lodge and others unhappy with the difficulty of duplicating the test asked repeatedly whether the deflection from Jupiter could be observed, instead of having to wait for a solar eclipse. (It could not.) Ludwik Silberstein argued that the plates showed significant nonradial displacements not predicted by Einstein and that an objective analysis unprejudiced by the theory would show that there was no radial effect.[143] The Joint Permanent Eclipse Committee had made copies of the eclipse plates available to anyone who wanted them (see fig. 3.4), and there was a frenzy of analysis across the astronomical community.[144] Silberstein's critique was refuted, and there was soon little doubt that there was a radial deflection of the star images.[145] Exactly what caused this displacement was still in contention. Refraction in the solar atmosphere was still the most popular alternative to Einstein's explanation, though there were others.[146] Lindemann again provided detailed calculations dismissing the possibility of refraction.[147]

Objections based on the aesthetic, conceptual, and mathematical unfamiliarity of relativity were legion but had surprisingly little effect on the debate among astronomers.[148] Careful analysis of the data had persuaded most of them that there was in fact a deflection. Charles St. John, himself still skeptical of the theory, wrote, "A deflection of light while passing through the near neighborhood of the Sun is now a fact of observation, destined either to influence profoundly our attitude toward the conceptions of space and time, as involved in the generalized theory of relativity, or to serve as a basis for the advancement of science in other directions."[149] Joseph Larmor took a similar position, calling the expedition a "very important astronomical determination that is to be regarded as a guide toward future theory rather than as the verification of the particular theory which suggested it."[150]

Eddington's assertion that Einstein's *law* and not necessarily his *theory* had been proved was an apt characterization of opinion among astronomers. This distinction was particularly useful for Eddington (he certainly did not think it true) in that he could publicize a confirmation of Einstein even in the absence of the spectroscopic results.[151] Opponents of the theory used the distinction as well: Silberstein, acceding to the unlikelihood of refraction, maintained that the deflection was an "isolated fact" and had no bearing on the truth of the theory. Others had similar thoughts: "The star images are certainly excellent and the photograph leaves little doubt that the deflection is close to the double amount. You and your colleagues are to be heartily congratulated on the results of the eclipse expeditions. I trust you are planning to observe the 1922 eclipse in the same way, as I am sure you will be among the first to agree that results of such importance should be thoroughly confirmed before we accept them as establishing Einstein's theory. I have a feeling that some explanation for the deflection will be found without resorting to non-Euclidean space."[152]

The lack of spectroscopic evidence meant many scientists were reserved on pronouncing relativity "proven," but astronomers were comfortable an-

FIGURE 3.4
A photographic plate from the expedition to Principe. Courtesy of the Science and Society
Picture Library.

nouncing that Einstein's quantitative prediction of the deflection of light
had been confirmed. Many astronomers, still much more comfortable with
positional astronomy than spectroscopy, found the combination of the deflection test and Mercury's orbit convincing.[153] Dyson, writing a few years later,
described the situation well: "But what appeals to me and the astronomers
and physicists I know is that Einstein's theory gives a formula—if not an

explanation—of a number of very difficult and unexplained observations and experiments. . . . Einstein's predictions have been verified and his Law of Gravitation is correct, and as far as I can see, there is no alternative law. It is possible to accept the law and reject the theory which led to it, but it is not an unreasonable view to regard the verification of the law as confirmation of the theory."[154] In short, the observations proved the law, and the law came from the theory. Many of the difficulties of the theory were far outside the expertise of astronomers, and its utility in investigating astronomical phenomena made it worth using. What was clear to astronomers was that there was a deflection, and that, coupled with Mercury's orbit, effectively verified Einstein's law of gravitation. The practical workings of gravity and celestial mechanics were by far the dominant concerns; esoteric matters of clocks running slowly and rulers shrinking were distinctly secondary. Even before the final spectroscopic results in favor of relativity were published, the "Oxford Note-Book" could comment casually that debate among astronomers had effectively ceased, with opinion firmly behind Einstein.[155]

The public response was, unsurprisingly, less measured. Among the audience at the joint meeting on November 6 were several members of the press, and the excitement of the scene quickly found its way into the public eye. The Friday, November 7, 1919, edition of the *Times* proclaimed the headline "Revolution in Science" (see fig. 3.5), which shared the page with an announcement of the first Armistice Day observance ("The Glorious Dead"). The correspondent reported that "the greatest possible interest had been aroused in scientific circles."[156] On Saturday the paper ran a follow-up article under the same banner, this time subtitled "Einstein v. Newton." Two paragraphs following the article introduced Einstein himself, who had barely appeared in the accounts of his theory. He was described as a Swiss Jew who had taught at Zurich and Prague but took a position in Berlin for a large salary. It noted that "during the war, as a man of liberal tendencies, he was one of the signatories to the protest against the German manifesto of the men of science who declared themselves in favour of Germany's part in the war."[157] Clearly, the correspondent was concerned to demonstrate that the originator of this new theory had nothing to do with the scientists that aided and abetted the war. A year after the Armistice, those British men of science who carried out the expedition and its measurements still needed protection against accusations of consorting with the enemy.

Einstein himself wrote an article for the *Times* three weeks later. In it he applauded the international character of the eclipse expedition: "After the lamentable breach in the former international relations existing among men

REVOLUTION IN SCIENCE.

NEW THEORY OF THE UNIVERSE.

NEWTONIAN IDEAS OVERTHROWN.

Yesterday afternoon in the rooms of the Royal Society, at a joint session of the Royal and Astronomical Societies, the results obtained by British observers of the total solar eclipse of May 29 were discussed.

The greatest possible interest had been aroused in scientific circles by the hope that rival theories of a fundamental physical problem would be put to the test, and there was a very large attendance of astronomers and physicists. It was generally accepted that the observations were decisive in the verifying of the prediction of the famous physicist, Einstein, stated by the President of the Royal Society as being the most remarkable scientific event since the discovery of the predicted existence of the planet Neptune. But there was difference of opinion as to whether science had to face merely a new and unexplained fact, or to reckon with a theory that would completely revolutionize the accepted fundamentals of physics.

SIR FRANK DYSON, the Astronomer Royal, described the work of the expeditions sent respectively to Sobral in North Brazil and the island of Principe, off the West Coast of Africa. At each of these places, if the weather were propitious on the day of the eclipse, it would be possible to take during totality a set of photographs of the obscured sun and of a number of bright stars which happened to be in its immediate vicinity. The desired object was to ascertain whether the light from these stars, as it passed the sun, came as directly towards us as if the sun were not there, or if there was a deflection due to its presence, and if the latter proved to be the case, what the amount of the deflection was. If deflection did occur, the stars would appear on the photographic plates at a measurable distance from their theoretical positions. He explained in detail the apparatus that had been employed, the corrections that had to be made for various disturbing factors, and the methods by which comparison between the theoretical and the observed positions had been made. He convinced the meeting that the results were definite and conclusive. Deflection did take place, and the measurements showed that the extent of the deflection was in close accord with the theoretical degree predicted by Einstein, as opposed to half that degree, the amount that would follow from the principles of Newton. It is interesting to recall that Sir Oliver Lodge, speaking at the Royal Institution last February, had also ventured on a prediction. He doubted if deflection would be observed, but was confident that if it did take place, it would follow the law of Newton and not that of Einstein.

DR. CROMMELIN and PROFESSOR EDDINGTON, two of the actual observers, followed the Astronomer-Royal, and gave interesting accounts of their work, in every way confirming the general conclusions that had been enunciated.

"MOMENTOUS PRONOUNCEMENT."

So far the matter was clear, but when the discussion began, it was plain that the scientific interest centred more in the theoretical bearings of the results than in the results themselves. Even the President of the Royal Society, in stating that they had just listened to "one of the most momentous, if not the most momentous, pronouncements of human thought," had to confess that no one had yet succeeded in stating in clear language what the theory of Einstein really was. It was accepted, how-

FIGURE 3.5

The *Times* headline announcing the eclipse results.

of science,... it was in accordance with the high and proud tradition of English science that English scientific men should have given their time and labour... to test a theory that had been completed and published in the country of their enemies in the midst of war."[158] Einstein could hardly have better expressed the hopes Eddington had for the eclipse observations. Primed with publicity from newspapers and magazines from across the United Kingdom, Eddington set out to spread the message of international science.[159] A representative passage from the opening paragraph of one of his articles on relativity: "The theoretical researches of Prof. Albert Einstein, of Berlin, now so strikingly confirmed by the British eclipse expeditions, involve a broadening of our views of external nature, comparable with, or perhaps, exceeding the advances associated with Copernicus, Newton and Darwin."[160] Here was the crux of his strategy: explicitly linking the wartime enemies through an epochal scientific advance.

Eddington's avalanche of lectures, public addresses, and articles on relativity reinforced the expedition's quality of adventure, as well (see fig. 3.6). For example, at an RAS dinner he presented the story of the expedition as a parody of the *Rubaiyat of Omar Khayyam*, full of excitement and the thrill of revealing the unknown. His best-selling *Space, Time and Gravitation* also gave the expedition a privileged place. Thus, Eddington did his part to mend international relations, while bringing home as invigorating a tale as the Friends who had been chased by bandits while feeding Russian refugees. Needless to say, Quakers in Britain were delighted by the expedition and Eddington's work to make it known. It was this work that brought him to prominence in the Society of Friends, and in later years he helped make the teaching of international relations part of science instruction in some Quaker schools.[161]

Eddington's correspondence with Einstein revealed his high hopes for the political and social effects of the expedition. In December 1919 he wrote, "All England has been talking about your theory.... There is no mistaking the genuine enthusiasm in scientific circles and perhaps more particularly in this University. It is the best possible thing that could have happened for scientific relations between England and Germany. I do not anticipate rapid progress toward official reunion, but there is a big advance toward a more reasonable frame of mind among scientific men, and that is even more important than the renewal of formal associations."[162] Like the Quakers working to rebuild Europe for a lasting peace, Eddington saw that rapprochement as a slow process, which began with personal relationships. His excited report of enthusiasm in Britain shows that he was concerned not only to humanize the Germans to the British, but also to show Einstein that there was more than ill will across the Channel. He went on to describe in some detail the interest the

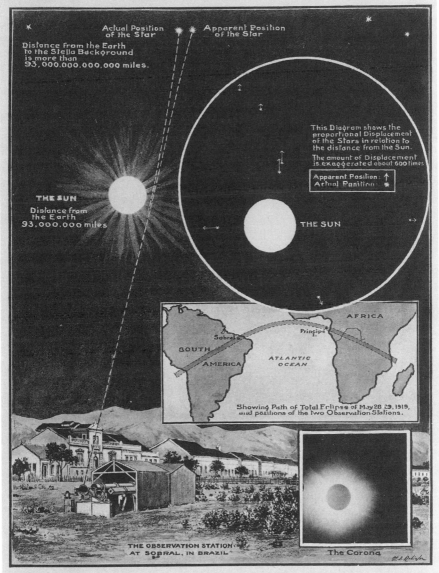

FIGURE 3.6

The *Illustrated London News* presentation of the 1919 eclipse expedition. *Illustrated London News*, November 22, 1919, 815.

expedition aroused: "I have been kept very busy lecturing and writing on your theory. My Report on Relativity is sold out and is being reprinted. That shows the zeal for knowledge on the subject; because it is not an easy book to tackle. I had a huge audience at the Cambridge Philosophical Society a few days ago, and hundreds were turned away unable to get near the room. . . . One feels that things have turned out very fortunately in giving this object-lesson of the solidarity of German and British science even in time of war."[163]

Despite the interest in relativity, Eddington remained the principal advocate of a return to normal scientific relations. When a group of German scientists organized a special meeting of the Astronomische Gesellschaft to discuss general relativity, the silence from British scientists was deafening. Writing to the head of the Gesellschaft, Strömgren, Eddington said: "I hope to show my interest in the Astronomische Gesellschaft by attending the next meeting— an individual step which no one has any right to object to. . . . International Science is bound to win and recent events—the verification of Einstein's theory—has made a tremendous difference in the past month."[164] Despite his clear implication that his British colleagues would object to his attendance, Eddington made plans for the trip to Germany. The meeting was held at Einstein's house and came to be known as the Potsdam Conference. Eddington was the sole British scientist present. He continued to participate in German science as if there had been no disruption from the war, even publishing a paper in the *Zeitschrift für Physik* despite his almost complete inability to write or read German.[165] The paper begins: "This paper is intended to give a full account on the theory of the radiative equilibrium of the stars. It is written primarily because the original papers are not easily accessible in Central Europe in present circumstances."[166]

Eddington's hopes for a reconciliation between the enemy nations was no doubt boosted at the November 14, 1919, meeting of the Council of the Royal Astronomical Society, when Einstein was nominated for the Gold Medal, the society's highest honor. He was in competition with H. N. Russell, for his theory of stellar evolution, and Annie Jump Cannon, for her catalogue of stellar spectra.[167] In December Einstein was chosen to receive the medal.[168] According to the Byzantine regulations surrounding the medal, the confirmation of the award would have to wait until the January meeting, but an excited Eddington decided to send word to Einstein even before the end of the year. Eddington's friend E. B. Ludlam was soon to go over to Germany as part of the Quaker Emergency Committee's war relief work among the former enemy countries. Ludlam had done research under Philipp Lenard at Kiel and had a great deal of experience with German language and culture. He met with Einstein personally to convey the news of his nomination.

Unfortunately, the confirmation of the award was far from certain; many were uncomfortable with the idea of giving an award to a citizen of a nation so recently at war with Britain. The minutes of the meeting only record that "the award of the Gold Medal to Professor Einstein was not confirmed."[169] But the message was clear: rather than give Einstein the award, the RAS elected to give no Gold Medal at all for the first time since 1891.[170] Eddington immediately sent an apology to Einstein:

> I am sorry to say an unexpected thing has happened and at the meeting on Jan. 9 the Council of the RAS rejected the award, which had been carried by quite a large majority at the previous meeting. The facts (which are confidential) are that three names were proposed for the Medal. You were selected by an overwhelming majority in December. Meanwhile the "irreconcilables" took alarm, mustered up their full forces in January, and managed to defeat the confirmation of the award in January. . . . I confess I was very much surprised when the motion was proposed and carried originally (it was proposed by two men who during the war have been violently "patriotic"). . . . I am sure that your disappointment will not be in any way personal; and that you will share with me the regret that this promising opening of a better international spirit has had a rebuff from reaction. Nevertheless I am sure the better spirit is making progress.[171]

He closed the letter expressing hope that Einstein would be able to visit England soon and even be present at a meeting of the RAS, although he admitted there might be "some awkwardness after what happened." Eddington clearly placed the blame for the rejection of the medal squarely on the shoulders of those who joined in with the wartime animosity. The facts of the election are unclear: the original nominations came from H. H. Turner and James Jeans, both of whom during the war had been extremely vocal opponents of any contact with German scientists. Whether they had a change of heart is unknown; they were absent at the January meeting that voted on the final award. In any case, the violence of anti-German rhetoric had not cooled in the few months since the Armistice, and it was likely that the attitudes voiced so loudly during the conflict weighed against Einstein in the end. E. B. Ludlam, writing on Emergency Committee stationery, apologized for his countrymen:

> I find it difficult to believe that English men of science can really be so narrow minded. I think one of the chief difficulties is that scientific men work so hard, and have so much to read, that they have not time to study the real facts in international affairs and accept too easily the opinions of the common press. Your visit to England may be postponed, but I hope not for very long, and it is more evident than ever that there is need of every effort to overcome these foolish and narrow-minded prejudices. . . . Perhaps, when you consider

the campaign of lies which has lasted for five years—in all countries—you will
not judge these poor islanders too harshly.[172]

For Ludlam and Eddington, the medal took on a symbolic political and
religious dimension as a tangible symbol of their work toward reconciliation
and the prevention of future hostility.

Eddington nominated Einstein for the Gold Medal again in 1920 but had
little support. Six years later, Einstein was finally awarded the medal. In a
letter, Eddington told Einstein, "I had not much to do with this decision."[173]
In a literal sense this was true, as Eddington was in Leiden when the vote
was taken. However, in a wider and very fundamental sense, Eddington was
directly responsible for helping create the conditions in which a former enemy
could be welcomed in Britain. The expedition's importance in this task was
noted by many contemporaries. "The fact that a theory formulated by a
German has been confirmed by observations on the part of Englishmen has
brought the possibility of cooperation between these two scientifically minded
nations much closer."[174] The conduct of Einstein's visit to England in 1921
showed both how much progress had been made toward reconciliation and
how much bitterness remained. He was welcomed by Lord Haldane, who
had himself been a victim of Germanophobia, and was feted at a whirlwind
series of dinner parties. There were serious misgivings, however, about his first
public appearance in a city where German science had even become an issue in
the recent election; one radical MP noted that his patriotic colleagues wanted
to "prevent the dumping of German science on these shores, and if [they
win, they] will preserve intact an all-British Law of Gravity."[175] In the end,
Einstein's lecture was well received, despite its being delivered in German.[176]

Conclusion

Scientists and laypersons alike marveled at the warm reception Einstein received despite the still manifest venom toward all things German. His talk
at the Royal Astronomical Society was especially remarkable given the hatred and dehumanization directed toward Germans and German thought in
the British scientific community, which at times appeared to overwhelm any
hope for postwar cooperation. The fact that Einstein's visit took place at all
was in large part due to Eddington's promulgation of Quaker idealism, placed
against patriotic anger. Eddington and the Friends both sought to alleviate the
suffering caused by the war as well as weaken the antagonism between nations
that made the war possible. Thus, Eddington played the same role within the
scientific community that his friends E. B. Ludlam and Anna Braithewaite

Thomas played in general British society: a reminder that the enemies were human and were themselves victims of war. The expedition provided a focal point for these efforts as a discrete and dramatic event that could be used to influence public opinion. Eddington presented the expedition as a milestone in both international relations and human thought, and it was his presentation that entered both the public imagination and historical record. It now appears in innumerable textbooks and popular histories, usually transparent of Eddington's Quaker values which brought it to prominence.

A crucial element of these values was the philosophy that responsibility for the war could not be hurled as a blanket aspersion against anyone in the enemy countries. Rather, responsibility was held up as an individual issue: this allowed Quakers to seek both to defeat German militarism and to ease the suffering of the German people. This distinction was flatly denied by the most patriotic supporters of the war, such as H. H. Turner, speaking through the mouthpiece of the "Oxford Note-Book." Such supporters argued that nationality alone was sufficient for incrimination and that the acts of the German government and army directly reflected the character of its citizens (and its scientists). Planning for postwar science had begun seriously in 1918, with a Royal Society draft of a statement on international science. The justification for excluding the Central Powers was quite clear:

> When more than four years ago the outbreak of war divided Europe into hostile camps, men of science were still able to hope that the conclusion of peace would join again the broken threads, and that the present enemies might then once more be able to meet in friendly conference, united in their efforts to advance the interests of science... [but] the Central Powers—and more particularly Germany—have broken the ordinances of civilization, disregarding all conventions, and unbridling the worst passions which the ferocity of war engenders. War is necessarily full of cruelties; individual acts of barbarity cannot be avoided and have to be borne. It is not of these we speak, but of the organized horrors encouraged and initiated from above with the sole object of terrorizing unoffending communities. The wanton destruction of property, the murder and outrage of innocent persons, the sinking of hospital ships, the insults and tortures inflicted on prisoners of war, have left a stain on the history of the guilty nations, which cannot be wiped out by their signatures attached to a treaty of peace.[177]

Their views held sway long after the war and gained validation with the establishment of the International Research Council and its subgroup the International Astronomical Union, which explicitly denied membership to scientists of the Central Powers.[178] The struggle for the future of international science would continue for nearly a decade after the end of the war, and its course would largely reflect the treatment of Germany in political relations, as it did during the war.

Eddington continued to work for peaceful internationalism both inside and outside science for the rest of his life. His activism was renewed with the rise of tensions leading up to the Second World War, when he joined a high-profile appeal for peace: "It is time, if we are not too late, that men of good will who value the fruits of civilization, who have no hatred or spirit of revenge in their hearts, and who desire in all sincerity to live on terms of friendship with their fellow men in every country, should speak across the frontiers to those who feel as they do, in order that they may use together their gifts of heart and mind to cooperate in preventing the supreme catastrophe and in breaking down the artificial barriers of hatred by which we are in danger of being divided."[179] The gathering clouds of war also spurred him to accept the position of president of the National Peace Council, an organization dedicated to a peaceful, decolonized postwar world.[180]

The scientific merit of the 1919 expedition continues to be controversial today. A typical accusation is that Eddington intentionally discarded or misinterpreted data so as to confirm Einstein's prediction.[181] The basis of this is the claim that there was no justification for viewing some of the data as more reliable than the rest. But as I have argued, the evidence shows that the quality and utility of the photographs were carefully considered by Eddington and Dyson, and the determination of the unreliability of the Sobral astrographic results was made in the field by the observers in Brazil (who did not include Eddington). Were these decisions difficult? Yes. Could they have been made only by trained and experienced observers? Yes. But the importance of this tacit knowledge does not mean that the results were untrustworthy. Indeed, since the community the actors needed to persuade (most directly, astronomers) was also well versed in this tacit knowledge, one cannot complain that it was used to obscure the basis of their choices.

None of this is to say that the results were precise and unarguable. The error was fairly large and would not be greatly improved on until the much later development of other techniques such as the Shapiro time delay. The bottom line, however, is that contemporary astronomers were persuaded there was a deflection and that it was most likely associated with Einstein's law of gravity. There was, of course, disagreement about the results throughout the 1920s and 1930s, but I know of no serious accusations of impropriety on Eddington's part. Why, then, has it now become common opinion that he fudged the results? Since this opinion exists largely as folklore, it lies beyond the realm of this book to investigate fully. However, I would like to suggest that it is the result of the severe blow Eddington took to his reputation as a result of his writings on mysticism and religion, as well as the spectacular rejection of his attempts to unify relativity and quantum mechanics. This later work of his, looked upon with disfavor by many of his colleagues, may have tarnished him sufficiently to cast doubt on his previous work as well. It has been my personal experience that physicists are much more willing to impugn him than astronomers, and I speculate that this is because of the former group's greater exposure to his unified field theory, while the latter remember his still-important work in astrophysics.

Despite frequent criticism of the results, astronomers and scientists still look back and point to the 1919 eclipse expedition as an example of how scientific internationalism could rise above any challenge. But the memory of the test of Einstein's theory as a straightforward and harmonious conjunction of the scientists of nations in conflict only came long after the end of the conflict. It was only with Eddington's deliberate presentation of the expedition as a milestone in international scientific relations that it came to have that meaning. To its contemporaries, the expedition was a symbol of highly contested visions of what it meant to do science in a world at war. Examining the expedition in the context of wartime Britain has shown us that it was not only a pivotal moment in scientific investigation, but also in the debate over the relationship between science, war, politics, and peace. For Eddington, Turner, and virtually every historical actor involved, science was an enterprise necessarily tied to some aspect of civilization, patriotism, or religion. Eddington's involvement with, and promotion of, Einstein and relativity therefore takes on the important and rich context of his role as the representative of Quaker pacifism in a scientific community grappling with the horrors of what was the greatest conflict yet in human history.

Pacifism
Confronting the State and Maintaining Identity in War

A. S. Eddington: "I am a conscientious objector."
Chairman of the Tribunal: "That question is not before us."
Cambridge Daily News, June 14, 1918

In June of 1918 A. S. Eddington stood in a crowded courtroom in Cambridgeshire. He faced a hostile tribunal determined to bring the full weight of the wartime state against his pacifist values. At stake were his career, his freedom, and possibly even his life. How did it come to be that Eddington, Fellow of the Royal Society, Secretary of the Royal Astronomical Society, holder of a centuries-old chair in Cambridge, and future knight of the realm was in danger of becoming a military prisoner and social pariah?

The scientific community was far from isolated from the social forces of the Great War. Chief among these forces was universal conscription, the introduction of which many Britons saw as a challenge to the very foundations of their liberal society. Eddington's Quaker co-religionists were among the first to experience the societal changes wrought by full mobilization: their peace testimony led them to refuse this new national obligation, and the full weight of British government and society was brought against them. On the other extreme of support for the war lay the scientific societies and their leaders, who sought to integrate British science as closely as possible into the war effort. They felt scientists could do as much to answer Lord Kitchener's call as anyone who could hold a rifle.

Eddington lay at the intersection of these two movements. The social place of the Quakers during the war was shaped by their refusal to participate, but the scientists were fashioning their own social role as indispensable elements of the war effort. Eddington's professional and personal identity meant he was pulled by both of these forces, and the tension was brought to a head by the machinery of conscription. Early in the war he was marked as a scientist and exempted from fighting by his apparent usefulness to the war as an astronomer. As the war progressed, however, the need for men became more desperate and his exemption for "work of national importance" was revoked. Eddington then stepped forward and declared his religious objection to the war. However, he was formally denied the opportunity to do so: his previous categorization as a scientist meant that, officially, *he could not also be religious*. The structure of the British conscription scheme and, by extension, much of British society and government, did not allow for the possibility of a religious scientist. Eddington's valence value of pacifism, however, did not allow him to capitulate one part of his identity in order to preserve the other. Valence values do not reside in only one aspect of a person; they contribute significantly to all of the realms they bond together. The difficulty for Eddington was how to maintain this symmetry when the power of the state was committed to disrupting it.

This chapter will examine the process by which Eddington came to contest the official understanding of the scientist and the Quaker. The process began with the introduction of conscription in 1916 and the reactions of the Society of Friends and the scientific societies. These events then formed the landscape through which Eddington had to navigate as a religious scientist and the conditions that shaped his experience of wartime Cambridge. Eddington's maintenance of his identity against dictated norms raised the possibility of specific and dramatic consequences: in a time of total war, it could lead to prison or even death.

The Emergence of Conscription

Even as the Great War revealed its brutal scale, it was widely expected that Britain would fight the entire conflict with the forces with which it began the war. It was thought that the average citizen's life would be little affected.[1] The entire Victorian state had been built on the assumption that personal freedom, without government interference, was the height of civilization and the root of British economic success. The historian A. J. P. Taylor claimed, with only slight exaggeration, that "until August 1914 a sensible, law-abiding Englishman could pass through life and hardly notice the existence of the state."[2]

So when the bloody stalemate of the first year of fighting made it clear that more forces would have to be raised, there was great hand-wringing in Whitehall. The obvious and practical solution was conscription, and, indeed, many public figures had been calling for universal military service on the Continental model for some years.[3] But this would be such a gross violation of what were seen as basic liberties that Prime Minister H. H. Asquith removed that option from the table. Many of the protests against conscription argued that it would be adopting the very Prussian militarism that Britain was fighting the war to destroy.[4] Lord Kitchener, the Secretary of State for War, was one of the few members of the government who expected a long war, and he immediately set out to raise troops by voluntary enlistment. At Kitchener's request, Parliament approved the induction of a hundred thousand men just a few days after the war started.[5] The first recruiting appeal read:

YOUR KING AND COUNTRY NEED YOU

A CALL TO ARMS

An addition of 100,000 men to His Majesty's Regular Army is immediately necessary in the present grave National Emergency. Lord Kitchener is confident that this appeal will be at once responded to by all who have the safety of our Empire at heart.[6]

There was a huge rush to enlist before the war was over, and the Army had no trouble filling its ranks. Valentine's Games even made a board game about enlistment called "Recruiting for Kitchener's Army," in which players try to pass the medical exam so they can serve king and country.[7] The Archbishop of Canterbury declined to make the Church an official recruiting arm, but the vast majority of individual clergy used their positions to make the case for the moral obligation to volunteer.[8] The defense of tiny Belgium was held up as a Christian duty.[9] Officers were in particular demand, and a commission was given to anyone with a public school or university background. Both Cambridge and Oxford were rapidly depleted of undergraduates.[10]

The ferocity of the fighting in France showed that this would not be enough, and Parliament approved another hundred thousand on August 28, and a half-million on September 15.[11] Recruiting remained strong until 1915, when there was a notable decrease in the numbers of young men coming forward. There was much grumbling about "slackers"—men of military age that were not enlisting of their own accord. There were more calls for conscription, but the prime minister resisted these and instead in July engineered the National Registration Act, which created a register of all persons (male and female) between age 15 and 65. Many people accused the government of preparing for conscription, but Asquith reassured them that he wanted to prosecute the war "along Liberal lines."[12] But it was clear to those running the war that something would need to be done soon.

The solution proposed the next month was the Derby scheme, "one of those shotgun weddings between the fair maid of Victorian liberalism and the Ogre of Tory militarism."[13] Lord Derby, Director-General of Recruiting, asked men to "attest" that they would serve when called upon. There was thus no formal compulsion, but a tremendous amount of moral and social pressure was placed upon men of all ages. One of the recruiting letters from Derby read, in part:

Mr. Asquith pledged this Country to support our Allies to the fullest extent in our power. It was a pledge given on behalf of the Nation and endorsed by all parties. Every man of military age and fitness must equally bear his share in redeeming it. May I, as Director-General of Recruiting, beg you to consider your own position? Ask yourself whether, in a country fighting as ours is for its very existence, you are doing all you can for its safety, and whether the reason you have hitherto held valid as one for not enlisting holds good at the present crisis. Lord Kitchener wants every man he can get. Will you not be one of those who respond to your Country's call?[14]

Those who attested were given an armlet to show that they were not shirking their duty, and not incidentally to free them from harassment by the groups of women organized by the Admiralty to shame men into enlisting.[15] The scheme, although of limited likely effectiveness, was a compromise that was acceptable to the entire political spectrum. The libertarians were happy because, if the scheme worked, the principle of voluntarism had been protected. The militarists were happy because, if the scheme failed, the case for universal conscription would be overwhelming.[16]

The first tribunal system was set up soon after the scheme was put in motion. Its original purpose was for employers to seek exemptions for their employees who had attested. For example, the owner of a munitions factory might ask to retain a particularly skilled worker. The questionnaire the employer needed to fill out was heavily weighted toward recruiting manpower: "What facilities have been afforded [in the workplace] to men to join the Forces? . . . How many men of military age are now employed at the same work as the man to whom this application relates and how many men were so employed before the War? How many of these joined the forces at the time they were employed by the firm? . . . [What makes this man] specially valuable and irreplaceable?"[17] Occasionally, men who had attested wished to withdraw from the scheme because of changed circumstances, such as an ill relative. Given the voluntary nature of the scheme, however, the work of the tribunals was both light and easy.[18]

After two months, the results from the Derby scheme were analyzed, and the news was not good. About half of the eligible men had attested, and the Army still found itself with a massive manpower deficiency.[19] Conscription was now inevitable, and the government introduced the Military Service Bill in January 1916. This imposed compulsion on all unmarried men between eighteen and forty-one and did not apply to Ireland. There was great protest from the left. Sir John Simon, the Home Secretary, declared that the bill was the beginning of an "immense change in the structure of our society."[20] He promptly resigned in protest. Many Labour MPs threatened to resign as well, although none actually did so. Debate was furious. The Quaker MPs Arnold S. Rowntree and T. Edmund Harvey argued for a clause that would provide conscientious objector status for those with principled objections to military service. Asquith supported this motion, not least because he hoped it would placate Simon into not organizing resistance in the House of Commons. He defended the principle of a conscience clause based on William Pitt's exemption of Quakers from service in the militia during the Napoleonic Wars.[21] One MP, Joynson Hicks, thought that any such conscience clause should apply only to Quakers and similar religious groups, lest it become a shield for cowards and "slackers." Rowntree saw that this would focus anger against the Friends and warned the House "against beginning a religious persecution." The clause was in the final version of the Military Service Act, but it was roundly condemned and commonly known as the "The Slackers' Charter."[22] The bill passed in the Commons by a vote of 431 to 39.[23]

The machinery of conscription was built upon the skeleton of the Derby scheme and the National Register. Recruiting offices smoothly became processing centers, and the existing tribunal system was given a host of new responsibilities and powers. Now that the forbidden step of compulsion had been taken, Parliament freely expanded upon it, though not always skillfully. In April a second bill seeking conscription of married men was introduced to a secret session of the House of Commons, but it was so poorly constructed that it was immediately rejected. A better crafted substitute was passed in May, and all men, married or single, of age in England, Scotland, and Wales could now be forced to serve in the Army.[24]

The Refusal to Fight

Organized resistance to conscription had been present long before the government began speaking of it as a possibility. The No-Conscription Fellowship was founded in October 1914 to prevent the adoption of military compulsion. Clifford Allen was the president, and the membership was drawn largely from the Society of Friends and the socialists. The NCF's position was a libertarian one. Their activities were widespread and influential and probably helped the introduction of the conscience clause in the Military Service Act. After the act became law, the NCF switched its goal to the repeal of conscription: "Fellow Citizens: Conscription is now law in this country of free traditions. Our hard-won liberties have been violated. Conscription means the desecration of principles that we have long held dear; it involves the subordination of civil liberties to military dictation; it imperils the freedom of individual conscience and establishes in our midst that militarism which menaces all social progress and divides the peoples of all nations."[25] They also produced useful leaflets such as "The Court-Martial Friend and Prison Guide." This sort of politically based pacifism was a new phenomenon, though, and British society still equated nonviolence with Nonconformist religion. When the government considered how to deal with conscientious objectors, they were considering how to deal with the Society of Friends.[26]

The Quakers began planning how to deal with conscription as soon as the National Register was formed. There was no question that they would refuse to fight on the front lines, and their internal discussions dealt with the morality of accepting alternative service (that is, noncombatant jobs within the military). John William Graham, a respected Quaker leader and Eddington's mentor from Manchester, was not sure where to draw the line. Hospital and auxiliary services were probably acceptable: it would not free a man to fight since such jobs would probably be filled with men unfit for the front anyway. In some sense, it was impossible not to help with the war in some way, just by being a British citizen.[27]

The Friends' Institute in Manchester quickly became a clearinghouse for information on the act and for resisting conscription. They held nightly sessions to orient conscientious objectors and help them fill out the forms. Mock tribunals were held to accustom young Friends to the sort of questioning they would receive.[28] Shortly after the passing of the act in January, the Quaker Yearly Meeting in London announced their official position on conscription: "We regard the central conception of the Act as imperiling the liberty of the individual conscience—which is the main hope of human progress—and entrenching more deeply that Militarism from which we all desire the world to be freed."[29] J. W. Graham wrote later, "You will not eliminate Militarism from Europe by opposing to a German Militarism a superior British-French-Russian Militarism. You will simply establish Militarism securely in England."[30] The Yearly Meeting concluded that Quakers must refuse alternative military service as well as combatant roles. Anything that helped oil the gears of war was contrary to the peace testimony. Militarism could not be resisted by taking part in its methods: "It is not by compromise with an evil thing, but by a passion of goodwill, that the war spirit must be met."[31]

Scientists Step Forward

The majority of the nation sought to help the war effort in any way they could. This included all of the significant scientific societies and organizations in Britain: the Royal Society, the Royal Astronomical Society, and the British Association for the Advancement of Science all made special efforts to put their skills at the disposal of the War Office. However, the government as a whole was not particularly interested in scientific and technical help, and the patriotic scientists were constantly forced to justify their importance.

In the fall of 1914 the Royal Society formed a War Committee to offer assistance to the government in scientific matters.[32] The committee reported favorable responses from the government offices it had contacted, and it was soon helping with wartime production difficulties.[33] Lord Haldane set up a Government Chemical Products Supply Committee when it was realized how much of its chemicals Britain had been importing from Germany.[34] There was shock in Parliament when it was suddenly confronted with how deficient Britain was in technical industries. The Royal Society's help proved invaluable in ameliorating this, as the chemical industry was making some of these chemicals for the first time ever in England.[35] It is important to note that the government committee did little in assisting research; the Royal Society generally dealt directly with the industries needing its help. The official attitude of Britain was "business as usual," and this meant no mobilization of science. The laissez-faire philosophy of Victorian and Edwardian times was still potent and had been the chief reason state support of science was so meager on the eve of war.[36] A further issue was the military's skepticism that intellectuals would have much to offer on matters such as munitions manufacture.[37]

Despite a hopeful beginning, the Royal Society became frustrated with the government's lack of active support for science's role in the war effort and in May 1915 made a public request along those lines. Whether the government was shamed or inspired is difficult to say (a highly visible campaign of letters to the editor may also have been influential), but at the end of July an Advisory Council was formed to deal with scientific and technical issues. It was given £25,000 for its first year and £40,000 for the second.[38] Despite this first good step, the council still did little to support research. It was overseen by the Lord President of the Privy Council, who had a stunning number of additional responsibilities. The committees that composed the Advisory Council rarely met.[39] The next year, many of the British scientific societies came together to form the Conjoint Board of Scientific Societies, intended to promote cooperation between the government and science.[40] This new board immediately issued a memorandum complaining of past and apparent continuing British official neglect of science, contrasted with the "recognition and encouragement" given to scientific men in Germany.[41] A Neglect of Science Committee was even formed, chaired by a figure of no less stature than Lord Rayleigh.[42]

Unfortunately, the limiting factor in the cooperation sought by the board was not the scientists, but the disorganization and lack of interest of the government. There were both cultural and structural problems. There was the general disinclination to work with scientists, a holdover from Victorian times that one historian describes as: "there was no question of the state, its officials and politicians placing any trust in the work and ideas of scientists."[43] Also, research efforts were fragmented between government offices. The Admiralty had set up its own Board of Invention and Research, as did the Ministry of Munitions, and the Medical Research Committee was an entirely independent entity formed in 1911.[44] The metamorphosis of the Advisory Council into the formal Department for Scientific and Industrial Research helped somewhat, and the government slowly began to synchronize its efforts with that of its resident scientists. It would eventually seek and support research on navigation, wireless transmission, and optical glass production.[45] Its efforts to increase scientific manpower through training were stymied by the huge number of science teachers and scientists that had enlisted or been conscripted.[46]

After the war, the Royal Society looked back sourly on its contribution to the war effort. In an internal document, it complained about the handicap Britain fought under early in the war. They directly fingered the government as the source of this handicap: "It was difficult to convince the Authorities that the application of science held out sufficient prospects of success to make it worth while to spend time and money on research. . . . Even where scientific methods were adopted, the work was often put into the hands of the wrong men or the wrong organization."[47] The document went on to list the benefits that the Germans and the French were able to reap from their more properly mobilized and appreciated scientists.[48]

British scientists thus had mixed success in defining their role in wartime society. Their help was appreciated but not generally sought. Most scientists who did war work did so as volunteers and never appeared on the government payroll.[49] They felt their work, including ballistics, acoustics, chemistry, meteorology, and optics, was crucial to the war effort but could not persuade the authorities of it.[50] This had important consequences after the introduction of conscription, when suddenly every man needed to justify his work as being of national value. The nation's ambivalence toward science and technology was clearly seen in the effect of recruitment on scientists and would be brought into particular relief when it collided with its official stance toward conscientious objectors in the case of Eddington.

Exemptions for Scientists

Even before conscription was a political reality, scientists across Britain worried about the effect of a general compulsion on their work. Partly this was a result of so many of their colleagues and assistants enlisting voluntarily, which had already begun to affect labs and universities. The Cambridge Observatory

Syndicate was quite concerned about the smooth functioning of their institution. Under the rules of the Derby scheme, they could file a request for exemptions for their employees who had attested, and they felt it was "their duty to apply that the scientific work of the Cambridge Observatory, which is one of the most important observatories in England, should be safe guarded as far as the national needs permit."[51] It was acknowledged that if Eddington was called up, the observatory would have to be shut down. The Syndicate drafted a note to the Vice-Chancellor of the University for further action on obtaining exemptions for its staff.[52] By March, Eddington was the only member of the observatory staff not serving with the military.

The Royal Society received several requests from university professors looking for help attaining exemptions for their assistants and technicians. The Society said it was sympathetic but did not feel justified in taking action on their behalf.[53] It was still committed to integrating science and scientists into the war effort and did not want to be part of anything that suggested scientists were not eager to fight. The death of the gifted young physicist H. G. Moseley at Gallipoli shook many scientists.[54] Despite the strong efforts of scientists to contribute to the war as scientists, the military was not sure whether their efforts would be best appreciated in the lab or in the trenches. Generally, when there was any doubt, the tribunals assumed every man would be best as a soldier. Part of the confusion was that the tribunals had tremendous latitude in deciding whether individuals in protected employment would actually be deferred: the MP Philip Snowden reminded applicants that "while certain trades are certified as indispensable, it does not follow that every individual working in such a trade will be regarded as indispensable."[55] Technical and scientific trades seemed to suffer from particular equivocation. As many pharmaceutical chemists were exempted as were conscripted; a bacteriologist's request was granted, but an industrial chemist's work was not considered to be of national importance.[56] A later clarification to the tribunals did not particularly help matters: "In some cases the most valuable national service a man can render to the State may prove to be the technical or professional work in which he is engaged. But generally it will be necessary to allocate him to some other approved service."[57] Those individuals who were declared to have work of national importance were entitled to wear a war service badge so they would not be mistaken for shirkers.

The details of the tribunals' operations were often ad hoc. The Military Service Act had been passed into law with little consideration of specifics. The first meeting of the Central Tribunal (headed by Lord Derby himself) decided farmers would not be "starred," that is, exempted, but the next meeting a week later announced that decisions on farmers would be made case by case.[58] The original starred occupations were production or transport of munitions, coal mining, agriculture, certain other mining related jobs, and railway maintenance and traffic.[59] This list was made up with no consultation of industry or employers, however, and it was quickly amended on an individual basis. The tribunals were made up of between five and twenty-five men (usually the minimum), with a Military Service Representative present. Typical membership was described as "elderly local magnates or tradesmen, often with a Labour man, known to be in favour of the war, added."[60] There was one tribunal roughly for every rural district, urban district, or town. An appeal tribunal was set up for each county, and the Central Tribunal was formally above of all of these, but only heard cases passed to it by an appeal tribunal.

As originally conceived, the purpose of the tribunals was to ensure that mass conscription did not adversely affect work of "national importance." But the government did not anticipate the difficulties the conscience clause would create. Instead of making more-or-less straightforward decisions about economic contribution, the tribunals found themselves spending most of their time judging the human soul and religious belief. When an applicant claimed they held a conscientious objection to war, the tribunals' responsibility was to decide whether the objection was *genuine*. It was assumed until proved otherwise that the claimant was a shirker attempting to evade their duties as a citizen.

Instructions to the tribunals read "While care must be taken that the man who shirks his duty does not find unworthy shelter behind this provision, *every consideration should be given to the man whose objection genuinely rests on religious and moral convictions.* Whatever may be the views of the members of the Tribunal, they must interpret the Act in an impartial and tolerant spirit. Difference of convictions must not bias judgment."[61] Unfortunately this idealism did not translate well into practice. One observer noted that the tribunals were "almost entirely composed of men openly antagonistic to pacifists."[62] They had few rules and regulations and their practices varied tremendously from place to place. The Military Service Representatives often dominated them. The tribunals saw their task as being to detect shirkers, not affirm religious belief. Because of this, the hearings were often more like interrogations. There was plenty of "petty bullying and even crass stupidity."[63] The tribunal members shared in the general social attitude of despising a conscientious objector (or, in the contemporary slang, "CO"), and such a person was considered to be "unpatriotic, a slacker, a weakling."[64] Objectors were subjected to aggressive questioning of their religious beliefs, and many men unused to such attacks simply crumpled under the treatment. The standard question was "What would you do if a German attacked your mother?" If the applicant responded that he would defend her, he should be sent to the front. If he responded that he would not, he was clearly lying and was labeled "not genuine." The tribunals did not care to hear nuanced answers differentiating between the policeman and the soldier and usually cut off any meaningful discussion. Verbal abuse of the applicant was common. One man was told: "You are exploiting God to save your own skin. You are nothing but a shivering mass of unwholesome fat."[65] This hatred of someone who refused to defend the liberties behind which they hid was virtually universal in Britain. The Earl of Malmesbury commented that the CO was "sailing dangerously near the very ugly word traitor."[66]

Some of the religious COs, denied the opportunity to assert their objection in their hearings, wrote moving tributes to their belief. Dr. Alfred Salter's essay "The Religion of the C.O." represented this group well. Here he responded to the common tribunal attack on the belief that the Christian religion was opposed to war: "Look! Christ in khaki, out in France thrusting His bayonet into the body of a German workman. . . . Hark! The Man of Sorrows in a cavalry charge, cutting, hacking, thrusting, crushing, cheering. No! No! That picture is an impossible one, *and we all know it.* That settles the matter for me." In response to the question of self-defense, he stated the position of non-resistance unequivocally: "I say deliberately that I am prepared to be shot rather than kill a German peasant with whom I have no conceivable quarrel. I will do nothing to kill a foe, directly or indirectly, by my own hand or by proxy. So help me, God."[67]

Despite testimony such as this, some tribunals declared they would not give absolute exemption to anyone. An exemption from all forms of service was extremely rare. Most of the time a "genuine" objector was granted exemption from combatant duties or conditional on finding work of "national importance." They then had the option of serving in the Non-Combatant Corps building roads and digging ditches in France or working in a munitions factory (or similar work). Those who refused even this as infringing upon their conscience were arrested, fined £2, and handed over to the military authorities. Their fate would be grim.

Cambridge

The end of February saw the first CO case before the Cambridge tribunal. The case was dismissed, but it probably should not be seen as typical, since the petitioner also asked for exemptions on the grounds of his business and a foot injury that had been bothering him.[68] It was clear that the conscience clause would be tested soon, though. On March 1 the *Cambridge Daily News* had a short discussion of conscientious objectors and their treatment:

As the Military Service Act provides for the exemption of conscientious objectors it is the disagreeable duty of the local tribunals to decide whether such applicants are really acting from conviction or from a desire to shirk their duty. In the case of Quakers, of course, the matter is simple enough: the doctrine of non-resistance is the distinguishing characteristic of their creed. It might have been wise of the Government definitely to have specified members of this body, and confined exemptions to them, but instead the door was opened to any religious fanatic and every shirker who is clever enough to hide behind this clause. It is a difficult matter for the plain man to appreciate the position of the man who accepts the protection of the Army and the Navy . . . yet refuses to take part in the protection of others. . . . To our mind, the willingness to fight for one's country is the proper test of citizenship, and might well form the basis of the franchise. To those who refuse to accept this principle we would present a free ticket to the United States or to any other country that would be willing to accept them on their own terms.[69]

The article noted that many CO applications also had another reason for exemption. Skepticism of applicants with multiple reasons for exemption was widespread, especially among the members of the tribunals. The categories of exemption were seen as self-contained and inflexible, and there was no consideration of overlap. An applicant often had to fill out a separate form for each requested exemption (one for employment, one for conscientious objection, etc.), making it difficult for them to make a unified case and easy for a tribunal to ignore or dismiss one of the petitions.[70]

As soon as conscription became law, Cambridge University obtained an exemption for Eddington based on his work at the observatory being of national importance. Partially this was simply to prevent losing him to the war, but it was well known that Eddington would have claimed CO status, and the university had no interest in the embarrassment that would have come with that. Eddington had actually applied for conscientious objector status, but the local tribunal did not even process it; the university had placed him in the category of "scientist," and it was impossible to address whether he should also be evaluated for the category of "Quaker." Indeed, there was no concept of an "also." There could be no overlap. Because the university applied as his employer, Eddington never had to appear before a tribunal. He was now officially a scientist doing work important for the war.

Everyday life in the Cambridge Observatory would have been a constant reminder that Eddington was refusing to serve his king and country. He and other COs were required to carry their exemption papers at all times. The War Office dictated that any place of employment that had an employee holding a national importance exemption hang a large poster detailing why those individuals were not at the front. It was headed "Defence of the Realm (Consolidation) Regulations: List of Male Employees between the Ages of 18 and 41. NB—this List must be posted in some conspicuous place on the premises in or about which the persons are employed." It listed the name, address, and job of the exempted employee and the nature of the document exempting the employee from military service.[71] The poster hanging in the Cambridge Observatory had just one name on it.

Life in wartime Cambridge was full of tension, even more so for Eddington in his awkward position. The University Sermons were enthusiastic about the war and the role Cambridge could play. One sermon said the war was God's way of bringing spiritual regeneration to England.[72] Men at the universities were particularly motivated as guardians of civilization by the destruction of Louvain. "He will carry with him into the field the memory of that martyred city, whose ashes cry aloud for the vindication of true culture against the barbarity made possible and said to be sanctioned by a false *Kultur.*"[73] Fear of invasion was an equally strong motivation, and a university senator argued that all men of military age should undergo training to repulse such an attack. Civil defense instructions in the case of an enemy landing were hung throughout the city (see fig. 4.1). The local newspapers regularly reported prosecutions for violating wartime blackout restrictions.[74]

The Officer Training Corps was a strong influence at Cambridge, as well as at Oxford and the newer universities.[75] One observer noted that Cambridge had been converted from academics to war preparation.[76] The biochemical, engineering, and Cavendish labs were commandeered for the billeting of troops, and after two in the afternoon, military duties were "the only thing." Bertrand Russell described it thus: "The melancholy of this place now-a-days is beyond endurance—the Colleges are dead, except for a few Indians and a few pale pacifists and bloodthirsty old men hobbling along victorious in the absence of youth. Soldiers are billeted in the courts and drill on the grass; bellicose parsons preach to them in stentorian tones from the steps of the Hall."[77] The halls were virtually empty. According to the *Cambridge Review,* attendance dropped from 3,263 in October 1913 to 398 in October 1917.[78]

Like the scientific societies, the university was eager to put its collective skills at the disposal of the country. The Vice-Chancellor proclaimed, "The policy of the university in this crisis has been to render all employment of its resources, material and intellectual, for the benefit of the country." Lloyd George formally invited Britain's educational institutions to help, and J. J. Thomson answered proudly for Trinity College. The Cavendish Laboratory had begun investigations into signaling, acoustics, wireless transmission, and high explosives. This and other lab work at the university was done with

COUNTY OF CAMBRIDGESHIRE

(Including the Isle of Ely and the Borough of Cambridge.)

Instructions for the Guidance of the Civil Population in the event of a landing by the Enemy in this Country.

Local Emergency Committees have been formed in all the Divisions of the County, acting under a Central Emergency Committee, appointed by Viscount Clifden, the Lord Lieutenant. Their duty is to act in co-operation with the Military Authority in case of an invasion.

In order to facilitate the operations of His Majesty's Forces and to hinder those of the Enemy, the following instructions have been given by the Central Emergency Committee and are to be followed by the Civil Population as soon as, but not before, the Military Authority declares that a State of Emergency has arisen. Their action will be taken by the Police Authorities assisted by Special Constables, working under instructions from the Military Authority.

Special Constables have the full powers of uniformed Police and must be obeyed.

HOUSEHOLDERS RECEIVING THIS NOTICE SHOULD MAKE ITS CONTENTS KNOWN TO ALL PERSONS LIVING IN THEIR HOUSES, AND EXPLAIN IT TO ANY NEIGHBOURS WHO CANNOT READ.

1. NOTICE TO THE PEOPLE.

Unless otherwise directed all persons are advised TO REMAIN QUIETLY IN THEIR HOMES, AND NOT TO MOVE WITHOUT ORDERS. IT IS STRONGLY IMPRESSED ON EVERY MAN THAT UNLESS IN UNIFORM AND ACTING UNDER ORDERS HE MUST NOT UNDER ANY CIRCUMSTANCES COMMIT ANY HOSTILE ACT AGAINST THE ENEMY OR BE IN POSSESSION OF FIREARMS, OR HE WILL EXPOSE HIM-SELF, FAMILY AND NEIGHBOURS TO CERTAIN REVENGE FROM THE ENEMY.

Persons in possession of FIREARMS may be required to hand them over to the Police.

2. TRANSPORT AND PETROL.

All horses, mules, donkeys, motors, bicycles, carts, carriages, and other vehicles, harness, petrol, launches, and lighters should be immediately moved to a preconcerted place, or as far as is practicable from the area of military operations.

FIGURE 4.1

A warning of enemy invasion.

the intent of helping the war effort but was largely the result of individual initiative and was not coordinated or sponsored by the government.[79] The Vice-Chancellor also asserted that pure culture such as literature and history was useful in war too: knowledge of out-of-the-way subjects and dialects could help with "propaganda, censorship or intelligence."[80]

Part of this enthusiasm for the war was a reaction to the growing unease regarding a blossoming peace movement at Cambridge. Particularly in contrast to patriotic Oxford, many saw it as a dangerous hotbed of pacifism. It had more peace societies than any other university, and they were becoming increasingly active.[81] The Cambridge University Friends Society (i.e., the Quakers) and the Cambridge University Socialists published a pamphlet detailing the close ties the senators were trying to make with the military. The cover illustration was a man in khaki scooping the brains out of armed men in academic gowns and dumping the contents into a sack marked "garbage."[82] The authors' intent was to mobilize the university for peace instead of war: "if [university men] resist a war they can very considerably hinder its progress."[83] Much high-profile pacifism also came from the *Cambridge Magazine*. Founded by C. K. Ogden in 1912, it was not avowedly against the war, but its inclination was clear.

During the national debate on the necessity and suitability of conscription, some of the Cambridge dons outdid the militarists by arguing for making military service compulsory for all graduates. The *University Socialist* responded with a piece intended to reassure the nation that Cambridge was not the forefront of militarism it seemed. The author ("a Cambridge B.A.") reported that the Union had been full of undergraduate debates on a possible rapprochement with Germany. He gave a nuanced view of this, saying that the details of the world situation must be considered carefully. On the issue of compulsion, he wanted the university to fulfill its intellectual role and "lead the nation" by considering "the need and effect of conscription."[84] Of course, this was exactly the sort of leadership with which the university administration did not want to be associated. When the Union for Democratic Control, an organization devoted to a negotiated peace, started meeting on campus, it was immediately banned.[85] Worried that its own Fellows were helping or organizing pacifist groups, the University Council tried to ban a Fellow from having a private meeting in his rooms. G. H. Hardy drafted a protest against this, aided by Eddington and several other Fellows.[86]

Eddington's workplace was thus a site of distinct tension. The authorities saw the university as a dangerous source of pacifism and were working to increase the support there for the war. The pacifists saw it as a dangerous source of militarism and were working to decrease support for the war. Polarization set in quickly: *pacifist* became applied to anyone who did not accept victory at all costs. "The word degenerated into a term of abuse."[87] Eddington's colleagues at Trinity, such as Thomson, were arguing that scientists must work for the war, especially if they already held national importance exemptions (as Eddington did). As the war progressed, what little tolerance there was for pacifists quickly evaporated. Pressure mounted for Eddington to conform to the ideal of the wartime scientist. Tensions among Fellows at Trinity grew to the point where they avoided each other.[88]

Trinity was not Eddington's only cultural location in Cambridge, however. The Friends Meeting House on Jesus Lane was a focus for his religious identity and practice and shaped his experience of the war. When conscription became a possibility, the Monthly Meeting (which Eddington, as head of several committees, probably would have attended) sought unity on the issue and came to the following statement: "This meeting has given very serious consideration to the subject of conscription and the attitude of friends in regard to it. Should a Bill to establish conscription be brought into Parliament, we earnestly desire that it should include exemption for all conscientious objectors, and if such a clause were not included we should work to obtain it."[89] If an exemption clause were obtained, they recommended young Friends take advantage of it even if it applied only to Friends. It was decided that the benefit of helping pave the way for other COs would outweigh the initial difficulty other objectors would have. As J. W. Graham did in Manchester, the Meeting appointed a committee to advise potential COs. Separating slightly from the Yearly Meeting, the Cambridge Quakers decided that the question of whether to accept alternative service instead of conscription "can only be settled by the individual conscience."[90] The choices of Quakers like Eddington, who had to decide whether accepting national importance exemptions violated their beliefs, was thus moved from the realm of the social group to that of the personal. This was in keeping with the most important Quaker traditions of respecting the Inner Light. When seen from outside the community, however, it likely appeared as inconsistency and wavering. Naturally, this would cause difficulty when Friends had to justify themselves before tribunals.

Anticipating confusion regarding their peace testimony (How could one be a pacifist and not be pro-German?), the Cambridge Friends attempted to publish a statement of peace in the local papers but were refused.[91] J. W. Graham submitted an article on the peace testimony to the magazine *Christian World*, but it was rejected because "it does not deal with the Quaker position in a way that is of much help at the present time." The editor went on:

> What you do—you will excuse me if I speak plainly—is to interpret the teaching of Christ for a world in which there are no marauding Germans. At a time

when all the liberties are at stake for which Quakers and others have contended for centuries, you argue that it would be wrong for you to take a hand in the rough work of preserving those liberties. You speak of one pacifist nation, but were not all the nations at any rate trying to become pacifist, except Germany? Again, as a body Quakers are not content with the common understanding that you are averse to war, recognised in the Militia Ballot Act, but you become a rallying centre for all those who cannot brace themselves up to a hard duty.[92]

Despite their efforts at explaining their position, local Quakers soon began to be persecuted for their pacifism. A Friends Meeting in Cambridgeshire reported that the authorities took over their Meeting House "for military purposes."[93]

The Cambridge Tribunal had been busy right up to the beginning of conscription hearing cases of attested men. On the final day of hearings before the Military Service Act came into effect, they dispensed with men wanting exemptions for important work (turned down), an ailing mother (approved), partners in a business (two months exemption for them to close it), and a student wanting to take the Classical Tripos exam (granted until May, though the tribunal was skeptical that classics was of importance to the country).[94] March brought the first Friends before the Cambridge Tribunal. Three of the assistant masters at the Friends School applied for conscientious objector status, and the tribunal heard all their cases as one. They were asked what they would do if a German ran them through with a bayonet, and would they be willing to live under German rule. The tribunal released them from combatant duties, but they filed an appeal based on unwillingness to serve in any capacity under military command.[95] The same issue of the *Cambridge Daily News* that reported their hearing also printed a letter arguing that the tribunals must decide not only whether a CO's beliefs were sincere, but also whether they were *valid*. Further, beliefs against taking life were said to be inconsistent, because sometimes one must take life to save it. In a metaphor that would prove popular with many tribunals, the letter writer compared the Germans to an armed lunatic running around the town who must be shot for the public good.

The Cambridge Tribunal seemed convinced that no one could legitimately hold to pacifism if presented with an outrageous enough scenario. One fine example, which the *Cambridge Magazine* described as the "burning question of the hour," was "If someone refused to sheathe his sword until he should have imbrued it with the blood of your deceased wife's sister—what would *you* do?"[96] Anger against the "shirker" COs was inflamed by rumors that attested married men would soon be called up and that all men in protected occupations under age thirty would be unstarred. A local newspaper printed a letter from a Cambridge undergraduate in the trenches who had been reading about the new CO hearings. Titled "Conscientious Objectors: What Those as the Front Think of Them," it read in part: "You have presumably heard of the proposal that conscientious objectors should be made to wear white armlets with a large red 'C' on them. They themselves can take 'C' to stand for 'conscience'; decent folk will consider it to stand for something else. There is a rumour that they are to be collected in a corps which will put up barbed wire between our trenches and those of the Huns. I have done that job myself, and they may have joy of it. They will not have much joy if they meet any of this battalion in days of peace."[97] His sentiment was warmly received.

There was a great deal of concern within the university community that its students, staff, and graduates were overly represented among COs. The *London Times* reported that there were three hundred conscience cases from the university, causing great anxiety in Cambridge. The mayor of the city quickly refuted the claim and showed that there were only seventy-two that could be associated with the university.[98] This sort of concern also drove verbal, and sometimes physical, abuse directed at pacifists. Local clergy were unsparing in their attacks on those who would not fight. One declared that "Liberals, Socialists and Pacifists [are] worse than Jews."[99]

The Cambridge Tribunal was the natural focal point of these sentiments, and proceedings were always reported in the local press. Hearings were an opportunity to publicly shame and humiliate pacifists and demonstrate the impossibility of their position. Young COs were treated particularly harshly and on several occasions were told that they were "too young to have a conscience."[100] It was common practice to have wounded soldiers present at hearings, presumably to shame those refusing to serve king and country.

Any applicants who departed even slightly from the mold of the religious objector to all violence were seized upon as shirkers. One Trinity undergraduate claimed a belief in passive resistance to Germany and was classified as nongenuine. A Quaker graduate of Trinity being trained as a schoolteacher asked for exemption on both national importance and religious grounds. In what would become an important pattern, the tribunal turned down anyone who claimed to be both valuable to the nation and to hold a religious objection. It is impossible to know their reasoning in this case, as they utilized their prerogative of silence. When directly asked why, they simply replied, "No, we don't give our reasons."[101] Friends were increasingly being pressed on their beliefs, and any departure from what was expected of a stereotypical Quaker endangered their claim to conscience.

Young men from Eddington's Friends Meeting were being called before the tribunal in increasing numbers. If they were birthright Quakers (that is, they came from Quaker families) they were usually recognized as having a

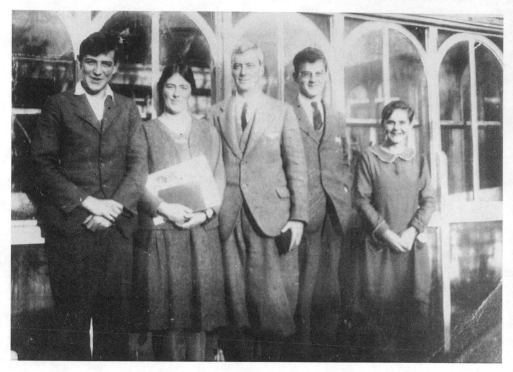

FIGURE 4.2

E. B. Ludlam (center) and family. Courtesy of the Cambridge County Record Office and Cambridge and
Peterborough Monthly Meeting of the Religious Society of Friends in Great Britain.

genuine conscientious objection and released from combatant duties. Some
then went to France to serve in the Friends Ambulance Unit, and only a few
agreed to dig ditches for the army in the Non-Combatant Corps (casually
known as the NCC). Those who refused these duties (a majority) were sent
to prison. Eddington would have gradually watched the numbers of Friends
dwindle on Sunday mornings as the community was culled by the state.

Scientists in the System

Eddington's Quaker colleagues at the university were generally better off, because the authorities there tried to use their influence to prevent embarrassing
CO hearings. Some COs refused the help that Eddington had tacitly accepted.
Ebenezer Cunningham, a mathematics lecturer at St. John's College, was ordered into the YMCA as a substitute for the NCC. The case of Ernest Ludlam,
a research chemist at the Cavendish Laboratory, is instructive (see fig. 4.2).
When the war began he was working on an ammonia production process,

and the Ministry of Munitions offered him a grant to support the research. He declined the grant, hoping his work would be applied to agricultural needs, but the university was able to secure him an exemption for his work anyway. Lloyd George's famous "knock-out blow" speech convinced Ludlam that his research would be put to military use regardless, and he quit the Cavendish to do relief work for the Quaker Emergency Committee. No longer protected by his national importance exemption and refusing to fight, he was arrested and sentenced to hard labor at the fierce Wormwood Scrubs prison.[102] Upon release, he was rearrested—under the same "cat-and-mouse" tactics the government had used against suffragettes—and sent back to prison.[103] After his second sentence to hard labor, his wife, Olive, wrote a letter to the Cambridge Friends Meeting thanking them for their support: "Ernest went to prison with a light heart, feeling it a privilege to suffer in such a righteous cause."[104] He was released several months after the armistice and immediately headed to Germany. He had studied there with Philipp Lenard for some years, and his knowledge of the language and culture were invaluable for the relief efforts. The timing was fortuitous, because he rejoined the Quaker relief workers just in time to help Eddington make contact with Einstein in the wake of the May eclipse.[105]

Ludlam's case is demonstrative in that it shows the same difficulties Eddington experienced, as well as the likely trajectory he would follow if he stepped beyond the protection offered by his scientific work. The treatment of COs like Ludlam in prison was a continuing source of horror for their friends and family left behind. The officers in charge of the COs seemed to be doing their best to fulfill Lloyd George's promise to make "the path of that class as hard as possible."[106] Solitary confinement was common. One imprisoned Quaker wrote: "Things are coming near the end this morning. I was taken up to a quiet place and simply 'pasted' until I couldn't stand and then they took me to the hospital and forcibly fed me.... The colonel was standing near me and thundered up and shouted 'What! You won't obey me?' I quietly answered 'I must obey the commands of my God, Sir.' 'Damn your God!'"[107] COs were technically conscripted members of the Army and therefore subject to military discipline.[108] Field punishments were often administered, including one known popularly as crucifixion, which involved being bound, standing, to some immovable object and left exposed to the elements. The worst abuse of military discipline involved a terrifying incident in which thirty-four COs were secretly taken from their prison cells and brought to the front in France. There they were given a military order. When they refused, an ersatz court-martial was set up. They were then sentenced to be shot, as was permissible for refusing an order in a combat situation. The men were allowed to absorb the implications of their sentence before being told it had been commuted to ten years hard labor.[109] Similarly, a CO, "on refusing to obey orders, was told that he was to be shot at dawn. In the morning a rifle was loaded in his presence, a [soldier] ordered to fire, and then he was told that he had been pardoned."[110] The Quaker MPs Harvey and Rowntree interceded on behalf of abused COs when they could, but such issues were of little importance in a wartime Parliament.[111]

Tensions at Trinity were inflamed by Bertrand Russell's pacifist stance and its legal complications.[112] Russell wrote a pamphlet publicizing the Ernest Everett case, a particularly notorious example of a CO imprisoned for his beliefs. The pamphlet was anonymous, but those distributing it were arrested. The pamphlet, along with much peace literature, was deemed to be harmful to recruiting and therefore illegal under the Defence of the Realm Act. Russell came forward and was arrested and tried. He was found guilty and given the choice of paying £100 or 61 days in prison. He chose the fine. The University Council in Cambridge was enraged over such a high-profile embarrassment and promptly stripped Russell of his position. He was not allowed to give his lectures nor even to access his rooms.[113] Twenty-two Fellows, including Eddington, protested the council's action in a petition: "The undersigned fellows of the College, while not proposing to take an action in the matter during the war, desire to place it on record that they are not satisfied with the action of the college in depriving Mr. Russell of his lectureship."[114] Eddington worked with Hardy and the other Fellows to prevent Russell's persecution for his pacifist views, but to little avail. Russell's legal problems only increased. He was banned from any county bordering the sea lest he communicate with the enemy, and the Foreign Office denied him a passport, thus preventing him from taking a position offered by Harvard.[115] He was hosted by J. W. Graham for a series of lectures in Manchester, but the first talk was broken up by police and recruiting officers looking for "deserters."[116]

By the end of 1916 the authorities at Trinity had no tolerance for pacifists in their midst and were terrified of being painted by the national press as a hotbed of conscientious objection. They were demanding that their staff and Fellows adhere strictly to the ideal of national importance: scientists and intellectuals should be valuable for their contributions to the war. Eddington therefore had this sword over his neck at the same time his friends and co-religionists were being imprisoned for the same beliefs he held. If his conscience had allowed it, he might have lasted the entire war in this state. However, external events drew him to a point where he would be expected to choose between his role as an astronomer and as a Quaker. This is a distinction he was unwilling to make.

Called Up

The unstable equilibrium Eddington had achieved in Cambridge was dis-
rupted in April 1918. A massive German offensive was in danger of completely
overrunning British and French forces, and the Lloyd George government found
itself in a deep crisis. General Haig's famous order of the day, "with our backs
to the wall, and believing in the justice of our cause, each one of us must fight
to the end," made it clear that Britain would fight on even as its armies were in
danger of breaking. Desperate for manpower, the government used the new
Military Service Act to revoke all occupational exemptions and raised the age
of compulsion to fifty-one. This new recruiting power had been passed in January, but had not yet been acted upon in earnest when the armies in France began to crumble. Immediately there were panicked calls for more soldiers, and
there were efforts to withdraw all exemptions, including conscientious ones.[117]

Because most exemptions had been issued for a specific period of time,
the easiest method for revoking them was simply to change the end date of
the exemption. Men suddenly found their two-year exemptions changed to
two months or even weeks. The Cambridge Tribunal reviewed Eddington's
certificate and decided to terminate his exemption as of April 30. We have no
record of his immediate reaction, but given his later decisions one can imagine
that it was a mix of fear and relief of long-held tension. The sufferings of his co-
religionists were well known, but he would no longer have to hide his deeply
held beliefs. The university, however, panicked. The Observatory Syndicate
tried once more to assert Eddington's importance: "In making application for
the exemption of the Director, Prof. A. S. Eddington, it should be stated that
in consequence of the death of the First Assistant [of the observatory] in the
explosion of the *Vanguard*, and of the death of the Second Assistant in action
in France, the Director is the sole remaining member of the Staff."[118] The
local tribunal, well used to working with the university by this point, decided
to extend Eddington's exemption by another three months. This probably
would have been enough to allow him to avoid the current manpower crisis,
but the National Service Representative on the tribunal (a Lt. Ollard, likely
not so beholden to local interests) appealed the decision. Eddington would
now have to appear before the tribunal to make his case. By this point he had
clearly resolved to claim a formal conscientious objection to the war, and his
colleagues and superiors at Trinity College (particularly Joseph Larmor and
H. F. Newall) were frustrated and angry. His Quaker values, though, would
no longer allow him to sit by idly.

Eddington tried to enlist allies for his case within the scientific community.
He wrote to Oliver Lodge asking for a letter of support:

I should explain first that I am a conscientious objector. (No doubt you will deplore that, but I can only say that it is a matter of lifelong conviction as a member of the Society of Friends from birth, and I have always taken a fairly active part in the affairs of the Society.) . . .

My position is that I should be willing to do work of that kind (not war-work) if ordered; but I find it difficult to believe that that would really be for the benefit of the world even from the most narrow point of view. . . .

One feels reluctant to make much fuss about a particular case like this, when so many obviously far harder cases are being ruthlessly dismissed every day in order to supply the army. Still I think I ought to make the attempt to continue my work, provided that it is in the national interest.

I shall quite understand if you think it best in your position not to appear to be mixed up with a conscientious objector's case; and you may be sure that I should not take a refusal amiss.[119]

It is interesting that Eddington thought he had a strong case for his scientific work being in the national interest and thought Lodge would agree. His tone was somewhat fatalistic; he thought it virtually certain that he would be forced into some kind of government work. The letter also reveals that he felt his case was of small moral importance compared to some other COs but that he needed to defend his case nonetheless. It is not known whether Lodge responded to Eddington. His reply, if it was ever made, has not survived. In any case, no letter from Lodge was offered by Eddington in his defense. Eddington clearly thought Lodge would be hesitant, and it seems he was correct.

Eddington came before the Appeal Tribunal on June 14. At the request of the Vice-Chancellor, Newall was present to speak for the university. The chair of the tribunal, a Major S. G. Howard, addressed the issue of Eddington's national importance, which was the official subject of the appeal. He "suggested that Prof. Eddington's ability might be better employed for the active prosecution of the war if placed at the disposal of the Government. He did not think Prof. Eddington would be taken as an ordinary soldier." Eddington stood and attempted to refocus the discussion. He declared, "I am a conscientious objector." Newall, no doubt concerned to separate the astronomy professor from the socialist riff-raff, hastily added that Eddington was a Quaker. The chairman refused to engage with the question of religion while the matter of science and the national interest was on the table: "That question is not before us." Frustrated, Eddington explained that he had made an application for CO status but that it had never been addressed because of his occupational exemption. From the start, he had defended his unified identity but been stymied by official decisions. The tribunal retired to consider the case in private. When they emerged, they announced that Eddington's services would be better rendered to the government than to the observatory. His exemption was now void. The chairman casually mentioned that they had "not considered the question of conscientious objection because it was not before them."[120] Science and religion had been neatly categorized by the government, and Eddington's objections to those categories could not or would not be addressed by the local authorities.

To pursue his claim to conscientious objection, Eddington now had to go back before the Cambridge Tribunal, as though he had never had an occupational exemption. And now, despite the university's best efforts, his case had come to public attention. The *Cambridge Daily News* topped its columns devoted to the hearing with the headline "Professor of Astronomy as CO." Eddington was described as "the Plumian Professor of Astronomy at the University, Director of the Observatory, and hon. Secretary of the Royal Society." When called forward, he at long last had a chance to declare his pacifism and objection to military compulsion. His statement was forthright:

> My objection to war is based on religious grounds. I cannot believe that God is calling me to go out and slaughter men, many of whom are animated by the same values of patriotism and supposed religious duty that have sent my countrymen into the field. To assert that it is our religious duty to cast off the moral progress of centuries and take part in the passions and barbarity of war is to contradict my whole conception of what the Christian religion means. Even if the abstention of conscientious objectors were to make the difference between victory and defeat, we cannot truly benefit the nation by willful disobedience to the divine will.

The Military Service Representative interjected that Eddington had no grounds for claiming CO status. The professor had accepted a national importance claim and had therefore forfeited his right to claim a conscientious objection. Eddington again explained his confusing path through the conscription system and repeatedly emphasized that he had attempted to secure exemption both as an astronomer and a Quaker. Representative Miller, in a tone not unlike an exasperated adult explaining the world to a child, asked Eddington to withdraw his religious claim, because it was clearly not consistent with his earlier claim of being a scientist. Eddington refused. The tribunal, reifying the separateness of the two claims, threw out his religious claim "until required." They then deliberated behind closed doors and apparently decided against him based on the validity of his national importance claim, instead of the conscientious one. The local paper reported that the tribunal "considered the case a very hard one—hard on Prof. Eddington." They took the unusual step of giving him until July 11 to get a cabinet-level intervention, which was no doubt an indication of the maneuvering being conducted by the university behind the scenes.[121]

While Eddington was moving through his hearings, Newall and Larmor worked furiously to arrange an exemption for him unrelated to his pacifism. Through unknown means, they managed to secure a new exemption. This was presumably a national importance exemption, although the details are lost. A letter was sent to Eddington, and all that was necessary was for him to sign and return it. Again, the exact text of the letter is not available, but given its origin within Trinity College, one can assume that it asserted Eddington's willingness to work in a field of national importance and made no mention of conscientious objection. Eddington signed it—and then added a postscript that he would pursue conscientious objector status regardless of the success of this new exemption. As he had from the beginning, Eddington refused to allow the mundane needs of the government to split the identity of the religious scientist that had proved productive and meaningful throughout his life. His action of course invalidated the offered exemption, and the college administrators were furious. But Eddington could see no reason for them to be; he felt had been nothing but honest and transparent in showing his loyalty to the Society of Friends. He would not abandon them now in the moment of their greatest crisis.[122]

Other Quakers were losing their exemptions as well. Ebenezer Cunningham, the lecturer at St. John's in mathematics, had been given conscientious exemption to work as a school teacher, but the National Service Representative appealed on the grounds that "Mr. Cunningham was not a fit person to teach children." The tribunal agreed and sent him to work in agriculture or minesweeping.[123] This strange pair of possibilities indicated some of the changes to treatment of COs since the introduction of conscription. Prison was still the destination of the absolute objector, but those COs willing to perform service outside the context and structure of the military now had other options. The Home Office Scheme and the Pelham Committee found work for them in agriculture or other fields of national importance—at least in theory. In reality, the work was generally menial, degrading, and of no importance whatsoever. The work was officially supposed to be of a "deterrent character," and stacking rocks certainly qualified.[124] Many participants asked to be sent back to prison. The tribunals continued to doubt the internal consistency of the concept of conscientious objection. The hearing after Eddington's dealt with a CO who had been working in an ambulance unit who wanted to move to combat service. According to the *Cambridge Daily News*, the chairman of the tribunal felt the case "showed very clearly that much of that trouble arose from associations. He was afraid that much of the conscientious objections came from an unhealthy influence in the objector's surroundings. That applied especially, he thought, to many conscientious objectors in Cambridge."[125]

Eddington's case came to a climax in the second week of July 1918. The Ministry of National Service had, in a significant departure from normal procedure, allowed Eddington's case to be heard again.[126] Frank Dyson, as Astronomer Royal, had become involved with his hearing, and it is difficult to know whether the change was due to his intervention or some figure in the university hierarchy. There was no question that Eddington preferred to accept Dyson's help. They were old colleagues and the Scottish astronomer had been invaluable in planning the 1919 eclipse expedition. A letter from Dyson was the first item presented by Eddington at his hearing. The letter was shrewdly written to appeal to the attitudes of the tribunal. It underlined the importance of keeping prestigious British scientists working to counter the dominance of enemy science:

> I should like to bring to the notice of the Tribunal the great value of Prof. Eddington's researches in astronomy, which are, in my opinion, to be ranked as highly as the work of his predecessors at Cambridge—Darwin, Ball, and Adams. They maintain the high position and traditions of British science at a time when it is very desirable that they be upheld, particularly in view of a widely spread but erroneous notion that the most important scientific researches are carried out in Germany. . . . I hope very strongly that the decision of the Tribunal will permit that important work to be continued.

After establishing Eddington's importance for British intellectual life, Dyson invoked a spectacular opportunity for astronomy that only Eddington could carry out:

> There is another point to which I would like to draw attention. The Joint Permanent Eclipse Committee, of which I am Chairman, has received a grant of £1000 for the observation of a total eclipse of the sun in May of next year, on account of exceptional importance. Under present conditions the eclipse will be observed by very few people. Prof. Eddington is peculiarly qualified to make these observations, and I hope the Tribunal will give him permission to undertake this task.

After presenting the evidence for his importance as a scientist, Eddington continued his argument for exemption by again describing his conscientious objection to the war. He said he had been a member of the Society of Friends since birth, which was often the key criterion for being seen as a genuine objector. When asked whether he would accept alternative service, Eddington took the most common position of COs. He refused to accept service under military auspices ("he did not think the War Office could guarantee that he be employed solely in saving life") but expressed willingness to work in other contexts. He suggested service in the Friends Ambulance Unit, the Red Cross, or helping with the harvest, "if it was thought he could be of more use to the nation in that way." The tribunal was not particularly interested in the details of his conscience, however, and wanted to know more about the eclipse. They interrogated him with regard to whether this eclipse was of particular importance, and Eddington assured them that it would allow observations that could not be made again for centuries. Without retiring to discuss his case, the tribunal announced that his work was of national importance and "therefore gave Prof. Eddington, in order to cover [the period of the eclipse], 12 months' exemption, on condition that he continued in his present work." They also, startlingly, said they were convinced he was a genuine CO. This was immediately forgotten, however, and the actual exemption was solely for national importance and made no mention of his conscientious objection. Eddington then left the hall, free to pursue his work.[127]

The ease of Eddington's hearing makes it clear that Dyson was able to influence the outcome through his contacts in the Admiralty. The end result (an exemption for national importance) was no different than if Eddington had accepted the deal arranged by Trinity. Why, then, did he accept the one arranged by Dyson and not the other? The difference was that the exemption received with Dyson's help allowed him to maintain a protest against the war that he saw as being as legitimate and important as the protests of his friends who were officially labeled as COs. He was unwilling to accept the Trinity exemption for his scientific work without also announcing his religious identity. Assuming he had no sea-change in his beliefs before his final hearing (and there is no reason to think there was one), there must have been some element in the final exemption that fulfilled his need to work for peace. This was the eclipse. As I argued in the previous chapter, Eddington saw the eclipse expedition and its associated publicity as an important pacifist opportunity. It was, therefore, in some sense "equivalent" to conscientious objection for him. What was important for him was to give expression to his values of pacifism. He was willing to do this by applying for an exemption on the grounds of both his scientific work and his religious beliefs (as he tried at first), attaching a statement of his conscientious objection to an exemption for his scientific work (as he would have done with the Trinity College exemption) or obtaining an exemption for scientific work that he saw as having value in line with the peace testimony (i.e., the eclipse and Einstein). That the peace value of the expedition was not obvious to the Cambridge Tribunal was irrelevant; this was a matter of conscience and of Eddington holding true to his values in a way that was, to him, both meaningful and productive.

The war exhausted itself before Eddington's twelve months was over, but the suffering of his CO friends and colleagues did not end on Armistice Day. Months after the war was over, many COs were still imprisoned, including a Cambridge lecturer in psychology.[128] The Cambridge Friends Meeting petitioned the government on their behalf to no avail: "This meeting enters its protest against the continued imprisonment of Conscientious Objectors, especially of two honoured members of our own Monthly Meeting, Ernest B. Ludlam and Robert G. Errington, both of whom hold the office of Overseer in our religious Society and are warmly loved and respected."[129] In the end, they were not released, but discharged, since they had been technically inducted into the Army. The grounds for discharge were stated as "misconduct."[130] Even the release of all the COs was no real victory, as across Britain ten had died in prison and sixty-one died soon after as a result of their treatment.[131] The Representation of the People Act, passed just after the war to finally extend universal suffrage to Britain, disenfranchised conscientious objectors for five years after the war.[132]

The implementation of conscription had been a debacle, and the War Office knew it. As early as June 1916 the Central Tribunal realized it had large numbers of men in prison that should have been exempted. Unfortunately, they could not be released without making it look as though the tribunals were not doing their job. A committee was appointed to look into the matter.[133] It was not until April 1918 that the War Office publicly admitted that "many hundreds of these men have been thrust into the Army whom the House of Commons never intended should become soldiers."[134] After the war was over, the government sent out a memorandum listing wartime records that could be thrown out. Records on conscientious objection were an exception—these needed to be destroyed. They were to be burned for privacy purposes.[135] Few of the official records survive today, although some sample papers from the Central Tribunal were retained for historical purposes.

Conclusion

Shortly after the Armistice, the chairman of the Middlesex Appeal Tribunal commented, "From the point of view of the civil population the Military Service Acts had a far more reaching effect than any legislation within the history of man."[136] In a sense, he was correct. Not since the Restoration had there been a political decision in Britain that had life or death consequences for virtually every family in the country. My study is concerned here with the peculiar feature of the Acts which dictated the categories of identity that would be considered acceptable in wartime. The extremes of war show that identity is not just an internal concept but can have concrete external consequences. Whether Eddington considered himself an astronomer or Quaker determined the state of his personal freedom. The war also showed that identity is also not only an internal construction. The mechanism of conscription created several types of personal identification that were considered legitimate. And those were not always in harmony with the self-identity held by the people who had to move through this mechanism. This is not to say that identity as understood by the individual and the state must necessarily conflict; there were many scientists who were perfectly happy acquiescing to the military's understanding of their value (e.g., H. G. Moseley).

However, as a society diversifies and allows more personal freedom and choice, it is inevitable that there be clashes.[137] Late Victorian and Edwardian Britain was undergoing just such a diversification with the lowering of the final barriers that kept Nonconformists such as Quakers isolated. British society had grown used to religious scientists on the model of the natural theologian, which always included a healthy dose of British patriotism and self-congratulation. Nonconformists, however, did not hold the same value structure, and the presence of a Quaker scientist was unexpected and baffling to the authorities. The conscription apparatus required the state to dictate the relationship between certain values, and in the Great War the British state declared normatively that values of pacifism *could not* intersect with values of national service, regardless of whether such intersections actually *did* occur.

The clash that resulted can be valuable for the historian, because it gives an opportunity to see exactly which values are in conflict. A superficial reading of Eddington's case might see it as a clash between science and religion (he was trying to unite two categories that should be kept separate) or a purely religious or purely scientific matter. A closer examination reveals that the conflict was between two different understandings of science clashing with two different understandings of religion. In this way, the abstract categories of science and religion are reduced to concrete decisions and actions: Eddington's self-identity as a religious scientist and the activities he felt it necessitated versus the decisions made by political and military officials regarding the relationship of scientists and Quakers to the state. Analyzing the substantive roots of conflict is extremely useful in translating putatively abstract issues like science and religion into concrete issues of action and belief.

Experience
From Relativity to Religion

The scientist and the religious teacher may well be content to agree that the value of any hypothesis extends just so far as it is verified by actual experience.

A. S. EDDINGTON, "The Domain of Physical Science"

Chapter 3 presented much of the story of Eddington's initial encounter with relativity, and here I will pursue the story synoptically. While earlier this book tracked the elements of his encounter that were salient for considerations of the wartime and postwar situation in Britain, I will now explore the development of his understanding of the *meaning* of relativity. By "the meaning of relativity" I indicate the ontological content and philosophical consequences of the theory. Separating the theory into "scientific meaning" and "philosophical meaning" would present a skewed picture of Eddington's approach. Instead, Eddington's considerations of relativity demonstrated an integrated approach to understanding the importance of a scientific theory. This integration required constant attention to the boundaries between and interdependence of physics, metaphysics, philosophy, and religion. In particular, Eddington's values supporting the importance of personal experience functioned as guides for the construction, maintenance, and demolition of such boundaries. I will examine his early work on relativity chronologically, through close readings of his texts, to show how such an approach developed and was received.

This chapter does not seek to present a general overview of the reception of general relativity in Britain; the reception of relativity has been studied in great detail elsewhere.[1] Rather, it will use Eddington's work as a thread with which to track the British religious and philosophical response to the theory. The specific issue here is how the religious implications of relativity were considered, and I will address this by examining both how one high-profile figure (Eddington) did this in concert with his technical work and how the wider community fit relativity into existing concerns in philosophy and religion. In particular, I will argue that Eddington's understanding and description of relativity fits squarely into the important contemporary religious movement known as liberal theology.

To borrow a metaphor from relativity itself, we can think of this analysis as exploring the different frames of reference from which relativity was studied. Philosophers, religious thinkers, and scientists were all situated in their own frames built on the coordinate systems of idealism/materialism, objectivity/subjectivity, or metaphysics/physics. The observers of early 1920s Britain read the location of relativity off these yardsticks, which had provided the points of reference for understanding developments in science, religion, and philosophy for decades. This is not to say that relativity was completely malleable in their hands; they understood that parts of the theory were "invariant," but those were not the parts with which they were concerned. They wanted to know how it factored into the debates in which they were already entangled. If they were interested in metaphysics, they typically were not interested in the materialist implications, for example. Eddington is a useful figure with which to trace relativity's trajectory through these cultural regions precisely because he *was* interested in the invariant aspects of relativity. To him, it demonstrated the fundamental truths of human experience that underlay all the enterprises of science, philosophy, and spirituality. His valence values supporting the irreducibility of experience made the approach he took seem essential. This led him to move through the different axes mentioned above—if he was to see how relativity had religious and physical significance, he had to engage with the theory in the contexts of both the church and the laboratory. He was intellectually and physically present in all the spheres in Britain where relativity was being engaged. Thus, this chapter will pursue Eddington's exploration of relativity as a unifier and take advantage of that to observe other actors' assessments of the theory against the existing axes of cultural and intellectual measurement.

The Early Exposition

Most scientists in Britain first encountered relativity from either de Sitter's articles in the *Monthly Notices of the Royal Astronomical Society* in 1916 or Eddington's informal expositions in the following year.[2] Much of the early debate in the astronomical community centered on the physical significance of some of Einstein's claims about time and space. A frequent concern was that the theory was too "metaphysical" and was not really physics. Even scientists with a somewhat favorable outlook toward relativity tried to downplay its statements about the nature of space-time. James Jeans warned that some presentations of the theory had clad it in a "metaphysical garment" that obscured its "more concrete part."[3] His position was that relativity made no claims that could be considered metaphysical: "Einstein's crumpling up of his four-dimensional space must, for the present, be considered to be just as fictitious as the crumpling of ordinary space represented by the metric transformation $ds' = \mu \, ds$ in optical theory."[4] This is, the bending of space-time was no more real than the bending of the image of a straw in a glass of water. To Jeans, such a technique could be useful but could not in any sense be considered an explanation for physical phenomena.

It should come as no surprise that Eddington joined this debate, and his response provides us with one of his earliest statements of his understanding of the meaning of relativity. He defended relativity's claims about space and time and denied that there was anything "mystical" about phenomena such as length contraction. Instead, he stressed the "essentially matter-of-fact character of Einstein's procedure."[5] The warping of space was simply a consequence of thinking rigorously about precise measurement. If asking questions like "What is measurement?" sounded metaphysical, so be it—that was the productive approach. In other words, if it helped do more physics, it was good. We can already see Eddington's willingness to blur the conventional boundaries of physics and philosophy. People like Jeans were concerned to show that relativity was not metaphysics, but Eddington was perfectly content to let previously philosophical categories such as time fall under the label of physics. Why was this? The key was the concept of measurement and what precisely that concept entailed. The evolution of Eddington's understanding of measurement will be the major factor in understanding the emergence of his religious relativity.

Eddington's first major publication on relativity was his *Report on the Relativity Theory of Gravitation*.[6] This was also the first substantial presentation of general relativity in English and formed the introduction to the theory for many British and American scientists. The *Report* followed up on Eddington's aggressive defense of Einstein in the previous two years, and particularly his debates with Oliver Lodge. Eddington relied heavily on the work of Continental scientists (particularly de Sitter) in writing it, but by the time he finished it he had certainly mastered the theory and felt comfortable speaking authoritatively on both its essentials and its meaning.[7]

The *Report* was written before the 1919 expedition, and the only empirical evidence Eddington had to present were the precession of Mercury's perihelion and the failure to detect uniform motion through the ether. His more striking claims for the importance of the theory were based on elegance of mathematical reasoning and novel predictions. In language similar to his defense of his astrophysical techniques, Eddington wrote:

Whether the theory ultimately proves to be correct or not, it claims attention
as one of the most beautiful examples of the power of general mathematical
reasoning. The nearest parallel to it is found in the applications of the second
law of thermo-dynamics, in which remarkable conclusions are deduced from
a single principle without any inquiry into the mechanism of the phenomena;
similarly, if the principle of equivalence is accepted, it is possible to stride
over the difficulties due to ignorance of the nature of gravitation and arrive
directly at physical results. Einstein's theory has been successful in explaining
the celebrated astronomical discordance of the perihelion of Mercury without
introducing any arbitrary constant; there is no trace of forced agreement about
this prediction. It further leads to interesting conclusions with regard to the
deflection of light by a gravitational field, and the displacement of spectral
lines by the sun, which may be tested by experiment.[8]

As with stellar physics, Eddington argued that relativity should not be discarded or accepted based solely on its correlation with known facts, but rather
on its ability to expand our sphere of scientific inquiry.

The technical elements of relativity were presented with great emphasis
on the role of the observer, and particularly the distillation of observations
to *relations*. Any experimental result was actually a set of relations between
material bodies (such as rigid rods and clocks) and an observer. As mere
intersections of "two entities in space and time," all astronomical and laboratory observations "rest finally on the coincidence of some indicator with a
division on a scale."[9] In a talk at the Royal Institution around the same time,
Eddington argued even more assertively, saying, "If we analyse any scientific
observation, distinguishing between what we see and what we merely infer, it
always resolves itself into a *coincidence* in space and time." That is, it is only
the coincidence of a pointer with the scale on an instrument that we really
see.[10] An observation that could not be reduced to such a coincidence could
not be a launching point for physics.

This instrumental approach allowed Eddington to define time and space
in the Einsteinian positivist style: time was the ticking of a clock, space was
marks on a rigid rod. He reassured his readers that this was indeed a legitimate
topic for physics: "The reader may not unnaturally suspect that there is an
admixture of metaphysics in a theory which thus reduces the gravitational field
to a modification of the metrical properties of space and time. This suspicion,
however, is a complete misapprehension, due to the confusion of space, as we
have defined it, with some transcendental and philosophical space."[11] What
made space accessible to physics was that it was measurable and analyzable.

Eddington used this reasoning to defend many of the apparent paradoxes
of relativity. So long as a claim produced a measurable result, it was scientific,

and valuable: "There is nothing metaphysical in the statement that under certain circumstances the measured circumference of a circle is less than pi times the measured diameter; it is purely a matter for experiment. . . . we certainly ought not to be accused of metaphysical speculation, since we confine ourselves to the geometry of measures which are strictly practical, if not strictly practicable."[12] The *meta-* in *metaphysics* was only a mark of discussion of the unmeasurable. To some contemporaries, Eddington seemed to be arguing for relativity's metaphysical significance, in that he claimed it gave us knowledge about the nature of time and space. But a more accurate assessment would be that he was proposing a new border for physics and metaphysics, as demonstrated by relativity. A critical corollary of this was that the effects and implications of relativity were *real* and omnipresent, and experiment dealt with nothing more than the measured relations among bodies.[13]

Eddington included a chapter on tensor theory to help ameliorate the difficulties of the mathematics in Einstein's theory. The field equations were presented so as to make their experimental predictions most clear, and Eddington briefly dealt with the continuing problem of the relationship between relativity and ether theory. Relativistic cosmology was still in its infancy, but the book devoted several pages to Einstein's and de Sitter's theories.[14] In conclusion, Eddington asserted that relativity was in essence a theory that illuminated our understanding of measurement, not of gravity. No "explanation" in the Victorian sense (that is, a mechanical model that provided ontological explanations) had been advanced by Einstein: "In this discussion of the law of gravitation, we have not sought, and we have not reached, any ultimate explanation of its cause. . . . We do not in these days seek to explain the behaviour of natural forces in terms of a mechanical model having the familiar characteristics of matter in bulk; we have to accept some mathematical expression as an axiomatic property which cannot be further analysed."[15] The question of whether general relativity "explained" gravity would remain controversial for some time, in large part due to Eddington's claims that relativity completely reshaped our understanding of physical explanation.

The *Report* was received with great interest and even some acclaim. One reviewer, who called it "the most remarkable publication during the war," continued to worry about relativity's lack of simplicity and its divorce from common sense. "Generalisation is the supreme intellectual achievement, but it may leave us thirsting for the particular and for simplicity."[16] One physicist glumly admitted, "It is not improbable that the statement which [Einstein] is alleged to have made to his editor, that only ten men in the world could understand his treatment of the subject, is true. I am fully prepared to believe it, and wish to add that I certainly am not one of the ten."[17] He did feel he could grasp the theory once he had time to adapt to the mathematics. Overall, the *Report* stimulated conversation about relativity not only because it provided a minimal education in the theory but also because it could be referred to as an authoritative baseline and therefore allowed more meaningful discussion.[18]

After the results of the eclipse expedition were made public, the tenor of conversations on relativity shifted, as has already been discussed in chapter 3. In the February following the presentation of the results, a lively discussion on relativity was held at the Royal Society, chaired by James H. Jeans. He compared the theory to conservation of energy and the second law of thermodynamics: "The three principles have in common that they do not explain how or why events happen; they merely limit the types of events which can happen. . . . All three principles deal with events, and not with the mechanism of events."[19] Relativity (like conservation of energy) was, therefore, a destructive, rather than constructive, theory. It *restricted* the behavior of nature, rather than *explaining* it in any way. Jeans held to his earlier position that relativity explained nothing, no doubt bolstered by Eddington's similar statements. Several astronomers and physicists, including Frank W. Dyson, addressed the status of the theory's tests. H. F. Newall, F. A. Lindemann, and Ludwik Silberstein continued to worry about the meaning of the fundamental principles of the theory and its application to the world. Eddington declined to discuss such matters and instead emphasized the empirical evidence for the theory.[20]

The growing scientific consensus on the theory created an increasingly safe intellectual and social space for the discussion of its meaning and foundations. Previously, concerns about the meaning and implications of relativity were mixed confusingly with questions about its validity and truth. Now, the latter issues faded in importance (though they did not disappear), allowing more vigorous pursuit of the lingering questions of how relativity should be integrated into the worldview of postwar Britain.

Engagement with Philosophy

Eddington's first public engagement with the philosophical concerns surrounding relativity appeared in the journal *Mind* in the spring of 1920.[21] It is not clear why Eddington chose to publish at this time, but it seems likely that he had already been pondering the issues and the timing was simply a matter of having sufficient time after his return from Africa. There is no reason to think he had a sudden inspiration to consider the issues in the article, as there are important continuities with his earlier writings on relativity. Rather, this appears to be a natural evolution of his understanding of relativity given his intellectual and religious heritage. I will approach this by first investigating

Eddington's early exposition of the meaning of relativity, and then exploring his influences and other interpreters of the theory.

The *Mind* article, "The Meaning of Matter and the Laws of Nature According to the Theory of Relativity," was designed to address many of the concerns that had been voiced with regard to relativity's implications. The title gave no clues as to the dramatic points Eddington would arrive at, however. On the first page, he revealed the conclusion that he felt was the inexorable result of relativity: "Whatever may be the true nature of matter, it is the *mind* which from the crude substratum constructs the familiar picture of a substantial world around us. On the present theory we seem able to discern something of the motives of the mind in selecting and endowing with substantiality one particular aspect of the world, and to see that practically no other choice was possible for a rational mind. It will appear in the discussion that many of the best-known laws of physics are not inherent in the natural world, but were automatically imposed by the world when it made the selection."[22] Eddington acknowledged that this was apparently in accord with some philosophical theories, particularly subjective idealism, but asserted that he was approaching the question from the point of view of physics. Of course, this necessarily integrated perspectives that some contemporaries would not have considered "scientific," but Eddington was genuine in saying that he treated this as a problem in physical science. The structure of the paper is similar to his *Report on the Relativity Theory of Gravitation*, and he uses equations, even the symbolism of differential geometry, quite often. Given his complaints in the article about not being able to make points clearly without mathematics, one senses that this was as often as his editors would allow.

Eddington began his argument with considerations of the general nature of physical theories. While the goal of physics is to explain experience, he said, everyday experience is too complex, so we must resolve it into new, simpler elements. This makes practical sense but has the strange side effect of requiring an explanation of the familiar in terms of the unfamiliar. Thus, the bases of physical theory are entities that have no meaning in ordinary experience and might not even be definable in a meaningful way: "There is no particular awkwardness in developing a mathematical theory in which the elementary constituents are undefined. But it is desirable that in some stage of the discussion we get to know what we are talking about; and this is achieved when we can identify one of the complex combinations of our undefinables with some object of experience recognised by the mind. Strange as it may seem, it is quite easy to overlook this necessity." The easiest way to achieve this identification, he said, is to find a way to measure it. The devices of measurement are elements of experience, and one needs to develop and extend the theory until it intersects with direct experience of those devices. There are many possible avenues to bridge the analytical world and the world of perception: experiments are the most common, but one can also "push" the theory until it resembles direct sense experience. Most productively, one develops a dynamic hybrid of experiment and mathematics (even if it involves unknown quantities) of the sort that led to Eddington's mass-luminosity relation.[23]

The undefinable quantity critical to relativity, he said, is the "point-event." The sum of all point-events may be known as "The World," which is expressible only as a four-dimensional mathematical structure that makes no appeal to intuitive understandings of space and time. The interval between any two events is measurable, but almost purely analytical. Despite the unknown nature of intervals and events, Einstein's law of gravity could be arrived at solely by manipulating their mathematical apparatus.[24]

Eddington pointed out that Einstein's law in the form $G_{\mu\nu} - \frac{1}{2}\, g_{\mu\nu} G = -8\pi T_{\mu\nu}$ was usually interpreted as showing that matter, on the right, disturbs the preexisting form of space-time, on the left.[25] But, he argued, there is something wrong in introducing an object of experience (matter) into a relation between purely analytical concepts (tensors). This should be rejected, he said, because it leads to dualism. Is there a way to interpret the equation that avoids this? His answer was that the equation should be thought of not as a law, but as a definition. It defined how we perceive the presence of matter: by the measured relations described by the left-hand tensors. "Matter does not cause an unevenness of the gravitational field; the unevenness of the field *is* matter." Einstein's equations were defining a mathematical, analytical quantity (the gravitational field) that was defined in a particular way with respect to an element of our experience—matter. The gravitational field did not *exist* in any ontological sense.[26] In the same way that we do not say heat is something that causes random molecular motions—we say that heat *is* those motions—we should not say that matter causes gravitational irregularities. Rather, we should say that matter *is* those irregularities. There was no need to introduce a new substance to "explain" a sense experience or measurement.

Matter, therefore, could not be said to exist apart from its effect on our measurements, which were themselves only the mind's perception of coincidences. "According to this view matter can scarcely be said to exist apart from mind. Matter is but one of a thousand relations between the constituents of the World, and it will be our task to show why one particular relation has a special value for the mind." We could gain a clue by noting that $G_{\mu\nu} - 1/2g_{\mu\nu} G$, in addition to describing gravity, has the property of conservation. This showed that if we *"measure space and time in one of a certain limited number of ways*, matter will be permanent." The four-dimensional World of discrete point-events could be observed in an infinite number of ways, only some of which allowed for the conservation of matter, energy, and momentum. Thus, there was something special about the fact that the law of gravity was linked to these conservation laws.[27]

Eddington said we did not need to know anything about the essences of the things that were conserved—the question was only what name the mind would assign to them. "For some unknown reason the mind appears to have a predilection for living in a more or less permanent universe. The idea of reality is at least closely associated with the idea of permanence." Therefore, the mind picked out a view of the World built from "permanent elements," in other words, matter, and thought this was real. This came from using just one method (of many) of seeing space and time, and once humans were conditioned to seeing things this way they forgot there were other ways. The laws of nature, including the conservation laws, gravity, and mechanics, thus came from how humans as observers chose to regard the world. They were not laws "of nature" in any sense. "The intervention of mind in the laws of nature is, I believe, more far-reaching than is usually supposed by physicists. I am almost inclined to attribute the whole responsibility for the laws of mechanics and gravitation to the mind, and deny the external world any share in them." Surely nature should have some responsibility for these laws? "I doubt it. So far as I can see, all that Nature was required to furnish was a four-dimensional aggregate of point-events . . . for the use made of point-events the mind alone is responsible."[28]

Recent work by Hermann Weyl suggested to Eddington that electromagnetism could soon be explained by the same method. The only phenomena of physics that seemed as though they might come from the external world were those of atomicity, in which Eddington included early quantum theory. He saw it as significant that physics found quantum phenomena bizarre and confusing: "Thus the domain, where the mind of the physicist has hitherto triumphed, comprises only those laws which have not their seat in the external world but spring ultimately from the mind. Will the human mind prove equal to formulating the genuine laws of a possibly irrational world, which it has no part in shaping?"[29]

Eddington closed his article with a summary of his argument and a cautionary note about the status of his claims:

> The physical theories which form the bases for this argument are still on trial, and I am far from asserting that this philosophy of matter is a necessary consequence of the laws of physics. It is sufficient that we have found one mode of thought tending toward the view that matter is a property of the

world singled out by the mind on account of its permanence, as the eye
ranging over the ocean singles out the wave-form for its permanence among
the moving waters; that the so-called laws of nature which have been definitely
formulated by physicists are implicitly contained in this identification, and
are therefore indirectly imposed by the mind; whereas the laws which have
hitherto been unable to fit into a rational scheme, are the true laws inherent
in the external world, and mind has had no chance of moulding them in
accordance with its own outlook.[30]

His warning that this was but one possible way of considering the nature
of matter and physical law, while familiar from his "seeking" methodology,
was speedily ignored by philosophers and physicists alike. Eddington had
apparently asserted that relativity required a radical change to the scientific
outlook, by changing the laws of nature to the laws of mind.

Eddington had a further opportunity to expound these views at the 1920
Oxford Congress of Philosophy that fall.[31] A special symposium on relativity
started the congress, presided over by A. N. Whitehead. Eddington began by
placing relativity with respect to physics and philosophy: "There is no doubt
that [relativity] was largely suggested by philosophical considerations, and it
leads to results hitherto regarded as lying in the domain of philosophy and
metaphysics. But the theory is not, in its nature or in its standards, essentially
different from other physical theories; it deals with experimental results and
theoretical deductions which arise from them." He played down the apparent revolutionary nature of Einstein's theory, saying there was no significant
breach of continuity with the history of science. Einstein's theory, like all scientific advances, was an incremental and exploratory step forward. Nonetheless,
it was still tremendously significant: "It would be rash to suppose [relativ-
ity] reaches finality; but it bears all the indications of being one of the more
permanent stages in the advance toward Truth." Its contribution was that it
was a major step in understanding how to correlate analytical science with
our ordinary world. In one passage, Eddington used "green" as an example
of the difficulty of linking qualitative, subjective sensations with quantitative,
rigorous concepts in physics—How does one match the appearance of grass
with electromagnetic oscillations without merely reducing one to the other?[32]

According to Eddington, relativity's enduring contribution was its ability
to deal with the absolute world of four dimensions and not just our "worm"-
like perception that divided up the world into space and time. Referring
to his *Mind* article, he reiterated that matter was not ontologically separate
from space-time geometry. Just as the mind could see constellations in the
stars, it could also see conservation in the universe. The laws of physics were

apparently implicit in the choice to observe matter, but the quantum laws were of a completely different character.[33]

The other speakers at the symposium help illustrate the aspects of relativity that were still found to be problematic. The moral philosopher W. D. Ross spoke strongly for the position that the philosophical foundations of relativity were flawed, and that a simple return to H. A. Lorentz's interpretation of the Michelson-Morley experiment would fix all outstanding issues. The apparent success of Einstein's theory was due to scientists being under "the influence of the glamour of relativity."[34]

C. D. Broad, a philosopher with much more experience in thinking about science, was convinced that relativity "or something like it must be true." He addressed Ross's many and varied misunderstandings and reacted to Eddington's paper. He argued that Eddington was unclear on the distinction between an observer as an instrument and an observer as a mind—the first is a part of nature, the second is not. Further, he asked whether all four-dimensional manifolds could be resolved into conservation laws. If so, were such laws not inherent in nature?[35] The Oxford physicist F. A. Lindemann closed the session with an evolutionary-epistemology explanation of relativity. Neither the absolute nor the relative world was "real," they were just different approaches. Human beings had evolved relativistic perceptions, and therefore it was easier for us to perceive the world in that way.[36]

It was reported by the idealist philosopher R. F. Alfred Hoernle that, of the entire congress, this symposium "excited the keenest anticipations and drew the largest audience. But it can hardly be said that the discussion resulted in the great clearing-up of ideas for which we had hoped." He demurred that this was not the fault of the scientists present, but rather the "inherent perplexities" of the subject. Even the experts seemed to have trouble expressing what relativity was about: "Most of [the scientists] are frankly puzzled, and the dogmatic ones are all at sixes and sevens."[37] With respect to the speakers: "It will be generally agreed that Professor Eddington is not lacking in speculative courage"; Ross's argument "seemed greatly to annoy" the other speakers; and Lindemann's "views were deservedly rejected by scientists and philosophers alike."[38]

Predecessors and Contemporaries

While Eddington's claims did appear radical to many physicists, they seemed quite familiar (perhaps even naively simple) to many philosophers. It was apparently a straightforward idealism of the sort that had dominated British philosophy at the end of the Victorian period.[39] However, it had many elements that were completely alien to traditional idealism, such as the emphasis on measurement and quantification. Was Eddington's interpretation of relativity a purely idiosyncratic construction, or can we trace a genealogy of his ideas?[40]

We can consider Eddington's intellectual influences prerelativity with some confidence, because his journal for that period records in detail what books he read.[41] Authors on whom he might have drawn to form his thinking on the philosophy of science and relativity include Bertrand Russell, A. N. Whitehead, W. K. Clifford, Karl Pearson, James Ward, and Henri Poincaré. I will not attempt a comprehensive examination of these authors; the intent of this section is solely to trace the origins of some of Eddington's perspective on relativity.

In 1906 Eddington discovered an interest in the philosophy of science and read both Pearson and Poincaré. It is possible that Eddington decided to read Poincaré after he became Chief Assistant at Greenwich, perhaps to acquaint himself with one of the titans of his new field of astronomy. If so, *Science and Hypothesis* may have been the spur of his sudden interest in philosophy. Poincaré's interrogation of the fundamental categories of physics (such as space and mathematics) may have been Eddington's first stimulus to approach those categories as subjects of open inquiry. Also, the French astronomer's insistence on the interwoven nature of observation, experiment, and theory had significant similarities to the dynamic theory-observation relationship that Eddington argued for in relativity and astrophysics.[42] Poincaré's relativity was, of course, very different from Einstein's, and Eddington does not mention the former in his work on general relativity. *Science and Hypothesis*'s importance was not an early warning about Einstein but rather an invitation to Eddington to explore the fundamental epistemologies and ontologies underlying space, time, physics, and mathematics.

Pearson's *Grammar of Science* had a dramatic impact on Eddington, and many of its perspectives can be found in his engagement with relativity. Pearson's insistence on the perpetual nature of the scientific project and the importance of science to sound citizenship resonated well with the lessons Eddington had learned in Manchester from Schuster and Graham.[43] *Grammar of Science*'s central argument was that science was built from sense impressions through the intervention of conscious minds. Pearson said the mind, like a central telephone exchange, received a variety of signals (sense-impressions) that the mind needed to organize into external objects based on past impressions. In this way the mind "constructed" both objects and natural laws. He denied that the mind could have any direct access to the substance of reality. Rather, it was restricted to inferences through sense-impressions. Humans tended to see design in nature because they had to make recourse to their own minds to make sense of things, and they correspondingly saw a reflection of themselves in the world.[44] Pearson's positivism was almost certainly Eddington's first exposure to these ideas, and they found fertile ground. While Eddington's later ideas were hardly simple copies of Pearson's, the *Grammar of Science* was one of the most formative sources for shaping his philosophical reaction to Einstein.

Eddington also read James Ward's Gifford Lectures, *Naturalism and Agnosticism*, in 1906, and perhaps had even met Ward while he was at Trinity. Ward seems a likely source for some of Eddington's fundamental understanding of the idealist/materialist dualism that would frame much of his philosophical work. Ward defined the materialist standpoint as the notion that mechanical laws were supreme and that they relied on quantitative consistency. This mechanical view subsumed mind under matter, making all mental and spiritual phenomena illusory.[45] Ward's position was hardly unusual, but as far as we know his was the only work of a major idealist philosopher that Eddington read. There is a short passage defending the exemption of mind from mechanical laws based on the unquantifiable aspects of mind, which certainly prefigures important aspects of Eddington's later views. Similarly, Ward stressed the abstract analytical elements of physics, and particularly how this prevented physics from speaking about the nature of matter. One of the critical parts of his argument was the danger of natural theology. According to Ward, invoking a "God of the Gaps" had only led to naturalism as those gaps were explained by science. Finally, he spent a great deal of energy considering the problem of determinism, and this was probably the first place Eddington encountered a rigorous examination of the problem.[46] Determinism, of course, would become an important issue for Eddington later (see chapter 6) but was not significant in his initial approach to relativity. Overall, Eddington may have taken from Ward general thoughts on materialism and determinism and perhaps a way of limiting physical science via its own abstraction and quantification (though these were not major parts of *Naturalism and Agnosticism*).

Apparently, it was not until the 1910s that Eddington paid any close attention to the work of Whitehead and Bertrand Russell. We know he attended Whitehead's lectures at Cambridge, but none of the mathematician's works appear on Eddington's reading list until 1912, when he read *An Introduction to Mathematics*.[47] The first chapter of *Introduction to Mathematics* focuses on the abstract nature of mathematics and physical science, which would certainly have been consonant with other sources, but overall the book had little that seems to have made a lasting impact. Eddington read his first book of Russell's (*The Problems of Philosophy*) the same year. This may have reinforced the rigorous and critical stance with respect to the nature of matter, space, and time that he learned from Poincaré.[48] Three years later he read Russell's *Scientific Method in Philosophy*, which also did not seem to have any direct influence, other than perhaps an optimism that science could solve many problems traditionally considered to be in the realm of philosophy.[49]

Many years later, however, Eddington credited Russell with being his most important philosophical influence.[50] Certainly Russell's emphasis on a structural approach and the significance of the relations between elements of logic, sense data, and language bears a great resemblance to Eddington's reduction of physics to relations.[51] The source for this structuralism was almost certainly Russell's *Introduction to Mathematical Philosophy*, which Eddington must have read just as he returned from the eclipse expedition.[52]

Russell argued strongly that the precise sort of knowledge that could be treated by mathematics and logic was only that of relations. He presented this idea in the context of language, through consideration of the problem of the possible meanings of a statement when the grammar is known but the vocabulary is not. That is, what happens when we understand the structure of a statement but not its meaning? To Russell this was not mere sophistry; it actually reflected our understanding of nature: "We know that certain scienific propositions—which, in the most advanced sciences, are expressed in mathmatical symbols—are more or less true of the world, but we are very much at sea as to the interpretation to be put upon the terms which occur in these propositions. We know much more about the *form* of nature than the *matter*."[53] Scientific statements could be made only about the relations between things, not the things themselves. Russell was not speaking about relativity in particular, but the similarity between his arguments and the German positivist interpretations of relativity (e.g., Moritz Schlick) must have struck Eddington as remarkable. A further connection with ideas Eddington already held was Russell's claim that mathematical analysis of incompletely known entities was perfectly acceptable and, indeed, should be expected. *Introduction to Mathematical Philosophy* stated that mathematicians never needed to know the "intrinsic nature" of mathematical elements (such as points and lines). As long as they fulfilled the mathematical requirements, they were serviceable. Thus, what was necessary in mathematics was not "true" knowledge, but the structural approach: "This is only an illustration of the general principle that what matters in mathematics, and to a very great extent in physical science, is not the intrinsic nature of our terms, but the logical nature of their interrelations." However, only mathematics and physics needed to hold only to structure; "pure philosophy" still considered intrinsic natures important.[54] Given Eddington's tribute to Russell's importance for him, the strong correlation between Russell's structuralism and Eddington's philosophy, and the timing (Eddington almost certainly read Russell as he was writing his *Mind* articles), we can safely attribute much of Eddington's emphasis on the structural elements of physics to Russell.

It does not appear that Eddington encountered W. K. Clifford's writings until he was preparing the *Report on the Relativity Theory of Gravitation*. Clifford is well known for having anticipated many of the novel elements of general relativity, most notably the idea that fields of force could be reproduced (or even created) by geometry of particular configurations. There does not seem to be anything in *The Common Sense of the Exact Sciences* or Clifford's papers that had a decisive impact on Eddington; he would have already accepted Clifford's most interesting hypotheses about the nature of space and force as essential elements of relativity.[55] He did, however, occasionally borrow Clifford's approach to these elements, as the mathematician's views often better expressed Eddington's thoughts on the inevitability and identicality of relativity.

There were, of course, expositions of the meaning of relativity before and at the same time Eddington was writing his *Mind* articles. Did he borrow anything significant from contemporary interpreters of relativity? We know Eddington read the writings on relativity of Ludwik Silberstein and Ebenezer Cunningham (and it would have been strange if he had not, because they were the most competent of the few British scientists who spent time on relativity pre-1916). Silberstein's textbook on relativity was straightforward and made no particular attempt to address the philosophical meaning of the theory.[56] Eddington did not think much of Silberstein's work and makes almost no reference to it. Cunningham was given more consideration, but his work was largely concerned with relativity's importance for ether theory.[57] Interestingly, both he and de Sitter emphasized the importance of instrumental aspects of relativity. Cunningham stated that "the only actual observation is . . . the *coincidence* of a pointer or of a spot of light with a certain mark on a scale."[58] Similarly, de Sitter equated all the observations in relativity with intersections of phenomena with measuring devices: "Now, what we *observe* are always *intersections* of world-lines. Take, *e.g.*, an observation of an occultation of a star by the moon, and let us imagine, to simplify the argument, that the face of the clock is illuminated by the light of the star. Then the world-line of a certain light-vibration starts from a point on the world-line of the star, it then intersects successively the world-line of a point on the moon's edge, that of the clock's hand, and that of a point on the clock's face."[59] And, of course, Einstein's own 1905 paper on special relativity used a highly instrumental positivism in its analysis.[60] Even though the formulation of general relativity often obscures its positivist roots, Eddington certainly had adequate resources for interpreting relativity as a whole as a theory dependent heavily on solid rods, clocks, and pointers.

Despite the similarity of these views to those of logical positivism, it is not known whether Eddington read any of the early writings of the figures who would eventually form the Vienna Circle. His ability to read German was, at best, poor, and his ability to write in the language was apparently nonexistent.[61] One source that he might have read was Moritz Schlick's *Space and Time in Contemporary Physics*. This was translated into English by Henry Brose for publication in 1920, so it is certainly possible that Eddington had read it sometime in 1919.[62] Schlick presented relativity in the positivist style, saying it liberated space and time from "metaphysical haziness." Reality was reduced to measurement and direct sense experience.[63] If Eddington encountered logical positivism in its early days, it was probably through Schlick's book. It would have only reinforced the instrumental interpretation to which he had already been exposed; direct interaction with the proto-Vienna Circle was not necessary for the development of his instrumentalism. There is little reason to think Eddington read or relied on Schlick, however, and it seems he independently anticipated many of the aspects of the positivism that would emerge from the writings of the Vienna Circle and, later, P. W. Bridgman.

Brose also translated Erwin Freundlich's *The Foundations of Einstein's Theory of Relativity* while interned in Germany during the war. Upon returning to England, he sought the help of H. H. Turner in publishing the booklet, to no avail. It was only after Eddington returned from Principe and became involved with the project that it was published.[64] Freundlich's treatment of relativity emphasized its origin as a theory of principles (causality and continuity, according to him) and stressed that only observations and "measure relations" could be treated. Any philosophical influence from this book on Eddington would have been minor (and along the lines of Schlick). It is likely that any influence went the other way—Brose quotes Eddington directly in order to "assist in making intelligible" some of the mysteries around relativity.[65]

One German, Hermann Weyl, provided extremely important resources for Eddington. Weyl's extension of Einstein's theory to electromagnetism had an inspirational effect on Eddington, who used it as the base for his own attempts to generalize the theory. He felt his efforts were so close to Weyl's that he "scarcely considered it a rival theory." This formulation was valuable because it emphasized the way the laws of physics fall naturally out of the definition of the interval: "I think that any other presentation obscures the simple meaning and inevitable character of the gravitational law."[66] Shortly after his *Mind* article, Eddington published a paper laying out his plans for using Weyl's insights to connect relativity with not only electromagnetism but also the properties of the electron.[67]

Space, Time and Gravitation

It is no exaggeration to say that Eddington's first book on relativity for a popular audience was eagerly anticipated.[68] Relativity had caught the public imagination, but confusion about the theory continued to reign. *Space, Time and Gravitation* appeared in 1920 (at about the same time as the Oxford Congress) as the first substantial, widely available resource in English for understanding Einstein's new view of the world. The fact that it was written by Eddington, the man who had "proved" relativity, made it even more authoritative. In it, Eddington developed his previous statements in much greater detail and rigor and made the philosophical implications of the theory much clearer.

The prologue to the book recounted a fictional conversation between an "experimental physicist," a "pure mathematician," and a "relativist." In this rather Galilean scene, the trio discussed the nature of geometry and space from the position of their respective disciplines.[69] The mathematician begins by saying he could not evaluate the truth of a geometric statement, because it depends on the truth of the axioms, which were selected arbitrarily. The physicist argues that the axioms of Euclid are true by measurement—it cannot be proven for all cases, but it is accepted to generalize in physics. The relativist corrects the physicist by pointing out that all he was really doing was using a measuring tool, and thus we must pay attention to the properties of the matter from which it was constructed. We could explore the "natural geometry" of space in the same way we explore a magnetic field.[70]

The relativist, of course, dominates the conversation, usually turning his colleagues' points back on them. When the physicist objects that our senses see a Euclidean geometry in the world and that non-Euclidean space is contrary to reason, the relativist replies that our senses are crude, and it is not contrary to reason but rather to common sense. When the mathematician points out that Poincaré had argued for space as a convention, the relativist agrees that geometry and physical laws are interdependent but argues that the correct meaning of that interdependence has never been explored.[71]

Here Eddington was preparing the reader for the strange features of the four-dimensional world, but he was also addressing his colleagues and Einstein's critics. His intent was to show that relativity did not so much overthrow the insights of mathematics and physics as it integrated and matured their results. In this conversation we can hear echoes of the disciplinary boundaries that Eddington defended during his debates over stellar physics and would later describe in *Internal Constitution of the Stars*. This preface is a rich source for seeing where Eddington located relativity with respect to mathematics and physics and where each discipline had something to contribute. The mathematician (sounding a bit like James Jeans) could only operate within a particular set of axioms. The physicist's naïveté about the nature of measurement had allowed him to be ensnared in the swamps of absolute time and space. But the relativist in the end needed to fuse the physicist's pragmatism and empiricism with the mathematician's deductive abilities. Relativity was the next step in simplifying and exploring our knowledge of the world and needed to rely on previous steps while simultaneously surpassing them.

The text of the book began with the FitzGerald contraction and other elementary consequences of the restricted principle of relativity (defined by Eddington as the inability to measure any uniform motion relative to the ether). Observers disagreeing on size and time made an appearance: "That is too absurd for fiction, and is an idea only to be found in the sober pages of science."[72] These were not simple tricks or errors in observation. Nor could Einstein's paradoxes be discarded as an ephemeral fashion of physics, as some critics had hoped: "But to those who think that the relativity theory is a passing phase of scientific thought, which may be reversed in the light of future experimental discoveries, we should point out that, though like other theories it may be developed and corrected, there is a certain minimum statement possible which represents irreversible progress."[73] Relativity's disproof of old hypotheses of space and time was irreversible, regardless of the truth of its propositions. The theory's great and fundamental accomplishment was discarding uncalled-for hypotheses that obscured the simplicity of nature. Like it or not, relativity was here to stay.

As in his *Mind* article, Eddington emphasized the crucial role of the observer in physics. The observer not only needed to be accounted for mathematically but also needed to be considered as the source of the experiences from which scientific inquiry began: "It is necessary to dive into this absolute [four-dimensional] world to seek the truth about nature; but the physicist's object is always to obtain knowledge which can be applied to the relative and familiar aspect of the world."[74] The ordinary elements of experience functioned as a boundary condition for the theories of physics; that is, analytical symbols and quantities needed always to return to recognizable human experience.

However, the only parts of experience accessible to relativity were those that dealt with the relations between two things. An observer's knowledge was limited to awareness of interactions: "If we examine the nature of our observations, distinguishing what is actually seen from what is merely inferred, we find that, at least in all exact measurements, our knowledge is primarily built up of intersections of worldlines of two or more entities, that is to say their coincidences." More specifically, our knowledge was built on intersections of measuring devices. Observations of current or temperature were actually the coincidences of scales and wires. The difficult part of the four-dimensional world was the correlation of these coincidences between different observers, which drove Eddington's presentation of an overview of tensors and their peculiar properties.[75] For a book aimed at a popular audience, the presentation was quite technical and must have been completely opaque to many readers.

The second half of the book developed the experimental tests of general relativity and its treatment of momentum, energy, and the space-time characteristics of the universe. Eddington also restated his hopes for using Weyl's theory to extend relativity's domain to include electricity and magnetism. The final chapter was his tour de force, however, and attracted the most notice. Titled "On the Nature of Things," it was a more mature and detailed expression of his views on the meaning of relativity.

The chapter's purpose was to deliver a straightforward answer to the question that had bothered the Anglophone world since 1919: What did relativity tell us about reality? Eddington argued that, at its foundations, relativity revealed that a true grasp of reality required some form of integration of all possible points of view and aspects of nature. Einstein had moved science dramatically closer to that grasp:

> Is the point now reached the ultimate goal? Have the points of view of all conceivable observers now been absorbed? We do not assert that they have. But it seems as though a definite task has been rounded off, and a natural halting place reached. So far as we know, the different possible impersonal points of view have been exhausted—those for which the observer can be regarded as a mechanical automaton, and can be replaced by scientific measuring-appliances. A variety of more personal points of view may indeed be needed for an ultimate reality; but they can scarcely be incorporated in a real world of physics. There is thus justification for stopping at this point but not for stopping earlier.[76]

Eddington's statement of relativity's great contribution also laid out its most important limitation. Only insofar as an observer can be replaced by "measuring-appliances" can physics claim to understand the world. He asserts that "more personal points of view" were necessary for the hoped-after ultimate reality, but he makes no note of what they might be. The critical point here was that physics' ability to represent the world ended with the measuring device, and something else had to take over after that. But this was not a flaw of science; indeed, it was only because relativity did this that it was successful in unifying the forces of nature. The acknowledgment of something beyond measurement was the foundation of the new physics.

The reduction of the physical world to coincidences in measuring devices led directly to ignorance of the essences of things. Eddington proposed the analogy of future archeologists trying to reconstruct the game of chess from a manual of strategies. They might be able to reproduce the structure of movements involved in a match, but they would be forever ignorant of the nature of the game itself. Similarly, physicists would never be able to understand what lay beyond measurements: "By no amount of study of the experiments can the absolute nature or appearance of these participants [electrical and gravitational fields, space-time intervals] be deduced; nor is this knowledge relevant, for without it we may yet learn 'the game' in all its intricacy."[77] Again, there was no shame in scientists' ignorance. It was only through deliberately ignoring essences that scientific advance was possible. The job of the physicist was to find coincidences and establish a symbolic structure to represent their relationships (i.e., a mathematical theory). Trying to comprehend what was below space-time coincidences resulted only in conceptual messes like the ether.

This discarding of essences was one of the most important consequences of relativity and, Eddington argued, led to a new understanding of matter. He recapitulated his novel theory of matter (that it was identical to, and not a cause of, space-time curvature) and the mind's role in finding continuity out of an unformed World. According to relativity, the sifting of space-time accomplished by the mind acted only on the relations between things. It was simply choosing a measurement scheme, not changing reality: "The relativity theory of physics reduces everything to relations; that is to say, it is structure, not material, which counts. The structure cannot be built up without material; but the nature of the material is of no importance."[78] Eddington pointed to Bertrand Russell's recent work (specifically *Introduction to Mathematical Philosophy*) as a further indication of the necessity of considering only structure. The structural approach showed that the laws of nature existed only in a latent sense, like walks on a moor. An infinite number existed, and a conscious choice was required to declare either a walk or a law to exist.

After these remarkable arguments, Eddington paused for breath and admitted that these views were not forced on us by relativity, but rather were made plausible: "This summary is intended to indicate the direction in which the views suggested by relativity theory appear to me to be tending, rather than to be a precise statement of what has been established. I am aware that there are at present many gaps in the argument. Indeed the whole of this part of the discussion should be regarded as suggestive rather than dogmatic."[79] Once again, Eddington was pursuing this as an inviting avenue of research, not as the Final Truth. He summed his argument in his famous concluding paragraphs:

The theory of relativity has passed in review the whole subject-matter of physics. It has unified the great laws, which by the precision of their formulation and the exactness of their application have won the proud place in human knowledge which physical science holds to-day. And yet, in regard to the nature of things, this knowledge is only an empty shell—a form of symbols. It is knowledge of structural form, and not knowledge of content. All through the physical world runs that unknown content, which must surely be the stuff of our consciousness. Here is a hint of aspects deep within the world of physics, and yet unattainable by the methods of physics. And, moreover, we have found that where science has progressed the farthest, the mind has but regained from nature that which the mind has put into nature.

We have found a strange foot-print on the shores of the unknown. We have devised profound theories, one after another, to account for its origin. At last, we have succeeded in reconstructing the creature that made the foot-print. And Lo! it is our own.[80]

Physical law and human mind were now intimately, and inextricably, linked.

Space, Time and Gravitation drew a great deal of attention and became a best-seller soon after its publication in 1920. It was expected to be the long-hoped-for exposition that would finally clarify the legendarily incomprehensible Einstein: "Professor Eddington has few rivals in the art of exposition . . . a more successful attempt to explain things of extreme difficulty has seldom if ever been made."[81] The review in *Nature* congratulated Eddington on his ability to make the theory comprehensible in concepts already possessed by those familiar with classical physics. The final chapter was noted to be "apparently addressed to philosophers, [and] we can see it raising many heated controversies." Eddington's assertion of the importance of mind and individual experience was seen to be a "disappointing conclusion, perhaps, to many."[82] The fears that relativity equaled relativism were perhaps realized.

A 1922 review by the philosopher H. R. Smart celebrated Eddington's willingness to be confusing. Most accounts of relativity, he said, suffered from "over-clearness," probably because the writers doubted the reader's ability. Admirably, Eddington was "more interested in provoking thought than in simplification of explanation" and providing a "healthy respect for the difficulties involved in the details of modern physical explanation." The last chapter was seen as lively and stimulatingly speculative and "all the more valuable because it does not leave the reader certain that he fully understands Einstein's great achievement." The book showed that "not even the initiated" fully understood the significance of relativity. Unlike *Nature*'s reviewer, Smart was quite pleased with Eddington's conclusions about the meaning of relativity: "One thing, however, is fairly certain, namely that there is an objective world outside us, which, while independent in a sense of purely subjective factors, nevertheless reflects the presence of a rational principle in the objective universe."[83] Scientists outside physics celebrated what they saw as Eddington's warning that physical science was not sufficient for understanding human life. Human experience could once again be described in terms recognizable to humans, instead of conglomerations of atoms. This gave "the courage to be anthropomorphic in describing man," instead of needing to use a "mechanical dummy which shall stand for man in our science."[84] Against all expectations, Eddington had apparently shown that relativity gave humanity a renewed place of importance in the scheme of reality.

Philosophers Grapple with Relativity

Space, Time and Gravitation marked the opening of relativity to serious philosophical investigation. A great deal of the popular British interest in relativity centered on its implications for religious thought, and I will focus on those issues here. The philosophical debates presented here provide the backdrop against which Eddington's ideas about the significance of relativity for religion were examined and understood. We need to explore the intellectual geography of British philosophy before placing Eddington within it.

There had been a few attempts to make a general assessment of relativity's philosophical and religious significance before Eddington's book, most notably *Space, Time, and Deity*, Samuel Alexander's Gifford Lectures for 1916–1918. Alexander, then teaching at Manchester, first encountered relativity in 1916. He noted that he did not feel qualified to address it—a remarkable statement given that he proceeded to give a lengthy series of lectures nominally based on the theory within the year. His lectures sought at first to establish a "metaphysical" theory of space and time that would arrive at the same results as the "mathematical" theory of Einstein and Hermann Minkowski.[85] He said he strove to ensure his space-time was "compatible" with theirs, but his analysis has virtually no reliance on relativity. Many more pages were devoted to Henri Bergson's theory of time than Einstein's. Alexander quickly departed from even a semblance of working from relativity and leapt into a conventional semi-idealist analysis of mind and "Deity" (an entity apparently different from, but virtually indistinguishable from, God). In the end his conclusions have nothing to do with Einstein's theory and could have come from any Gifford Lecture in the previous two decades.

Most of the British philosophical inquiry around relativity came in the years after Eddington's early work and needed to respond to his position. A high-profile and popular book was Viscount Haldane's *The Reign of Relativity*. As an indication of the interest in relativity at the time, his book went through three editions in four months in 1921. According to Haldane, Einstein had propelled philosophy in the direction it had already been going, namely, toward a general relativity of knowledge. Haldane presented himself as representing a widespread concern in Britain in the first years after the war that the nation was being dominated by unreflective emotion instead of the faith that had made it great.[86] His agenda was to show that relativity (of the Einsteinian sort and otherwise) brought together people and God and would help restore British civilization to greatness.

For Haldane, relativity demonstrated the individual character of knowledge and the variation of perception with circumstance. His approach was as a philosopher. While admitting that relativity was mathematical, he said that philosophers occasionally had to help mathematicians remember the important things. He was well read in relativity and addressed most of the major interpreters of the time. Schlick was criticized for continuing to treat space and time as separate entities, and Whitehead was applauded for refusing to follow the "Victorian view" of dividing the world into subjective and absolute elements. Whitehead was further praised for demonstrating effectively that relativity required us to move beyond the normal boundaries of science. Eddington's structural approach and definition of matter were discussed in some detail, as was his assertion that mind determines the laws of nature. According to Haldane, Eddington was in the "metaphysical borderland of mathematics." He called Eddington an "acute and courageous thinker" and seemed to accept much of what he had to say.[87] Einstein himself received little attention; Haldane was unaware of any philosophical pronouncements from him.

The relativity of knowledge that Haldane developed looked to unify all dualities: observer/observed, spiritual/material, whole/part, and so on: "From this point of view the theory of the relativity of knowledge derives a meaning wider than that which the physicists give to it.... For it shows us that the material and the spiritual are not separate and self-subsisting facts, but are illustrations of different fashions in which reality presents itself when regarded from standpoints divergent in the logical character of their methods." This resolved problems of personality, determinism, and government and of morality in general. God was to be found in the ultimate knowledge of the world and, specifically, in the creation of knowledge of the universe. Haldane then used Eddington's argument for the mind's creation of natural laws as evidence for the ability of human minds to approach the divine through the creative act.[88]

The god that came out of Haldane's reasoning bore little resemblance to the traditional Christian deity and only the most liberal of believers would have been satisfied with it. And while his ideas were based (at least somewhat) on Einstein's physics, critics were quick to point out that *The Reign of Relativity* arrived at results virtually identical to Haldane's earlier philosophical work.[89] It was widely read, however, and quickly helped ally relativity with liberal theology, a move that helped soothe fears that the theory had displaced all traditional values.

The relationship between relativity and relativism was the fulcrum for much of the early discussion of the theory's meaning. Oliver Lodge, as an exceptionally visible critic of relativity, continued to complain about the implications of Einstein's geometric approach even after he had more-or-less accepted relativity's empirical confirmation: "In such a system there is no need for Reality; only Phenomena can be observed or verified: absolute fact is inaccessible. We have no criterion for truth; all appearances are equally valid; physical explanations are neither forthcoming nor required; there need be no electrical or any other theory of the constitution of matter. Matter is, indeed, a mentally constructed illusion generated by local peculiarities of space."[90]

Had Einstein rendered truth inaccessible? What was left of the scientific project of understanding the world? The philosophers H. Wildon Carr and J. E. Turner carried out a lively debate in *Mind* that helps illuminate these issues—their concerns and misunderstandings were shared widely. Carr's 1920 book on relativity was not well received, at least partially because of its numerous technical errors.[91] In response, Turner attacked both Carr and Haldane for their attempts to link Einstein's theory with traditional understandings of relativism.[92] Turner equivocated on whether relativity had philosophical implications; he thought Einstein had stated there were none, but he happily seized on Eddington's work to demonstrate that there were still absolutes in the world.[93] Turner wanted to isolate the results of relativity from epistemology in general. While the results of relativity, he said, "certainly affect most materially the detailed content of knowledge, they can have no bearing whatever on the *nature* of knowledge, or of experience, or of reality, in general and as such; unless, *i.e.*, we are going to base our epistemology on the rate at which we happen to be moving, or on gravitational potentials."[94] Carr was criticized for credulously applying the properties of matter in motion to experience as a whole. But like Haldane, Carr defended his relativism as being really a defense of absolutes; his absolute was simply not independent of experience. He attacked Turner as being a dogmatic philosopher unwilling to accept the new world revealed by physics: "It is not a little curious to contrast this marked indifference of a philosopher to the new scientific discovery with the profound consciousness the mathematicians express of its fundamental philosophical significance. I have in mind particularly Eddington, Weyl,

Thirring and Einstein himself, to mention only a few."[95] Scientific facts had turned philosophy on its head, and critics of relativistic philosophy were no better than the geocentrists who persecuted the Copernicans.

All of these players were present at a session of the 1922 meeting of the Aristotelian Society, which discussed relativity and particularly its relationship to idealism.[96] Einstein's theory was a strong presence even outside this session; most of the discussions not devoted to classics or oriental thought discussed relativity in some way. The session on idealism was overseen by Viscount Haldane. H. W. Carr started the conversation and surprised no one with his claim that relativity was "in complete accord" with neo-idealism and "in complete disaccord" with neo-realism.[97] The other speakers were quick to refute his alliance of Einstein and the idealists. T. P. Nunn said that what Carr had really shown was that materialism and relativity were not compatible, but that had no significance for realism and his overall weakness was his appeal to overly vague notions of experience.[98] Dorothy Wrinch argued that relativity had no special philosophical implications different from physics in general, and physics in general was based on facts, which were real and independent of mind. Mind had no privileged place in physics or relativity.[99] The discussion was essentially an extension of the realism-idealism debate, with relativity simply being the new point of contention.

One of the contributors to the session was A. N. Whitehead. He also rejected Carr's simplistic idealist interpretation, though he had no particular stake in the realism-idealism issue. If Carr were to prove his case that mind was at the root of relativity, he said, it must be shown that "a man in love necessarily measures space and time differently from a man given over to avarice."[100] Whitehead maintained that relativity did have important philosophical implications, though they were outside the issue at hand. Perhaps to separate himself from the preexisting debate, Whitehead also contributed a paper on relativity outside the session on idealism. In that second paper, he argued that relativity's significance was that it showed the importance of relatedness and also new understandings of space and time. The fusion of space and time made the ultimate fact of the universe a kind of "process."[101]

Whitehead's contributions to the interpretation of relativity had begun some years before, at the 1916 meeting of the Aristotelian Society. He developed his relational theory of space and time as an outgrowth of his work on the logical foundations of epistemology and mathematics. He restructured relativity theory to rely on a system of time more in concert with experiential (rather than operational) notions of time. His "pan-physics" and "extensive abstraction" grew out of this during the ten years following 1915, but his philosophy of relativity was not widely read and appreciated until his Lowell

Lectures. They were published as *Science and the Modern World* in 1925 and were tremendously influential in forming what would eventually become process theology.[102] Interestingly, the most lasting contribution to the philosophy of religion to come out of relativity has often forgotten its Einsteinian roots.

The early 1920s saw many attempts to fuse relativity with idealism and metaphysics in general. Einstein's willingness to talk about the nature of time and space was apparently a spur to metaphysicians like Alexander who saw it as a sort of approval for their work. After centuries of discord, it seemed that philosophy finally had a permanent place in physics.[103] Generally, philosophers' concerns with relativity's connection to metaphysics and idealism were simply extensions of existing debates. Similarly, there were attempts to extend the issue of materialism to Einstein's new theory. Carr claimed that the role of mind in relativity (that is, the necessity for an observer) showed that materialism was no longer tenable: "It may not be obvious at once that the mere rejection of the Newtonian concept of absolute space and time and the substitution of Einstein's space-time is the death-knell of materialism, but reflection will show that is must be so. [In relativity] the existence of mind is essential." Materialism was held to be essentially a monistic and atomistic conception of reality in which matter was primordial and mind was derived. But now mind needed to exist before there could be any awareness of the material world; thus, the crude world of matter had been left behind.[104] Hugh Elliot immediately wrote to refute him: "Prof. Wildon Carr has for a number of years been busily engaged in ringing the death-knell of materialism. I was therefore not a little surprised to read in *Nature* his statement that Einstein's theory was the 'death-knell of materialism' . . . we are led to the conviction that materialism must have very singular properties to survive so many tragic executions." Elliot defended materialism as the basis of science, and he rejected the view that relativity disproved it as a confusion of things (which exist apart from mind) with concepts (which do not).[105] Materialism in modern physics would become a much more significant issue in the later 1920s, and the cultural storm around it will be dealt with in the next chapter.

By the time of the 1922 Aristotelian Society meeting, discussions about relativity usually accepted its truth, at least in some sense. Eddington's *Space, Time and Gravitation* not only established a common set of fundamentals for discussion, it also marked philosophical inquiry about relativity as acceptable. The near-ubiquitous references to it (with occasional exceptions for Weyl, Schlick, or Freundlich) in philosophical considerations of relativity show that it was effectively *the* text on relativity for Anglophones. Eddington was seen as the major authority on relativity immediately after Einstein, and the latter's silence on the philosophical meaning of relativity meant those questions defaulted to the former. While not everyone agreed with Eddington's interpretation, his mere act of philosophical speculation made further speculation acceptable—if Britain's technical expert on relativity thought relativity was worthy of philosophical explanation, the task was worth doing. The fact that Eddington's analysis was in harmony with many tenets of traditional idealism was simply a bonus. His philosophy per se was not yet particularly discussed, but his book as a foundation for philosophy was everywhere.

Mathematical Theory of Relativity

After years of presenting relativity in various forms truncated by his own technical inability (*Report on Relativity*, 1918) or the inability of his readers (*Space, Time and Gravitation*, 1920), Eddington finally had the opportunity to present the theory in what he considered to be its true form. This was his *Mathematical Theory of Relativity*, which became one of the most important textbooks of interwar physics. It should come as no surprise that he used the full mathematical apparatus of differential geometry, but it is intriguing to note how much of his perspective on the *meaning* of relativity was included. Indeed, he warns that the reader should already "have a general grasp of the revolution in thought associated with the theory of relativity" before approaching it "along the narrow lines of strict mathematical deduction." That is, they should have already read his other writings on the subject, particularly *Space, Time and Gravitation*. Although *Mathematical Theory of Relativity* was effectively a text for scientists, its goal was to bestow more than simple technical competence: "But for those who have caught the spirit of the new ideas the observational predictions form only a minor part of the subject. It is claimed for the theory that it leads to an understanding of the world of physics clearer and more penetrating than that previously attained, and it has been my aim to develop the theory in a form which throws most light on the origin and significance of the great laws of physics."[106] Relativity had uncovered the fundamental structure of the universe, and Eddington explicitly presented it in a way that best revealed its implications for mind, matter, and reality.

The book was comprehensive in presenting the technical elements of relativity, which differed only in detail and clarity from previous expositions. By the time *Mathematical Theory of Relativity* was published in 1923, however, relativity had been under discussion for nearly four years since the 1919 expedition, and Eddington needed to position the theory with respect to the various attempts to refute and interpret it. One of the dominant themes of the book was where relativity fit in the scheme of mathematics and physics. This reflected Eddington's awareness of the debates about relativity in the previous few years. Claims that the effects predicted by relativity (such as the crumpling of space-time) were merely mathematical fictions or that a theory based simply on geometry could not represent reality needed to be addressed. Similarly, arguments by philosophers that relativistic effects dealt with only ephemeral concepts and not the physical world were a serious misunderstanding that required remedy. Thus, despite the title, Eddington made a sustained argument that relativity was *physics* and that its claims therefore needed to be treated as real.

Physics could be distinguished from mathematics by its use of "physical quantities" (such as length, work, and current) that were defined primarily by the way we observed them in the world. These observations were defined by a procedure of observation and were, therefore, a "manufactured article." The connection between these physical quantities and the "existent world-condition" was only through the numerical measurements created by the observation procedures. This gave science only an indirect knowledge of the world. And a physical quantity had no reality other than "the series of operators and calculations of which it is the result." How do we find the definitions of physical quantities? We simply watch physicists work. There should be no assertion that these procedures are right or wrong; the task is simply to ensure that the definition actually reflects the measuring practice.[107]

Relativity began its reconstruction of physics with the realization that "any operation of measurement involves a comparison between a measuring-appliance and the thing measured." The crucial breakthrough was the recognition that physical quantities are not part of objects but only the relations between these objects and something else. Physics gathered the results of "measuring-appliances" and searched for uniformities, which were then called laws of nature. Eddington stressed that it was not necessary to immediately grasp all the operations; physicists could simply work as though they understood them and proceed until problems arose. "We cannot be forever examining our foundations; we look particularly to those places where it is reported to us that they are insecure."[108] In order to do physics, physicists had an obligation to move forward even when not all axioms had been proven.

But this was a *mathematical* theory being presented. What role did mathematics play, and when could the physicists hand their quantities over to the logicians? Remarkably, Eddington claimed this could be done as soon as a quantity yielding measurement was defined. This measurement then needed to be expressed in the symbolic language of tensors, unusual mathematical objects that allowed the measurement to be consistent regardless of coordinate system (i.e., regardless of observer). Eddington thought this methodology, at the heart of general relativity, had proved itself to be of fundamental importance for all science: "I do not think it is too extravagant to claim that the method of the tensor calculus . . . is the only possible means of studying the conditions of the world which are at the basis of physical phenomena."[109] Any theory had to take into account all observers, and tensors were the only way to do this.

Eddington argued that the definition of the space-time interval in tensor form, the principle of covariance, and the absence of gravity in empty space were essentially all that was needed to begin mathematical deduction. Mathematics provided a way of expressing and analyzing the two-way connection between physical observations and the structure of the universe that allowed observations to be made. The mathematician had two jobs:

> to examine how we may test the truth of these postulates, and to discover how the laws which they express originate in the structure of the world. We cannot neglect either of these aims; and perhaps an ideal logical discussion would be divided into two parts, the one showing the gradual ascent from experimental evidence to the finally adopted specification of the structure of the world, the other starting with this specification and deducing all observational phenomena. The latter part is specially attractive to the mathematician for the proof may be made rigorous; whereas at each stage in the ascent some new inference or generalisation is introduced which, however plausible, can scarcely be considered incontrovertible. We can show that a certain structure will explain all the phenomena; we cannot show that nothing else will.[110]

Relativity could be understood either as the construction of a theory from empirical evidence or as an a priori theory that predicted observational phenomena from basic postulates of measurement. Eddington's position was that physics was top-down *and* bottom-up: one could not understand the nature of physics without giving equal epistemological weight to both observation and theory. It did not matter which approach was taken, because the rigor of mathematics was needed alongside the perils of induction.

Eddington chose to begin from the basic postulates of observation and sought to engage in deductive "world-building," because he felt this method best expressed the implications of the theory. The deductions were an exercise in pure mathematics that attempted to build a world out of tensors that functioned the same way as the empirical world.[111] To Eddington, it was staggering that this project was possible at all. It was even more staggering that it worked and that the laws of mechanics and gravity actually did fall neatly from the definitions of measurement. Weyl had shown a method for generalizing this technique to include electromagnetism as well, and Eddington found this to be "unquestionably the greatest advance in relativity theory after Einstein's work."[112] He had been developing this gauge transformation technique since he first read Weyl years before and thought it to be the best prospect for unifying all physics in deductions from the postulates of relativity.[113]

Eddington prefaced the book with the claim that this method would reveal the "origin and significance" of the laws of physics. And what was the conclusion? "The investigation of the external world in physics is a quest for *structure* rather than *substance*. A structure can best be represented as a complex of relations and relata; and in conformity with this we endeavor to reduce the phenomena to their expressions in terms of the relations which we call intervals and the relata which we call events."[114] He had made his argument for the structural basis of physics before, but this time it was inextricably woven into his physical reasoning and mathematical deductions. It was not merely an interpretation layered on top of physics, it was wholly integral to the science. Structural considerations made the laws of physics inevitable:

> If we could see that there was the same inevitability in Maxwell's laws and in the law of gravitation that there is in the laws of gases, we should have reached an explanation far more complete than an ultimate arbitrary differential equation. This suggests striving for an ideal—to show, not that the laws of nature come from a special construction of the ultimate basis of everything, but that the same laws of nature would prevail for the widest possible variety of structure of that basis. The complete ideal is probably unattainable and certainly unattained; nevertheless we shall be influenced by it in our discussion, and it appears that considerable progress in this direction is possible.[115]

Eddington rejected Victorian mechanical models as tenuous and arbitrary but celebrated relativity as an irresistible deduction from the recognition that the physical world was only a superstructure of relations. Just as kinetic theory showed that all the macroscopic properties of gases followed from the simple microscopic motion of matter, the laws of mechanics followed from the simple operation of measurement. The power of this approach was in the possibilities for progress, not in what it had already achieved.

Anything more than an analysis of structure was, therefore, an error. Presumably, things in the world had both substance and structure, but substance was inaccessible. Eddington summed up the philosophy of science that he argued was inseparable from relativity:

> The physicist who explores nature conducts experiments. He handles material structures, sends rays of light from point to point, marks coincidences,

and performs mathematical operations on the numbers which he obtains. His result is a physical quantity, which, he believes, stands for something in the condition of the world. In a sense this is true, for whatever is actually occurring in the outside world is only accessible to our knowledge in so far as it helps to determine the results of these experimental operations. But we must not suppose that a law obeyed by the physical quantity necessarily has its seat in the world-condition which that quantity "stands for"; its origin may be disclosed by unravelling the series of operations of which the physical quantity is the result. Results of measurement are the subject-matter of physics; and the moral of the theory of relativity is that we can only comprehend what the physical quantities *stand for* if we first comprehend what they *are*.[116]

But the laws of physics followed inevitably only from measurement with the addition of one further element. This was the mind of the observer, who searched for permanence among the cloud of four-dimensional relations that made up the world. Eddington's reasoning around the importance of mind was fundamentally identical to that in his *Mind* articles, but here it was threaded throughout the text. As with structural considerations, Eddington presented the significance of mind as something that was addressed not after the physics, but as something that must be addressed within the process of physics. $G_{\mu\nu}$ could not be understood without reference to the human mind.[117]

   Mathematical Theory of Relativity was received with great enthusiasm in the scientific community (its readership was limited for obvious reasons). Some reviewers noted that its name was a misnomer and that it was written as physics. Complaints about the difficult mathematics continued, but Eddington was praised for having produced a book that was "quite clearly written (a quality too rare in texts on relativity) and is to be heartily recommended to any one seeking a comprehensive grasp of the theory."[118] The *Philosophical Magazine* was firm in warning readers that the book needed to be read with *Space, Time and Gravitation* close at hand. The reviewer praised the brief "quasi-mathematical, quasi-philosophical" introduction but insisted Eddington's thoughts on relativity could not be done justice in anything short of a full-length treatment.[119] There was surprisingly little comment from reviewers on Eddington's interpretation of the theory; it may be that a "mathematical" book's audience was uninterested in philosophical matters or that (like the above reviewer) readers took seriously Eddington's warning that this was only a companion to his previous book. Issues of Eddington's readership and their disciplinary identity would only become more complex after *Mathematical Theory of Relativity*, as this was nearly the last of his books that could be easily categorized as mathematics, physics, or philosophy.

Reality

In 1925 a volume appeared titled *Science, Religion and Reality*. Edited by Joseph Needham, it brought together an eclectic mix of philosophers, scientists, and historians to consider the relationships of religion and science from the perspectives of their disciplines. Needham sought to help round out human character by having "some feeling" for each of the forms of human experience, including religion and science.[120] One chapter, "The Domain of Physical Science," was authored by Eddington.[121] His contribution represented the culmination of his thinking on relativity by itself (the development of quantum mechanics would provide a new and important spur) and was his first explicit public statement on the relation of science and religion. However, despite the apparent introduction of religious elements into his philosophy, it was in no way a departure from the ideas he had been developing. Indeed, "The Domain of Physical Science" was a natural outgrowth and likely put forth ideas that he had held for some time.

Eddington's argument here will seem familiar by this point, so I will pay special attention to the loci where he brings that argument into contact with spiritual values. The core of the essay was the establishment of the legitimate boundaries of the scientific conception, illustrated with the image of the simple act of stepping into a room. The common man unhesitatingly strides forward, but the physicist hesitates: the floor's solidity is an illusion, and in truth it is an insubstantial web of hurtling atoms and shifting force fields. This parable does double duty for Eddington. First, it shows the dramatic divorce between the scientific and everyday conceptions of the world. Second, it shows the limitations of the use of the scientific conception. We must not insist on it constantly or we cannot function. Instead, expediency should be our guide in its application.[122]

Given this status of the scientific worldview, what grounds were there for believing in an external world at all? According to Eddington, it was our experience of the existence of other conscious beings. We can compare experiences and find that our experience is not independent, and the common elements can be synthesized and called the external world. "The purpose of conceiving an external world is to obtain a conception which could be shared by beings in any other physical circumstances whatever." This was what Einstein did, by his "refusal to reject any natural point of view as a 'wrong' one." His synthesis also had the spectacular by-product of explaining gravity. "Real" was no more than an accurate description of those experiences we have in common. Scientific statements must be judged right or wrong based on correspondence with this external world.[123] There is some circularity here: the external world comes from comparing experiences, but the purpose of conceiving a world is so experiences may be compared.

Eddington positioned himself against the vision that exact science was sufficient to explain all the phenomena of life and experience. He refuted this with relativity's insight that the subject matter of physics was more restricted than earlier thought.[124] The idea that physics could explain consciousness and religion was based on the view that the microscopic view (the shifting floor) was more correct and more fundamental than ordinary experience (the solid plank). But there was now powerful evidence that this view was profoundly flawed. Physics could speak of nothing beyond the "pointer readings" produced by measurements. Any measurement of any quantity reduced to a pointer on a scale, yielding a number, which could then be fed into "the machine of scientific calculation."[125] As he had argued before, physicists had to give up the notion that they have knowledge of things beyond pointer readings, and it was this that underlay much of the continuing resistance to relativity. There were limits to the sphere of exact science.

And yet, there were human responses that could not be reduced to pointer readings, such as affection and love. The traditional response to this was the split between the material and the spiritual, but Eddington reconceptualized this based on the structural nature of relativity: "I venture to say that the division of the external world into a material world and a spiritual world is superficial, and that the deep line of cleavage is between the metrical and the non-metrical aspects of the world." Was one of these real and the other false? No—both the metrical and the nonmetrical began with experience and therefore were the bases for knowledge of the external world. The non-metrical had a slight advantage in that it included the experience of our own consciousness, which was direct knowledge, while metrical knowledge was always inferred through a chain of pointer readings and calculations. "The spiritual phenomenon of consciousness is the one thing of which our knowledge is immediate and unchallengeable. It seems to be the most undoubtedly real thing we are aware of."[126]

Eddington then explained relativity as succinctly as possible to show the reality it attributes to mind. He introduced a new scheme for describing this, the circular nature of physics (see fig. 5.1). The circle worked like this: Einstein's law of gravitation was based on a series of potentials, which were derived from space-time intervals, which were descriptions of measurements made by scales and clocks, which were made of matter, which determined the mass-momentum-stress tensor, which determined the gravitational potentials. Physics happily moved around this cycle and thus created a unified, self-contained metrical world. But there was a way to escape from this cycle, through the

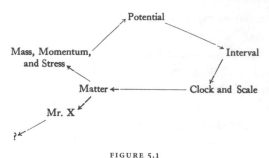

FIGURE 5.1
The closed cycle of physics, from Eddington, "Domain of Physical Science," 209.

observer who perceived matter (and, following Eddington's earlier argument, in a sense created it). The cycle exists independently of consciousness, because physics can operate only on metrical things, which are independent of consciousness.[127] This cyclical world was selected from the elements of the world amenable to this treatment but was solely structural. It said nothing about the nature of its elements.

The World was effectively a chaotic flux of space-time intervals and events that could take many different forms, with many different laws. Why did it look the way it did? As Eddington argued in earlier work, the conservation laws came from the mind's expectation of permanence. Thus, the world had "actuality" (i.e., it actually does look a particular way) that depends on consciousness: "We cannot describe the difference [between possible worlds] without referring to a mind. The actuality of the world is a spiritual value. The physical world at some point (or indeed throughout) impinges on the spiritual world and derives its actuality solely from this contact."[128] That the world took any coherent form was evidence for a world beyond the metrical: the spiritual. The spiritual and the metrical shared roots in experience and consciousness.

Consciousness was the source of all values, aesthetic values as well as physical ones like mass and force. The mind *chooses* to address certain values (by looking for permanence, etc.), and this determines the nature of the "superstructure" that theoretical physics manipulates.[129] The laws of nature produced by our consciousness were called identical laws (because they emerged from mathematical identities), but there were also statistical and transcendental laws, which did not appear to have their origin in mind. They were wholly other, and scientists should not expect to have the same success in explaining them that had been achieved with identical laws like electromagnetism. This distinction among the types of laws provided Eddington with his first opportunity to formally address an issue that would become a major project: free will. He argued that, like all phenomena, we can have confidence in the reality of free will because we experience it as real. The question was then, what was the efficient mechanism for the ability of the human will to influence the physical world? He took an agnostic position. It was clear from the nature of identical laws that they could not be modified, so he said that will must somehow influence statistical or transcendental laws.[130] But the laws of thermodynamics had been seen to operate in animals, so how could this influence exist?

Eddington acknowledged that this was a problem, but instead of cobbling together an explanation or abandoning apparently conflicting evidence, he used this as an opportunity to address the nature of scientific advance. He described science as a jigsaw puzzle: sometimes pieces do not fit, but the investigator thinking of the entire picture does not despair. The assembler must accept that there are connecting pieces that have yet to be found. But sometimes pieces fit together so well that there is no need to worry about future disruptions. Revolutions in thought like relativity dealt with the overall picture being formed by the puzzle.[131] Scientists do not worry about these changes, because they are confident in the growing portions of the completed puzzle. In this way Eddington defended the implications of relativity as being of lasting value. Einstein had assembled substantial portions of the puzzle that any future revolutions would need to take into account.

The completed portions of the puzzle had particular importance for religion. Above all, the model of a soulless mechanical universe was gone forever: The "unqualified materialism of the last century is not to-day the most inviting bypath.... Our thesis has been that the recent tendencies of scientific thought lead to the belief that mind is a greater instrument than was formerly recognized.... In exploring his own territory the physicist comes up against the influence of that wider reality which he cannot altogether shut out." Our minds are creative instruments, sparks of the divine Logos that created the world as a whole. For Eddington, mind and consciousness were identical with spiritual values, and the recognition of the former was a recognition of the latter. In Quaker style, he suggested that the divine sparks of human minds pointed to the presence of a greater Mind. But he had no illusions that physics alone had somehow replicated all the essential features of traditional religion:

> It will not be expected that science should indicate how this colourless pantheism is to be made into a vital religion. Science does not indicate whether the world-spirit is good or evil; but it does perhaps justify us in applying the adjective "creative." It is for other considerations to examine the daring hypothesis that the spirit in whom we have our being—our actuality—is approachable to us; that He is to us the beneficent Father, without which, it seems to me, the question of the theoretical existence of God has little significance. This

image of the divine nature is not a convenient fiction [to be discarded when needed].... If this hypothesis is correct, it signifies a direct relation of spirit to spirit which can scarcely be made clearer [by using mathematical physics].[132]

Relativity provided the critical first step through its recognition of mind, but further exploration into physics would give no more religious insight. That could come only through direct spiritual experience, in other words, mysticism. Eddington closed with an affirmation of the common empirical roots of both religion and science and an implicit warning about the provisional status of all knowledge: "The scientist and the religious teacher may well be content to agree that the value of any hypothesis extends just so far as it is verified by actual experience."[133]

Eddington's "Domain of Physical Science" sought to show how physics, and relativity in particular, allowed a new approach to reconciling religion and science. His argument was, concisely, that relativity showed that physics could address only the metrical aspects of reality. These metrical aspects needed to come from mind(s), whose experiences formed reality. Neither metrical nor nonmetrical experiences were privileged, so the nonmetrical spiritual perspective formed a picture of reality just as important as the metrical picture of physics. These spiritual experiences could then form a legitimate way to begin to think about the divine. Eddington in no way claimed that relativity *proved* any truths about religion, only that it removed some of the major objections to it (e.g., materialism). Once those obstacles were removed, it was up to the individual to follow their experiences. The argument defended the reality of aesthetics as much as it defended the reality of God. Natural theology was nowhere present, and there was no certainty to be found. The burden of discovering God was on the believer, not the scientist. Relativity, as Eddington, saw it, gave no solace to the religious faithful if experience was not central to their practice. A consequence of this was that when Eddington said "religion," he really meant "religious experience." The applicability of his ideas to any given religious tradition was directly proportional to its experiential basis, which both cemented his popularity with liberal theologians and angered traditionalists.

Eddington's contribution to Needham's book followed naturally from his work on relativity since 1920. He seems to have formed the fundamental aspects of his interpretation of relativity early in his encounter with Einstein's theory. Indeed, the majority of "The Domain of Physical Science" could have been written at the same time as his *Mind* articles. The work that relativity did in 1925 for Eddington in the religion-science interface consisted of two demonstrations: the restriction of physics to metrical phenomena and the dependence of physical law on the mind. *Both* were present in Eddington's earliest writings on the interpretation of relativity. There are two important points that follow from this. First, Eddington almost certainly saw the religious implications of relativity early on, if not in a sophisticated way. Everything he needed was present, and his valence value of religious experience would have made religious conclusions natural. Second, these two demonstrations appeared in varied forums: philosophical (*Mind*), technical (*Mathematical Theory of Relativity, Proceedings of the Royal Society*), and religious (*Science, Religion, and Reality*). The arguments he made for the further technical development of relativity and the arguments he made for the reality of religion were all based on these same demonstrations. Both arguments integrated his value of experience as a real and efficacious part of human life. As a Quaker, it was clear to him that experience was necessary for properly understanding and advancing religion. As a scientist, it was clear to him that relativity had shown that experience was also necessary for properly understanding physics. Einstein had been triumphant in unifying and simplifying physics because his theory (as Eddington understood it) used experience and consciousness as a fundamental premise. An experiential and mental outlook on reality, as demanded by relativity and quantum mechanics, required one to discard a materialist worldview. Thus, the values of mystical religion were of great benefit to the physicist: "The anti-materialistic attitude of religion would certainly be an advantage in modern science. It would help a man to see certain possibilities, to entertain certain speculations, which the old materialists, who regarded the universe as composed of little billiard balls, would have difficulty in grasping."[134]

In the domains of both religion and science, Eddington saw that experiential values were critical. This had multiple effects. Partially it was a matter of theory choice. In other words, he was more willing to accept relativity as valid, because it integrated certain values. But his values also shaped the way Eddington interpreted and used the theory. His discourses on the meaning of relativity inevitably centered on the significance of consciousness, as did his arguments for the new understanding of the nature of matter. Weyl's methods for unifying gravitation and electromagnetism were chosen for future investigation because they took relativity's emphasis on metricality even further and thus emphasized the role of the conscious observer. To Eddington, this set of values not only moved back and forth between religion and science, they were efficacious in both realms. Physics and mysticism were bound together, and neither could move forward without experience.

Conclusion

The Quaker value here is clear: reliance on individual, nondogmatic experience. It is useful at this point to step back and consider the trends in the larger religious community of early 1920s Britain. Eddington's argument, while Quaker, would have sounded familiar and persuasive to large numbers of Christians involved in the growing movement toward liberal theology. This movement was a turn away from evangelical emphases on scripture and doctrine and toward acceptance that sources other than the traditional Christian account (modern history, experience, science) could and should have authority in religion.[135] The liberals capitalized on the prewar sense of progress, which still had some life in it immediately after the war, to defend tolerance and broad religious views. Religious experience had been singled out as the essential element of faith in the influential works of Rudolph Otto, William James, and Friedrich Schleiermacher, and it had become possible to seriously discuss religion in the encompassing, ecumenical terms used by Eddington. Religion still had a great deal of vitality in this period, though many worried that the liberals were discarding essential parts of Christianity.[136] Rational approaches to religion, such as Dean Inge's writings, were acknowledged as important but dangerous. "That very spirit of broad-mindedness, and toleration, which is generally approved as consonant with the best democratic spirit, may soon prove to be a greater menace to the integrity and organized unity of the church than the fiercest dogmatic bigotry." Traditional parties within the Church of England asserted that attempts to bring together religion and science would result in a rationalization and therefore a destruction of religion. Science could never provide the eternal and universal truths needed.[137]

The liberal movement continued to draw closer to science and history, however, and *Science, Religion, and Reality* played an important role in the 1920s by helping create "a climate of opinion in scientific circles which was far more tolerant of religion."[138] Since the early days of liberal thinking, theologians had sought science and modern thought as an ally instead of an opponent, and the *Times* reviewer delighted that Eddington had shown that religious experience could no longer be "explained away": "If [these essays] do no more than enable one to take the mystical experience seriously, we think these speculations are of value. The true mystic can probably afford to ignore science, but for most of us it is a relief to know that science and mysticism cannot, by the nature of things, conflict. . . . On the old theory of naturalism, the old scientific outlook, the mystical experience must necessarily be an illusion. But we have seen that science no longer claims to have the kind of knowledge it once thought it possessed."[139] The Manchester Conference of

1895 was a perfect example of this hope (see chapter 1), and many religious figures thought they had been successful in turning science to the cause of religion without diluting either.[140] *Nature* reported on the writings of mathematicians E. W. Barnes and J. M. Wilson on science and religion, celebrating the "scientific view of religion, now accepted by men of science and Churchmen alike, [which] is that religion is the spiritual life of the individual, and subject to development."[141]

For a time it seemed that relativity, with its apparent kinship with relativism, was poised to upset all that. After the 1919 eclipse, there was a surge in concerns that the theory had upset centuries of natural theology. The *Times* received letters worrying about the theory's subjectivity and its lack of an "absolute framework."[142] Einstein seemed no better than Descartes. Interestingly, this anxiety did not seem to upset the representatives of organized religion. The *Church Times* did not think relativity had any particular significance for Christianity: "The question may arise in the minds of our readers, What bearing has this theory upon the spiritual conception of the universe? For ourselves we fail to see that it can in any way modify our present outlook. . . . Einstein has made no difference to religion; the permanent value of his work is in the domain of physics. . . . He has opened out to us new vistas of thought, and new possibilities in intellectual progress."[143] What worry there was about relativity dispelling religion seems to have been a small part of larger immediate postwar concerns about the failure of the Church. There was a widespread sense that the Church "had undoubtedly . . . failed her people during the days of crisis."[144] The spiritual crises of the war had proved to be too much for an institution with a confused and unclear mission and an increasingly skeptical and incredulous flock. There was debate about the best way to fix this problem, but the liberal approach had a great deal of support. Christianity needed to be adjusted to focus more on personal spiritual life and less on dogma. The Church must progress "hand in hand" with science.[145]

And yet, later retrospectives and recollections of the 1920s in Britain often presented Einstein as a virtual harbinger of destruction of traditional values. For example, *The Long Week-end* described relativity as "a terrible blow for elderly mathematics masters and mistresses and for all who had held fast to Euclidean truth as the one practical certainty in a weltering world."[146] Truth was no longer absolute: "The failure of the Churches to hold younger and more progressive-minded people even in weekly lip-service to the notion of Christendom; the knocking away by Einstein and his popularists of the lynchpin of geometric absoluteness which held the conventional universe together; all this was thoroughly unsettling."[147] This vastly oversells both the immediate crisis in the Church and the immediate impact of relativity on religious culture. Graves and Hodge were hardly alone in doing so—why was this a reasonable thing to do? Some clues can be found in the similar claims made in John Galsworthy's *A Modern Comedy*: "Everything being now relative, there is no longer absolute dependence on God, Free Trade, Marriage, Consols, Coal or Caste."[148] Galsworthy's linking of relativity to the industrial uncertainty, labor unrest, and economic shifts of the interwar period is significant, because those events did not appear until some years after 1919. It was only as authors looked back on the chaos of the interwar period that relativity seemed to be the beginning of the end of the good years. Einstein became blended with the General Strike and the "intractable million" as symbols of all that had gone wrong with Britain, even though contemporaries saw relativity as more intriguing than unsettling.[149]

The theory was in contention among philosophers and scientists for some time (as has been discussed), but among religious thinkers there was a general expectation that it would be handled like any other part of modern thought. They were not particularly anxious, and it did not stand out in discussions of science and religion. The religious implications of Einstein's theory were subsumed under general concerns with the state of religion and the state of science. Relativity became a serious religious issue only in the late 1920s and 1930s, when practicing, cutting-edge scientists like Eddington, Jeans, J. D. Bernal, and Joseph Needham became major players in the debate about religion and science. The next chapter will deal with these issues in detail, particularly the way they became fused with the political concerns of the period.

In this chapter Eddington has functioned as a trace for understanding the British approach to the meaning of relativity. Questions of philosophical and religious meaning were addressed immediately but were largely simple extensions of existing debates around idealism and materialism. Philosophical discussion about relativity was generally confused and made little reference to the theory itself until an authoritative source was available. Eddington's early writings (particularly *Space, Time and Gravitation*) both provided this source and encouraged interpretation through its own philosophical explorations. The divisions between the physical, mathematical, philosophical, and metaphysical aspects of relativity were of paramount concern to all interpreters and function as a useful way for tracking the theory's movement through existing concerns in those fields. Similarly, Eddington's interpretation of relativity provides insight into the values he thought relevant for the progress of both physics and religion.

Just as he did with the values of mysticism and pacifism, Eddington saw human experience as a valence value linking religion and science. Further, this bond was to be beneficial to both; he thought it to be no coincidence that the value of experience was recognized by the leading edge of science (relativity) and religion (liberal theology). Because he saw them linked in this way, Eddington's work on relativity and its reception gives us the double benefit of being able to track both an important portion of the response to relativity and shifts in religious thought.

Religion in Modern Life
Science, Philosophy, and Liberal Theology in Interwar Britain

The challenge now comes not from the scientific materialism which professes to seek a natural explanation of spiritual power, but from the deadlier moral materialism which despises it.... But is it true in history that material forces have been the most potent factors? Call it God, of the Devil, fanaticism, unreason; but do not underrate the power of the mystic. Mysticism may be fought as error or believed as inspired, but it is no matter for easy tolerance.

A. S. EDDINGTON, *Nature of the Physical World*, 1928

Eddington is often referred to as one of the great popularizers of science of the 1920s and 1930s, usually in the same breath with James Jeans.[1] His writings were astoundingly widely read, especially given the difficulty of topics such as quantum mechanics and relativity.[2] Indeed, his ability to make such abstruse topics comprehensible was a major selling point. He and his ideas were sufficiently diffused in British culture that the mystery writer Dorothy Sayers could comfortably have one character admonish an overanalytical colleague: "For heaven's sake, don't go all Eddington."[3] He had already gained a public reputation from the 1919 eclipse and his expositions of Einstein, which likely gave him a ready audience willing to grapple with difficult issues. His 1927 *Stars and Atoms*, taken from a public address, was further evidence that his conversational lecturing style made for good reading and interested audiences, and he was a commonsense choice to give the famous Gifford Lectures.[4] But it was not just his scientific exposition that made him a significant figure; rather, it was how his popular science was embedded in the dominant social and economic issues of the interwar period, particularly Britain's grappling with the question of socialism. Eddington's popularizations were not just tools for education, they were weapons in the battle to defend traditional values—an obituary described him as "one of mankind's most reassuring cosmic thinkers."[5] It is this aspect of Eddington as a *reassuring* scientist that made him so beloved. In a century during which science and technology seemed to be an increasingly mixed blessing, he persuaded his readers that they did not have to leave their lifelong beliefs behind to embrace the new.

One of the important issues for the interwar period in Britain was the simple question of what *kind* of people scientists were. Good? Evil? What were their moral intentions? The widespread tendency to blame the horrors of the Great War and the subsequent economic instability on science and technology meant scientists had a great deal at stake in portraying themselves as morally upstanding citizens who were aware of the larger implications of their work.[6] The particular question of religion was ubiquitous in both the United States and Great Britain; in the former largely because of increasing tensions around evolution, in the latter largely because of anxiety about declining church attendance and Christian belief. Popular science, and Eddington's work, played an important role in addressing these issues.

An important inflection point in the history of science popularization was the shift from a natural theological to a scientific naturalist form of communication between scientists and the public.[7] Britain, unsurprisingly, provides a particularly vivid and informative location for observing this shift and what it implies for the change in the cultural standing of science and the role of the scientist as public intellectual. The British culture that produced William Paley and William Whewell had never quite come to accept that it also produced John Tyndall and T. H. Huxley, and the tensions between the naturalistic and natural theological perspectives on science remained in force when Eddington became a household name. The banner of materialist science was brought vigorously forward in Britain in the late 1920s by communists and socialists inspired by the example of the new Soviet Union and its application of science to all aspects of life.[8]

This group would eventually become the social relations of science movement. The materialist and left wing scientists that would fuel this movement clustered at Cambridge, and Eddington was literally in the epicenter of a resurgent materialist community that sought to apply their beliefs to all of society, including science, work, and belief. This community chose the public understanding of science as one of their prime battlefields; they would change the nature of science by persuading the common person. Popularization was a primary weapon in the revolution.[9]

British science popularization had traditionally relied heavily on natural theology, particularly in the prestigious Gifford Lectures. They were typically a soapbox for proclaiming the harmony of current science with an ecumenical Christianity. Eddington, of course, was no exception to this, and his emphasis on approaching science through liberal theology makes him an excellent example of the religious tenor of the time. But he also had a particular kind of science popularization in mind to refute. His writings were explicitly aimed against both a philosophical materialism and a moral materialism of the sort that many of his Cambridge colleagues were espousing. Coming out of his experiences during the war, Eddington felt a moral responsibility to try and improve Britain as a society. He had worked for internationalism, and this was his opportunity to work closer to home. The Gifford Lectures, Eddington's initial broadside against the sort of materialism he saw ruining Britain, developed into a decade-long battle between differing views of science for the public's allegiance. The ensuing debates are extremely valuable for teasing out the subtle and varied meanings given to the concepts of science, religion, materialism, and idealism between the wars. Further, we can see the high stakes placed on the resolution of these varied forces—public leaders inside and outside the scientific community felt they were battling for the very future of the British state and British culture.

Becoming a Spokesman for Science

Eddington's 1927 Gifford Lectures, delivered in Edinburgh, were published as *The Nature of the Physical World*.[10] This book would become one of the most influential popular books on science published between the wars and would go on to sell 72,000 copies in Britain by 1943 (it was translated into French, Swedish, German, Japanese, Polish, Italian, Spanish, and Hebrew).[11] His humor, literary allusions, and engaging style were highly effective at creating an immersive experience for readers. Eddington described his lectures as treating of "the philosophical outcome of the great changes of scientific thought." His goal was to show the "scientific view of the world as it stands at the present day, and, where it is incomplete, to judge the direction in which modern ideas appear to be tending." He saw the lectures as contributing "new material" for the philosopher; the philosophical consequences presented were supposed to flow directly from the physics. For example, he described his own idealist philosophy as having come from his research in relativity. While he felt the material presented was critical for modern philosophy, he did display some reservations about pushing beyond his own discipline's boundaries. "From the beginning I have been doubtful whether it was desirable for a scientist to venture so far into extra-scientific territory. The primary justification for such an expedition is that it may afford a better view of his own scientific domain." Elements that professional philosophers might find suspicious, then, were to be eventually justified on the grounds of science, not philosophy.[12]

The lectures and the book were not identical; Eddington rewrote the text in August 1928 for two reasons. The first was to incorporate the implications of Heisenberg's indeterminacy principle (later and more widely known as the uncertainty principle). The second was to trim the freer speculations of the lectures. He described a public lecture as a forum in which unrigorous speculation was acceptable, but a publication needed to be on firmer ground lest philosophical critics take advantage. Wide-ranging speculation was more acceptable in a transient lecture than in a permanent book.[13] For the remainder of this chapter, I will reference the book directly, rather than the lectures, as the book was the form that was most widely known.

The majority of *Nature of the Physical World* was structured around expositions of the relativity and quantum theories, with Eddington's philosophical interpretation integrated directly. The structuralism he had been developing since *Space, Time and Gravitation* was one of the main themes. He described the two worlds, the symbolic/scientific and the commonsense. Eddington said the "aloof" world of scientific symbols was one of the most important advances in modern physics, but he counted himself unusual as a scientist in considering the philosophical meaning of that: "The frank realisation that physical science is concerned with a world of shadows is one of the most significant of recent advances. I do not mean that physicists are to any extent preoccupied with the philosophical implications of this. From their point of view it is not so much a withdrawal of untenable claims as an assertion of freedom for autonomous development." Scientists were concerned with the use and development of theories like relativity and quantum physics, not their essential truth. The significance attached to symbols was solely a product of the alchemy of the mind and had no direct link to science itself. He re-presented his tautological interpretation of Einstein's physics, famously asserting, "The law of gravitation is—a put-up job."[14]

Eddington wrote his Gifford Lectures soon after he read Werner Heisenberg's famous 1925 paper, and he eagerly drafted the new quantum mechanics to support his structural physics. He argued that that paper (and particularly Paul Dirac's interpretation of p and q numbers) showed that individual symbols had no actual numbers behind them, but became physical numbers only in combination. This suggested that, as Eddington had been arguing, the symbols themselves were not adequate for a description of the world. As Einstein had shown a decade earlier, *relations* between symbols were necessary before the mind could turn them into something recognizable. Eddington acknowledged that the quantum theory certainly did not demand this interpretation; rather, he said the suggestion of this implication should be taken seriously because it fit so closely with the implications of relativity. The building material of the world was again shown to be relations and symbols, not crude matter. Physics had accepted that it could attain great progress without any real knowledge of the nature of entities. Of course, the quantum theory was quite new; the scientific worldview would likely be modified in the future, but the interpretive exercise nonetheless "widened our minds to the possibilities" and had given science a new sense of physical law.[15]

Eddington continued to defend his philosophical interpretation as springing straight from scientific practice: "I should like to make it clear that the limitation of the scope of physics to pointer readings and the like is not a philosophical craze of my own but is essentially the current scientific doctrine. It is the outcome of a tendency discernable far back in the last century but only formulated comprehensively with the advent of the relativity theory."[16] There were conclusions that were philosophical, and then there were conclusions that were "definitely scientific." Among these was the conclusion that science no longer identified the real solely with the concrete and the corollary that those things that lack concreteness need no longer be automatically condemned. This liberated science from having to deal with parts of the world that did not submit easily to its treatment:

> The cleavage between the scientific and the extra-scientific domain of experience is, I believe, not a cleavage between the concrete and the transcendental but between the metrical and the non-metrical. I am at one with the materialist in feeling a repugnance toward any kind of pseudo-science of the extra-scientific territory. Science is not to be condemned as narrow because it refuses to deal with elements of experience which are unadapted to its own highly organised method; nor can it be blamed for looking superciliously on the comparative disorganisation of our knowledge and methods of reasoning about the non-metrical parts of experience. But I think we have not been guilty of pseudo-science in our attempt to show in the last two chapters how it comes about that within the whole domain of experience a selected portion is capable of that exact metrical representation which is a requisite for development by the scientific method.[17]

The religious subtext is clear, but this was presented as a benefit for science. This was not disingenuous on Eddington's part. He felt and argued passionately that recognizing the nonmetrical aspects of human experience was essential to both progress in science and correct insight in religion. The value of experience did not belong to one realm of thought; it underlay all.

Eddington warned his readers when he thought he was veering from strict science. They were venturing "into the deep waters of philosophy; and if I rashly plunge into them, it is not because I have confidence in my powers of swimming, but to try and show that the water is really deep." His attribution of mental states as the foundation of observation, which he credited to Bertrand Russell, meant the world was built from the "mind-stuff" of relations.[18] This conclusion rested, fundamentally, on the internal recognition that mind was the first and most direct thing in experience.

The structuralism presented in *Nature of the Physical World* was more developed than that presented earlier by Eddington and had the added support of Heisenberg's matrix mechanics. The novel addition to his philosophy was the analysis of quantum indeterminism. He had considered the classical statistical laws of thermodynamics before, but he saw the uncertainty principle as a wholly new way to approach the problem (the addition of the implications of the principle was one of his primary tasks in revising his manuscript). Like relativity, the uncertainty principle related directly to measurement techniques. It was a perfect embodiment of positivist, instrumentalist physics. "It reminds us once again that the world of physics is a world contemplated from within, surveyed by appliances which are part of it and subject to its laws. What the world might be deemed like if probed in some supernatural manner by appliances not furnished by itself we do not profess to know." The fact that it was an immensely powerful tool of physics was further evidence for the significance of identical and transcendental laws. And, like relativity, it added nothing new. It instead "represents the abandonment of a mistaken assumption which we never had sufficient reason for making."[19] In the light of the uncertainty principle, scientists must discard analogies between microscopic elements of the world and gross particles, in the same way that scientists had discarded their primitive theories about ether.

Heisenberg's uncertainty principle denied classical determinism because it eliminated the foundational elements of the Laplacian calculator: precise measurements of position and velocity. Eddington described his personal struggle with the question of determinism. While he knew in his heart that predestination was not right, he could not conceive of a kind of scientific law other than deterministic. This all changed with quantum mechanics, because "*physics is no longer pledged to a scheme of deterministic law.*"[20] Thus, Eddington's and the reader's innate intuition that they have free will needed no special defense. Instead, the traditional objection (the tyranny of deterministic physics) was simply gone, so, for the first time since Descartes, volitionists started on an equal footing with determinists. "We may note that science thereby withdraws its moral opposition to freewill. Those who maintain a deterministic theory of mental activity must do so as the outcome of their study of the mind itself and not with the idea that they are thereby making it more conformable with our experimental knowledge of the laws of inorganic nature." This "emancipation" from determinism affected both the mind and the physical world, and the philosopher and the psychologist must now play their parts and consider the freedom of the human mind and spirit as an elementary datum. Eddington admitted that it was unclear how this freedom was to be reconciled with the statistical laws of quantum mechanics and that quantum indeterminacy was "only a partial step towards freeing our actions from deterministic control."[21]

The defense of mysticism Eddington conducted in *Science, Religion, and Reality* was stronger than ever in his Gifford Lectures, but he made a special effort to link it to the practice of science. As I discussed in chapter 2, this alliance of the mystical outlook with scientific practice was evident in Eddington's day-to-day activity, but this was one of his first substantial presentations of it in a formal fashion. Scientists were driven to look for truth regardless of the symbolic nature of their field: "The path of science must be pursued for its own sake, irrespective of the views it may afford of a wider landscape; in this spirit we must follow the path whether it leads to the hill of vision or the tunnel of obscurity."[22] The justification of science as an entity was to be found in the spiritual realm because "the impulse to this quest is part of our very nature." Eddington was concerned to refute those scientists who claimed science's justification was solely in its material benefits, as he felt many of his Marxist colleagues were arguing. He said no scientist would "allow his subject to be shoved aside in a symposium on truth."[23] Science, then, had something beyond the material world, and this something was an inherent seeking in human nature. This was also a response to those critics who said Eddington's tautological approach to fundamental laws made science meaningless, a mere mathematical game of symbols. *It did not matter* whether scientific symbols connected to a real world beneath, Eddington replied, because what motivated the scientist was internal and independent of any particular subject.

To Eddington, the spiritual life was ubiquitous whether it was recognized or not. The indication of a spiritual world was simple—it was that which was "good enough to live in." This was contrasted with the scientific world of symbols, which cannot be inhabited in any meaningful sense: "My conception of my spiritual environment is not to be compared with your scientific world of pointer readings; it is an everyday world to be compared with the material world of familiar experience. I claim it as no more real and no less real than that. Primarily it is not a world to be analysed, but a world to be lived in."[24] Everyone lived in the spiritual world, whether they knew it or acknowledged it. Mysticism was simply the embrace of this reality and the use of it for productive ends. This was the fundamental premise of the Quaker Renaissance: spirituality was everywhere, and its importance was in how it changed people, not in how it fit a sectarian doctrine.

The target of Eddington's arguments was specific and significant. Even beyond those who attacked religion as untrue, there were those who discarded it as irrelevant or unworthy of attention. The materialists who dismissed religion based on physics had been disposed of by his structuralism, but the

Marxist view of history needed to be dealt with as well. The Marxists were clearly in the cross-hairs:

> The challenge now comes not from the scientific materialism which professes to seek a natural explanation of spiritual power, but from the deadlier moral materialism which despises it. Few deliberately hold the philosophy that the forces of progress are related only to the material side of our environment, but few can claim that they are not more or less under its sway.... But is it true in history that material forces have been the most potent factors? Call it of God, of the Devil, fanaticism, unreason; but do not underrate the power of the mystic. Mysticism may be fought as error or believed as inspired, but it is no matter for easy tolerance.[25]

Eddington saw the forces of materialism at work in British society, probably in the wake of the General Strike and the subsequent agitation that stoked fears of communism across the country.[26] He was hardly alone in seeing a genuine threat from Marxism, and one of the overriding agendas of *Nature of the Physical World* was a refutation of the ideology that would turn Britain into another Soviet Union.

Eddington was firm that his role was a defender of religious experience, not an evangelist who sought to prove a theological conclusion:

> This must be emphasised because appeal to intuitive conviction of this kind has been the foundation of religion through all ages and I do not wish to give the impression that we have now found something new and more scientific to substitute. I repudiate the idea of proving the distinctive beliefs of religion either from the data of physical science or by the methods of physical science. Presupposing a mystical religion based not on science but (rightly or wrongly) on a self-known experience accepted as fundamental, we can proceed to discuss the various criticisms which science might bring against it or the possible conflict with scientific views of the nature of experience equally originating from self-known data.[27]

He harshly dismissed the "naïve" idea that thermodynamics could be used to demonstrate Creation and warned that religious believers should be apprehensive of any "intention to reduce God to a system of differential equations, like the other agents which at various times have been introduced to restore order in the physical scheme. That fiasco at any rate is avoided." The changing conclusions of scientific knowledge made it far less reliable as a basis of religion than personal experience. "The lack of finality of scientific theories would be a very serious limitation of our argument, if we had staked much on their permanence. The religious reader may well be content that I have not offered him a God revealed by the quantum theory, and therefore liable to be swept away in the next scientific revolution." He acknowledged that many in his audience wanted a silver bullet to be used against unbelievers, but he demurred that "I could no more ram religious conviction into an atheist than I could ram a joke into [a] Scotchman."[28]

Eddington argued that in any case the desire for a proof of religion was a misplaced value. Proof had no use or meaning outside tortured systems of logic. Religion, like physics, had to make do with what was known based on transitory evidence based on experience. "*Proof* is an idol before whom the pure mathematician tortures himself. In physics we are generally content to sacrifice before the lesser shrine of *Plausibility*."[29] He was concerned that he not be misread as saying that religion was only justified by the new physics: "It will perhaps be said that the conclusion to be drawn from these arguments from modern science, is that religion first became possible for a reasonable scientific man about the year 1927. If we must consider that most tiresome person, the consistently reasonable man, we may point out that not merely religion but most of the ordinary aspects of life first became possible for him in that year."[30] Religion was in the same category of experience as aesthetics, happiness, and love. All had clearly been in evidence (and therefore de facto possible) since the emergence of humanity; modern physics only resolved a criticism that had been leveled against them by misguided materialists. Someone who truly lived the spiritual life already knew these things were true, and Eddington's job was just to reassure them that they were right.

Nature of the Physical World's impact was dramatic. It was seized upon by journalists, theologians, and community religious leaders across the English-speaking world. Readers were impressed with the clarity of the scientific exposition, but it was the religious ideas that made it a cultural phenomenon. Many influential theologians stated that Eddington was essential reading.[31]

It quickly became a standard against which to judge the currency of arguments on science and religion. A *Hibbert Journal* review of a book on the free will problem dismissed the work as out of date because it made no reference to Eddington. His demonstration of the collapse of determinism in physics was said to qualitatively change the landscape of discussions of determinism in philosophy, and his ideas about free will were even used to justify particular interpretations of political events.[32] Similarly, Gifford Lectures in subsequent years were inevitably compared to Eddington's.[33]

The *Times* reviewer declared that "it is a book which every one interested in the modern developments of science should procure and study."[34] One review of *Nature of the Physical World* congratulated Eddington on the book's success, calling its great sales "highly creditable to the intelligence of the British reading public." The book was called "brilliant, thoroughly expert," but its ideas "require careful re-examination and co-ordination in order to form the basis for anything which could be called a new philosophic view of the universe embracing the latest conclusions of physics." The sections on mysticism were singled out as being of special interest to the general reader.[35] A less positive article in the same issue gave Eddington credit for recognizing "that the garment of truth is seamless" and being willing to speak about religion and philosophy in concert with physics. But the author, the Oxford philosopher H. W. B. Joseph, was very concerned to defend the necessity of a proper philosophical investigation. He criticized Eddington for making imprecise analogies and (indirectly) for using his fame as a scientist to support unsubstantiated ideas.[36]

The literary skill evident in *Nature of the Physical World* cemented Eddington's standing as a public spokesman for science. This book and his carefully prepared lectures and radio broadcasts smoothly moved between technical matters, literary allusions, and whimsical humor. Lecturing in person, he seamlessly brought together verbal deftness, mathematical equations, and carefully chosen props to fully engage his audience.[37] Eddington constantly sought out opportunities to address the public, even regularly taking invitations from undergraduate societies, but his smooth public lectures often contrasted with his famously difficult style in the classroom and shy personal manner. Sometimes hesitant to the point of incommunicability in ordinary conversation, Eddington became witty, verbose, and charming when on prepared ground or when discussing a subject of personal interest.[38] The many honors he received in the following years (including a knighthood and the Order of Merit) invariably referred to his writing talents alongside his scientific achievements. When Eddington received the freedom of his hometown, Kendal, J. J. Thomson paid the following tribute: "He has by his eloquence, clearness, and literary power, persuaded multitudes of people in this country and in America that they understand what relativity really means. . . . Sir Arthur is one of those rare cases where great literary ability is combined with great scientific ability."[39] A later book was described as having "a beauty little short of the sublime."[40] Soon Eddington had become a celebrity whose name could sell books on its own—for the "cheap edition" of *Nature of the Physical World*, his name was unusually set above the title, indicating that *he* was the draw.[41]

Smooth prose and strong reviews were not enough to deflect his critics, however. Sir Oliver Lodge's public response to Eddington, in the journal *Nineteenth Century*, expressed concern that *Nature of the Physical World*'s restriction of science would serve science's enemies, and "it undoubtedly has enemies."[42] He wrote privately to Eddington to express concern about the emphasis on pointer readings, echoing similar thoughts held by many in the scientific community. Lodge was concerned about the implications this had for a full "philosophical outlook." He had two points. First, he was skeptical about extending the abstraction of physics to all aspects of physical reality. He worried that this would have deleterious effects on religion by abstracting away "transcendental realities" like the Deity. Second, he took issue with Eddington's assertion that measurement was the basis of science:

> I don't feel sure that I agree with all you say about pointer-readings. I half-feel this is confusing the measure of the thing with the thing itself. I am not ready to admit that science deals only with the metrical aspect of things. If that were seriously true it would indeed have been presumptuous in taking any philosophic outlook at all. It is well for science to be modest and decry its own competence; but there is a tendency among philosophic, theological, and artistic people to hope that the scientific treatment, which they don't understand, is far less important than it is, and that their ignorance of it is not deleterious.

He quoted his own son, a literary and artistic critic, as saying "How then can the soul of man and the universe be dealt with by people whose only real concern is pointer-readings?" Fundamentally, he was concerned that Eddington's emphasis on metricality, however well intended, would prevent physicists from saying anything meaningful about the world at all.[43] In an important sense, this was Eddington's agenda: to prevent physicists from linking the substance of their science to the *spiritual* world. But Lodge was continuing to defend a position in which physics was supposed to directly defend the truth of a worldview. This was a direct clash between the orthodox and liberal outlooks on both religion and science: the former outlook required universal proof and the authority to speak about proof in all contexts, and the latter depended on notions of progress, change, and personal experience.

One reviewer called the lectures "brilliant" but "dangerous," because they could make complex ideas seem more intelligible than they really were. Old-fashioned idealist philosophers were said to be pleased by Eddington, and old-fashioned scientists were suspicious. The real danger came from making unqualified people feel like they could understand what was being discussed: "Meanwhile the very width of [Eddington's] sympathies and the catholic range of his thought invite comments upon his argument from those who are least qualified to follow it in its more technical aspects. And this implied invitation is hereby pleaded as the excuse for what is here written by one who cannot claim even an elementary knowledge of those special subjects in which Prof. Eddington is a recognised authority."[44] He applauded Eddington's efforts to provide "a fresh and powerful vindication of this faith which to so many to-day appears a pathetically groundless paradox." Eddington was criticized for focusing too much on the mind's intimate knowledge of itself. He was also skeptical that Eddington's theory of mind and matter, even if true, actually made any contribution to idealism.[45]

While the reception of Eddington's ideas was to a certain degree smoothed by the long tradition of British religious scientists, it was complicated by the expectation that those scientists would be discussing some flavor of natural theology. Confusion about whether Eddington was a natural theologian was further engendered by the entry of his old rival James Jeans into the arena of popular science. Cambridge University Press was excited by the success of *Nature of the Physical World* and approached Jeans about writing a similar volume. Jeans's contact at the press suggested that he accepted the offer at least partly to maintain his "friendly rivalry" with Eddington.[46] His *Universe around Us* was a straightforward popular astronomy book and did modestly well, but it was his Rede Lecture, published as *The Mysterious Universe* in 1930, that put Jeans in the same public category as Eddington. The university press planned carefully to ensure the success of *The Mysterious Universe*, including distributing large numbers of advance copies to the press and coordinating publicity.[47]

Most of the book was straightforward exposition of scientific ideas, but the last chapter, titled "Into the Deep Waters," made the dramatic claim that "a scientific study of the action of the universe has suggested a conclusion which may be summed up, though very crudely and quite inadequately, because we have no language at our command except that derived from our terrestrial concepts and experiences, in the statement that the universe appears to have been designed by a pure mathematician."[48] This is as straightforward a natural theological claim as one could hope for: science has shown the existence of a "Great Architect of the Universe" and reveals His character ("a pure mathematician").[49] It should be no surprise that Jeans originally wanted to title the book "Religio Physici."[50] His book was quite comfortably in the tradition of natural theology, and *Mysterious Universe* sold a huge number of copies.[51] Nor was he alone. Linking science and religion in this way was still a popular (if increasingly divergent) project at the time, and books by Lodge and others filled the shelves along with Jeans's.[52]

Jeans's emphasis on the importance of mind and "pure thought" in the universe and its creation no doubt helped link him with Eddington in the thoughts of many readers and observers (including, as I will soon discuss, Bertrand Russell). But it is crucial to note that they differ on the most important point of natural theology: whether any point of religion can be proven via science. Jeans unequivocally said yes; Eddington unequivocally said no.

Both were seen (correctly) as defending religion and traditional values, however, and their important differences were frequently glossed over in favor of "Eddington-Jeans" idealism by both supporters and enemies.

Indeterminism

Eddington's next major project in philosophy and popularization was a refinement of his interpretation of indeterminism. His contribution to the Aristotelian Society meeting in 1931, his presidential address to the Mathematical Association in 1932, and his address to the British Institute of Philosophy in the same year all focused on the precise meaning and implications of the collapse of determinism in physics.[53] Eddington described his contribution to the Aristotelian Society meeting as his fullest treatment of indeterminism, and I will rely most heavily on it for an examination of his views. This 1931 meeting brought together the Aristotelian Society and the Mind Association for a discussion of "Indeterminism, Formalism, and Value." The symposium on "Indeterminacy and Indeterminism" consisted of C. D. Broad, R. B. Braithwaite, and Eddington. All three agreed that determinism as a concept was either no longer useful or no longer true, although they disagreed on precisely what determinism meant. (Was it causality, predestination, distinction between past and future, etc.?)

Broad's analysis was conventionally philosophical in strategy and terminology, and Eddington explicitly declined to follow his example. Eddington instead stated that he wanted to express the ideas that best fit his outlook. By this he seemed to mean an emphasis on scientific practice and personal experience; he noted that Broad's definition of determinism was one "only an expert could employ."[54] The basis of old physical determinism was the straightforward concept of "predictability," which was simply the ability to predict the state of the world (including human behavior) at any future time. The Heisenberg principle of indeterminism (a name which Eddington noted he was apparently responsible for) did not introduce indeterminism into this scheme, but it did make it impossible to ignore. Determinism had left physics, and no amount of wishful thinking by prominent physicists could bring it back. Simply hoping that there was some underlying property of radium that would explain its decay was "a frivolous conjecture."[55] Neither indeterminism nor determinism could be proven, and Eddington argued that in science belief implied active assent—so someone talking about determinism at all was making a "groundless assertion." The onus of proof is on the person making a positive claim, in other words, the determinist. Disproof of strict causality was not necessary. Like the idea of the moon being made of green cheese, determinism had been dismissed but not disproved.[56] Now that determinism had been dethroned in physics, it made no sense to retain it in other categories, and free will became a reasonable idea. Eddington speculated that there was clearly some kind of trigger mechanism in our bodies that eventually depended on indeterminate phenomena, but thought it would happen on the cellular scale rather than depending on a single quantum jump. The important consequence was that indeterminism in mind meant it was reasonable to attach importance and significance to bodily motions, because one was no longer obligated to explain thoughts and emotions away as being other than what they seem.[57] Our daily experience of control over our actions and interest in the outcome no longer needed to be an outrageous illusion. As he put it in a later interview: "Determinism is opposed both to our intuitions and to the evidence. Why not drop it?"[58]

Eddington's audience in Trinity College that July was comparatively receptive, but sharks lay in wait. The Liberal MP, socialist, and social reformer Sir Herbert Samuel wrote a critique of Eddington's indeterminism that was the sharpest and most vigorous yet. The stakes for Eddington's arguments were said to be the very highest—the future of British society. Eddington and his colleague in popularization James Jeans were described as living in "an underworld" where shady physicists and philosophers met to discuss the most abstract of ideas. Ordinarily the average person would have no interest in this, but "suddenly they may be stirred out of their complacency by reading accounts of a sensational raid by intellectual bandits who, after stupefying their victims with mathematical formulae, try to rob them of their most valuable beliefs. There has, in fact, been lately such a raid, led by that very distinguished scientist Sir Arthur Eddington. It is high time he and his accomplices were arrested!"[59] This highly polemical article spared no energy in accusing Eddington of undermining the very foundation of society: the faith that events are the result of causes. Samuel was dedicated to the idea of social improvement through deliberate action, a philosophy he would later call "meliorism." He saw Eddington as endangering the possibility of national progress: "This, then, is no remote or unimportant discussion. It touches the very springs of thought and action in contemporary life." Eddington had led people to think they could believe anything they wanted to, a conclusion which Samuel called "perverted."[60] Albert Einstein, Max Planck, and Ernest Rutherford were all invoked as definitive authorities in physics who still held to determinism and thus refuted Eddington. Further, determinism was the basis of the productive everyday activities of chemists, engineers, and even politicians. Those interested in politics should be particularly concerned, because their project of making the world better based on reason would be made meaningless in an indeterminate universe. If it were true that there were no causes or effects, "why trouble with our vast organisations for child welfare, education, sanitation, penal reform, anti-war propaganda or all the rest? You can have no ground for thinking that, at the end, things will be any different from what they would have been without them." Thus, indeterminism was demonstrably false because attempts at improving the world actually do make it better—and to give up on cause and effect was to prevent people from doing so.[61]

Samuel's article was one of the few attacks on his work to which Eddington responded directly and in a substantial way. Why might this have been the case? Samuel's critique was not particularly robust or threatening from an intellectual standpoint, but it did call out the political implications directly. And these sociopolitical implications were one of the major factors that Eddington was particularly concerned to address with his popularization. Specifically, he wanted to restore the importance of personal conviction to questions of society and politics. As a Quaker scientist, he felt he had a moral responsibility to defend liberal values.

Eddington responded to Samuel first by clarifying his position on causality. He said he did not deny that events followed from causes. Rather, he denied the universal predetermination that followed from an infinite chain of causality. Further, he argued that he had no need to prove indeterminism; the onus of proof was on the determinists. It was Samuel and other "popular science writers" who relied on a principle of causality "which has not been experimentally verified." Eddington pointed out that even Einstein admitted determinism was a positive principle. He did not shy away from Samuel's appeal to authority and particularly attacked Planck's claim that the statistical regularities of nature showed there was nothing like free will at work. Eddington retorted that life insurance companies would be shocked to learn that the free will of humans invalidated the statistics that generated their profits. He was firm that his position was not merely the idiosyncrasies of a particular scientist: "The law of causality does not exist in science to-day." The hope of Einstein and others that determinism would someday be restored was irrelevant, since interpreters in physics must base their interpretations on current theory, not on a theory that Einstein might someday produce.[62] Eddington's payoff to his clarifications was that Samuel's worry about the "social and political consequences" was invalid—the exceedingly probable physics produced by indeterminism was identical with classical certainty for day-to-day work. Samuel was therefore wrong not just on the phenomenological front, but also in his understanding of the basis of social progress. Eddington denied the socialist principle that he saw Samuel espousing (that determinism of thought and action was necessary to improve the general welfare) and said a future that was not prearranged by physical law was incomparably better. In such a universe one can have confidence that humans can change the characteristics of the world in which they live and have good reason for thinking their actions can have a demonstrable effect. Thus, indeterminism reproduced the best circumstances possible: physical science was useful in the world while still allowing the individual volition and responsibility that were essential to social stability and progress.[63] Eddington was arguing not just for a philosophical principle. His values of experience not only gave justification for indeterminism, they also invested the principle with tremendous social significance. To Eddington, the stability of a future Britain could be built only on a recognition of the reality of human experience, and his popularization and philosophy would return to this goal again and again. The foundations of physics would become the touchstone for the progress of the liberal state.

Different Forums

The success of *Nature of the Physical World* brought Eddington special recognition within the Quaker community. He was invited to give that year's Swarthmore Lecture, an annual address to the London Yearly Meeting meant to present new ideas about the meaning and role of Quakerism. His fame as a scientist and attention as a popularizer made him probably the best-known Quaker in Britain (even if most of the public was unaware of his Dissenting identity). The invitation for him to give the prestigious lectures was a measure of how well the Society of Friends thought he had represented their community through personal example. His address was published almost immediately in a slim volume as *Science and the Unseen World* and was extremely popular, going through three printings in four months.[64]

He covered many of the same themes as in his Gifford Lectures, but he was clearly thoughtful about his audience. His emphasis on what he saw as the most important themes for his fellow Quakers reveals two useful features. First, he directly revealed what issues in the community he thought most pressing or fruitful to address. Second, seeing how he discussed topics also presented in *Nature of the Physical World* shows us how the Quaker labels had often been removed from those topics in the interest of ecumenicism.

Eddington's choice to begin with a poetic exposition of the evolution of the universe and human beings was interesting, as it was rather unrelated to the larger topic. It was perhaps aimed at the remnants of that older generation of evangelical Quakers who still held to scriptural literalism, or perhaps he was anticipating an outside audience with such views. Having obliquely addressed an expected conflict between science and religion, he reframed the problem in what he saw as a more Quaker fashion. Eddington stated that the issue at hand was experience, defined as the interaction of the self with the environment. If science claimed any authority, it was because it dealt with experience, and religion dealt with experience or it "is not the kind of religion which our Society stands for."[65]

With a rare scriptural reference, he explained his move away from natural theology in favor of a mystical approach to the divine:

> "And behold the Lord passed by, and a great and strong wind rent the mountains, and brake in pieces the rocks before the Lord; but the Lord was not in the wind: and after the wind an earthquake; but the Lord was not in the earthquake: and after the earthquake a fire; but the Lord was not in the fire: and after the fire a still small voice. . . . And behold there came a voice unto him, and said, What doest thou here, Elijah?"
>
> Wind, earthquake, fire—meteorology, seismology, physics—pass in review . . . the Lord was not in them. Afterwards, a stirring, an awakening in the organ of the brain, a voice which asks "What doest thou here?"[66]

The lecture focused on the mystical, seeking outlook that was the result of this focus on experience. This was present in *Nature of the Physical World* as well, but it is clear that Eddington wished to stress this aspect of science when addressing the Friends. He was validating the pursuit of scientific truth as something worthy of a religious man. His words would have been familiar to anyone in attendance at the 1895 Manchester Conference: "We seek the truth; but if some voice told us that a few years more would see the end of our journey, that the clouds of uncertainty would be dispersed, and that we should perceive the whole truth about the physical universe, the tidings would be by no means joyful. In science as in religion the truth shines ahead as a beacon showing us the path; we do not ask to attain it; it is better far that we be permitted to seek." And later: "Quakerism in dispensing with creeds holds out a hand to the scientist. . . . The spirit of seeking which animates us refuses to regard any kind of creed as its goal."[67] He sought to evoke as much kinship as possible between the religious and scientific quests.

Eddington's familiar antimaterialist position (the symbolic nature of physics, etc.) appeared without much detail, and his efforts to streamline his argument help reveal what he felt to be the essential elements. In particular, he was concerned to describe the ideas he was battling: "Let us for a moment consider the most crudely materialistic view of [the connection between mind and matter]. It would be that the dance of atoms in the brain really constitutes the thought, that in our search for reality we should replace the thinking mind by a system of physical objects and forces, and that by so

doing we strip away an illusory part of our experience and reveal the essential truth which it so strangely disguises."[68] This is what he calls elsewhere
"billiard ball" materialism, and he admitted that this view perhaps was no
longer widely held. But he used a rhetorical strategy whereby he took the
weaknesses and unpleasantness of this philosophical viewpoint and married
them to a more sophisticated materialist position, without reevaluating the
new position's implications for mind. He smoothly made the struggle against
this outmoded materialism synonymous with any view that placed science or
matter paramount: "It is this belief in the universal dominance of scientific
law which is nowadays meant by materialism."[69]

To illustrate the profound limits of materialist physical explanations, Eddington spun a fable based around a moment of great emotional significance
for all his readers: the annual minutes of silence commemorating the end of
the Great War. "Let us suppose that on November 11th a visitor from another
planet comes to the Earth in order to observe scientifically the phenomena
occurring here. He is especially interested in the phenomena of sound, and at
the moment he is occupied in observing the rise and fall of the roar of traffic in
a great city. Suddenly the noise ceases, and for the space of two minutes there
is the utmost stillness; then the roar begins again."[70] The visitor was a trained,
materialist scientist and looked for an explanation for this in terms of forces
and matter. There did not seem to be any reason why such an explanation
would not be possible—the silence was the result of a clear series of physical
events, such as feet pushing on brake pedals, and so forth. There was no supernaturalism evident. Each event came from a chain of physical antecedents
ending in the human brain, but if the visitor knew the location and movement
of all the atoms in the brains of all the people there, would he *understand*
Armistice Day? "He understands perfectly why there is a two-minute silence;
it is a natural and calculable result of the motion of a number of atoms and
electrons following Maxwell's equations and the laws of conservation.... Our
visitor has apprehended the reality underlying the silence, so far as reality is
a matter of atoms and electrons. But he is unaware that the silence has also
a significance.... The more complete the scientific explanation of the silence
the more irrelevant that explanation becomes to our experience." The Mar-
tian materialist could not understand the silence, because the significance of
Armistice Day could not be explained in terms of matter and motion; the significance came from human values, experiences, and emotions. It was the
human reaction to the tragedy of the slaughter in the trenches that caused the
silence, and that human reaction could not be captured in materialist terms.
Thus, human consciousness caused, outside of the scheme of physics, a *real
event*. "If God is as real as the shadow of the Great War on Armistice Day,

need we seek further reason for making a place for God in our thoughts and lives?"[71] No one in the generation that had lived through the war would deny the significance of that moment of silence, and Eddington was challenging them to consider their experience of God with the same seriousness.

Comparing *Science and the Unseen World* to *Nature of the Physical World* provides a clear example of Eddington's sense of shared values between religion and science. The everyday world of solid tables and waves on the ocean described in his Gifford Lectures became, in Friends House, the *spiritual* world. What was spiritual about the everyday world? It was based on values of practicality, efficiency, and utility—the very bases of the Renaissance Quaker. Eddington had argued in many scientific situations the importance of these values, and his Swarthmore Lecture gave him the first opportunity to explicitly celebrate their importance for a Quaker life as well.

Science and the Unseen World received wide circulation despite its small printing and was popular perhaps because of its more succinct presentation of the ideas found in the somewhat lengthy *Nature of the Physical World*. The emphasis of *Science and the Unseen World* on personality was attractive to many who sought to defend idealism, as was Eddington's refusal to base religion on scientific discovery.[72]

An edited version of *Science and the Unseen World* was even published by a freethinker press in the United States. Following in a long-established pamphleteering tradition, Emanuel Haldeman-Julius of Girard, Kansas, printed an edited *Science and the Unseen World* so he could append his own rebuttal to Eddington.[73] Haldeman-Julius (born Julius, he adopted his wife's name) was an extremely successful publisher focusing on self-help, education, and freethinking material. His "Little Blue Books" were a tremendous marketing innovation, selling over 100 million copies in a decade. The books, among the first mass paperbacks, sold for a nickel and were widely read.[74]

Haldeman-Julius originally entered publishing through a socialist newspaper, and his books provided a forum for many atheist and anti-Catholic writers. His reply to *Science and the Unseen World* elaborates the antireligion materialist position as it stood in the middle of the interwar period. Its similarity in America and Britain, despite the significant differences in the Marxist movement in those countries, indicates that the literati of socialism were perhaps more homogeneous across borders than might be expected.

His attack on Eddington aimed at the very point that *Science and the Unseen World* was most concerned to defend: the validity of the identity of the religious scientist. His jeremiad opened with incredulity at the entire concept: "The combination of a scientist and mystic must ever seem too curious for belief. . . . Consistency is an obstacle that can be evaded in many adroit—or even unconscious—ways. . . . Perhaps it would be unfair to say that Prof. Eddington is trying hard to make himself believe something which his reason cannot very plausibly or firmly endorse; yet that is the impression which is somehow left, after a perusal of his feeble sermon. . . . Undoubtedly he is sincere. It seems he really finds satisfaction in his incongruous twofold character of scientist and mystic. But how little it takes to satisfy him!"[75] As with many other polemics against Eddington, this one portrayed him as a good-natured but feebleminded dupe being manipulated by the forces of oppressive theology. Haldeman-Julius expressed relief that few scientists led such a "double life." Eddington's reliance on spiritual experience as evidence was portrayed as a pathetic retreat to primitivism. "But essentially there is no difference between Prof. Eddington, talking about what the spirit feels of God and the 'unseen world' and a Tennessee yokel dancing and yelling under the influence of the Holy Ghost. I say [this] quite seriously as defining the sort of 'experience' which Prof. Eddington would have us look upon respectfully, aye believingly." If we were to accept that kind of evidence, Haldeman-Julius argued, we would have to be charitable to assume it had been arrived at "without the aid of whiskey or dope."[76] He insisted that materialism was wholly competent to explain not just religion, but also the aesthetics to which Eddington appealed. Eddington's attempt at being a mystic-scientist was a futile attempt to defend the indefensible.

Pamphlets like this were part of an explosion of popular writing on science and religion, of which *Nature of the Physical World* was part of the vanguard.[77] These issues found their way into new media as well; in late 1930 the recently established BBC aired a symposium on science and religion.[78] Public interest had been whetted, and there was demand to hear the opinions of the experts.

The broadcasts were made between September and December, and their stated purpose was to make available "a personal interpretation of the relation of science to religion by speakers eminent as churchmen, as scientists, and as philosophers; and to determine, in the light of their varied and extensive knowledge, to what degree the conclusions of modern science affect religious dogma and the fundamental tenets of Christian belief."[79] The disciplines of participants were varied, though philosophy was heavily represented and a few religious leaders seem to have been brought in solely to assuage fears of a disrespectful discussion.

J. S. Haldane and Eddington were among the main scientists invited to speak, and they represented two of the bodies of thought on religion that were becoming influential in the scientific community. Eddington, of course, described the symbolic skeleton of physics and the eclipse of determinism. He made it clear that this was all he could say as a scientist, and he appealed to personal intuitions of spirituality as necessary to move beyond. His presentation of the spiritual world was broadly conceived and included "a sense of beauty, of morality, and finally at the root of all spiritual religion an experience which we describe as the presence of God." As always, he was concerned to connect religious experience with human experience as a whole, to integrate religion with life as deeply as possible. He made his rejection of the need or possibility of the proof of religion a major part of his broadcast; he was likely responding to the many reviewers of *Nature of the Physical World* who either critiqued him for not providing proof or thought he was providing proof.[80]

Haldane argued against the preeminence of the physical science that supported many religious perspectives like Eddington's. Holism was the answer to science's problems, and physics was nothing but trouble: "Neither biology nor philosophy can afford to cringe before the physically interpreted or mathematically formulated universe." The exactitude of the physical sciences came only from their idealized view of the universe, and this idealized view had become a "nightmare" that the world was trapped in. Biology and its implied holism was needed, first, to save science from itself and, then, to provide a justification for concepts like personality and God. If holism was accepted, Haldane reasoned, then we would need to recognize things that have no material existence as emergent properties of those that do. The human mind was a holistic property of the brain, and God was a holistic property of the universe.[81]

The theologians and philosophers present had a mixed reaction to Eddington's ideas. Samuel Alexander disputed that we could know our own mental states better than the material world, while Dean Inge was suspicious of attempts to strictly demarcate intellectual territory.[82] Inge was not so impressed by claims that science and religion were no longer in conflict, but was intrigued that so many eminent scientists seemed to have been "driven" to theism.[83] He was pleased that Eddington was defending a liberal theological position and rejecting syllogistic proof in favor of the unanimity of mysticism. L. P. Jacks agreed that Eddington's and Haldane's stress of personality was more in line with contemporary theology; Eddington's striving after truth was more like Rudolph Otto's *mysterium tremendum* than was Alexander's metaphysically motivated deity.[84] Jacks called Eddington's contribution "the pivot of the symposium," in that it best expressed the values that liberal theology was championing: the centrality of the person, experience, and the human desire for truth.[85]

The symposium was a fascinating snapshot of British viewpoints on science and religion. The theologians and religious leaders included both the old guard and the new, and the style of their contributions is a useful barometer for the cultural position of their religious outlooks. The orthodox High Church representatives' contributions were virtually indistinguishable from what would have been written in a similar discussion fifty (or even a hundred) years before. To them, science either had not fundamentally changed since then or its change was of no significance. Their arguments about religion and science were in principle identical to Paley's and Whewell's. This reflected the state of their religious community as well; they had been resisting change strongly and generally reacted to attempts at modernization by a hardening line.

The representatives of the liberal religious community, however, had a completely different character to their contributions. They were daring, ambitious, and aggressively sought to bring the most modern of knowledge (scientific or otherwise) into play. They were optimistic about the notion of both improving the state of religion and using religion to improve society at large. To them, progress was key to resolving all issues of religion, especially those bordering on science. They thought of themselves as defending a particularly modern approach to religion. One spokesman declared that it was "the conviction of the Modernist that religion meets an indispensable human need. The Modernist desires to preserve and conserve all those values in our personal and social life." Attachment to scripture was discarded, and the liberal approach was seen as a way to restore religion to a leading role in British society: "[Modernism's] mission [is to] seek to educate the alienated masses in moral and essential truth."[86] The liberals' embrace of innovation and improvement was fundamental to their arguments, and the confidence of those arguments in the BBC symposium was a striking contrast to the bland conservatism of the orthodox religionists. The British religious community was at an inflection point where the tension between the liberal and orthodox outlooks was shaping the course of the nation's religiosity. Eddington had chosen the liberal side and, indeed, was embraced by them. Eddington's ideas, particularly those in *Nature of the Physical World*, were seen as important resources for the liberal theological outlook and provided an influential framework for progressive thinking on science and religion. His valence values allied him with the forces of social liberalism, but those bonds were seen by many as dangerous violations of the boundaries of proper science. The same contributions to liberal thought that made Eddington such a celebrated public figure also made him a target for academics with a different social agenda.

When Philosophers Attack

One of Eddington's best-read critics, and one of his credible rivals as a science popularizer, was Bertrand Russell. By the 1920s, Russell was already equally famous, both for his contributions to logic and his political agitation. Largely fueled by his need for financial stability, his writings became gradually more accessible after 1920 or so, and the combination of his razor-sharp intellect and his role as unending social disruption helped attract public interest.[87] His reaction to Eddington's philosophy provides an interesting case study: Russell performed a nearly complete about-face with respect to Eddington's capabilities after reading *Nature of the Physical World*. In this he is an excellent example of how both critics and supporters thought about Eddington. Once Eddington made clear his thinking on spirituality, his ideas became an inextricable part of the debate about the nature and role of religion in the modern world.

Russell's 1926 Tarner Lectures eventually became his widely read *The Analysis of Matter*.[88] Here he investigated the philosophical outcome of modern physics. He was particularly concerned with how logical analysis could be brought to bear on the problems of matter, causality, and natural law. A significant amount of the book was spent addressing the impact of relativity on the philosophy of science, and Russell returned again and again to Eddington's writings on the subject. He did not agree with all of Eddington's ideas, but he was clear that Eddington's proposals (such as tautological and epistemological laws) needed to be treated carefully and seriously. He said that Eddington's methodology for unifying relativity with laws of nature might appear unorthodox but was essentially no different from conventional scientific methods. Criticisms of Eddington were respectful and often were simple requests for Eddington to clarify or further develop his ideas.[89] In treating Eddington's theory of matter, he did not accept the theory, but he used it as a model of "the *sort* of definition to which modern physics is bound to be led."[90] Russell even credited Eddington with being his guide in thinking about the philosophy of relativity and explicitly said that *Mathematical Theory of Relativity* and *Space, Time and Gravitation* were important philosophical resources: "The theory of relativity, to my mind, is most remarkable when considered as a logical deductive system. That is the reason, or one of the reasons, why I have found occasion to allude so constantly to Eddington. He, more than Einstein or Weyl, has expounded the theory in the form most apt for the purposes of the philosopher."[91] At all times, he treated Eddington respectfully and as a colleague in philosophy. Eddington was portrayed in *Analysis of Matter* as someone who knew what he was doing and who could make meaningful contributions to the philosophy of physics.

This was about a year before Eddington's Gifford Lectures, which completely changed Russell's treatment of Eddington and his ideas. *The Scientific Outlook* was Bertrand Russell's first major response to Eddington's popularizations. Eddington received credit for his expository abilities, but that was nearly all.[92] The book was highly critical of scientists speaking outside their fields of expertise, and it is easy to see that the targets are Eddington and Jeans even before their names appear. Russell railed against any careful scientist who expressed "wholly untested opinions with a dogmatism which he would never display in regard to the well-founded results of his laboratory experiments." Scientific versus religious knowledge was an important theme of the book, with the former being associated with induction and the latter with deduction. His example of this conflict in the trial of Galileo made his thoughts on their relative value quite clear, and pure deduction was singled out for as much bile as theology.[93]

Eddington was explicitly targeted as a danger to the very existence of science. "Whoever wishes to know how and why scientific faith is decaying cannot do better than read Eddington's Gifford lectures." Eddington's idea about identical laws, which in *Analysis of Matter* was treated as worthy of serious investigation, was now dismissed with: "respect for Eddington prevents me from saying it is untrue."[94] Russell was angry with Eddington's reservation of a nonscientific part of human experience: "Eddington proceeds to base optimistic and pleasant conclusions upon the scientific nescience which he has expounded in previous pages. This optimism is based upon the time-honoured principle that anything which cannot be proved untrue may be assumed to be true, a principle whose falsehood is proved by the fortunes of bookmakers. If we discard this principle it is difficult to see what ground for cheerfulness modern physics provides." Eddington's claim that religion was somehow aided by modern physics was dismissed as fantasy. Further, he wondered whether Eddington's "scientific scepticism" would bring about the "collapse of the scientific era."[95] Russell speculated that the recent tendency for intellectuals to profess reconciliation of science and religion was only coming about because bishops saw Bolshevism as a common enemy with science. "It follows, of course, that science, if pursued with sufficient profundity, reveals the existence of God." Eddington and Jeans were giving up science's claims to total knowledge and were approaching the old order and apologizing for science's past arrogance. "In return, the established order showers knighthoods and fortunes upon the men of science, who become more and more determined supporters of the injustice and obscurantism upon which our social system is based."[96]

Russell grew increasingly cynical through the course of the book, and his prose was liberally scattered with accusations and insinuations that scientists like Eddington were simply capitulating to a corrupt, aristocratic, and capitalist system: "what they have said in the way of support for traditional religious beliefs has been said by them not in their cautious, scientific capacity, but rather in their capacity of good citizens, anxious to defend virtue and property." The war and the Russian Revolution had made all timid men conservative, and all professors were said to be timid men.[97] He seemed particularly bitter about James Jeans's knighthood.[98] There was plenty of scorn for the religious institutions as well: "In recent times, the bulk of eminent physicists and a number of eminent biologists have made pronouncements stating that recent advances in science have disproved the older materialism, and have tended to re-establish the truths of religion. The statements of the scientists have as a rule been somewhat tentative and indefinite, but the theologians have seized upon them and extended them, while the newspapers in turn have reported the more sensational accounts of the theologians, so that the general public has derived the impression that physics confirms practically the whole of the Book of Genesis."[99]

Eddington's interpretation of quantum indeterminism in defense of free will was attacked from every possible angle. Russell denied that mind could influence matter in any way, that atomic behavior was in any way indeterminate, that the uncertainty principle implied any kind of scientific ignorance, and that quantum indeterminacy would be a permanent part of physics. He also denied Eddington's defense of free will as something that we can experience, invoking Pavlov's experiments as evidence that conscious choice was subject to empirical laws.[100]

Jeans was dismissed as a modern-day Bishop Berkeley, and Eddington as a modern day Descartes. Russell was particularly irritated that the two professed completely incompatible views about the universe, but were both seized upon by religious leaders: "Eddington deduces religion from the fact that atoms do not obey the laws of mathematics. Jeans deduces it from the fact that they do. Both these arguments have been accepted with equal enthusiasm by the theologians, who hold, apparently, that the demand for consistency belongs to the cold reason and must not interfere with our deeper religious feelings." He marveled that theologians could be enthused at all about the sort of God given to them by modern physics: the deity was either a supremely distant mathematician or quantum fluctuations, neither of which Russell thought would provide much comfort.[101]

Russell said that Eddington's attack on scientific knowledge was coming at a time when humanity was having great difficulty integrating science into modern society, and Eddington was just making it more difficult. Crises such as the Great War indicated there were serious dangers, but Russell warned that going back to the "infantile fantasies" of religion was not going to help. To lose faith in our ability to have knowledge was to "lose faith in the best of men's capacities."[102]

The difference between Russell's treatment of Eddington in 1926 and 1931 is stark. Ideas that were formerly considered to be worthy of serious investigation were now dismissed with barely any consideration. More importantly, Eddington was no longer being treated as a professional colleague who could be disagreed with respectfully. Instead of being an important ally in examining the philosophical implications of modern physics, he was a lackey to theologians and knightly honors. Not only was he wrong, he was *completely* wrong, with every statement or suggestion being useless. Why did this shift occur?

Despite his claim otherwise, Russell clearly did not think intellectuals should never speak outside their direct area of expertise—the last section of *The Scientific Outlook* was devoted to his ideas on social engineering.[103] It is clear that what changed Russell's view of Eddington was specifically the issue of religion. When Russell could read Eddington's philosophy as being solely motivated by physical science, it was useful and worthy of study. But when *Nature of the Physical World* revealed that Eddington's religious values were involved significantly in his philosophical thought, the whole structure of Eddington's work underwent redefinition from scientific to religious. Russell's outlook on science and religion was wholly binary. Science was inductive, rigorous, empirical, and the future of humanity. Religion was scholastic, dogmatic, and a throwback to primitive times. Something could only be properly scientific if it had no trace of roots in or support for religion.

Of course Eddington's religious values were involved in *Nature of the Physical World*; it could not have been written without them. Russell's allergic reaction to those values, and his shifting of the categorization of Eddington's ideas, made it impossible for him to see what the actual project of the book was. Even though Eddington explicitly rejected any notion of proving the existence of God or the truth of religion, he was still grouped together with Jeans in that task. Both were reduced to the role of stooge to the religious establishment, because that was the only role Russell could imagine for a scientist writing amicably about religion.

It should be no surprise that Eddington's and Jeans's philosophical writings were more widely read than those by almost all other "professional" philosophers. Many philosophers were frustrated not only that the physicists did not seem to be doing very good philosophy, but also that the two were representing the discipline of philosophy of science to the larger public. One influential philosopher who tried to remedy this was L. Susan Stebbing, a professor at the University of London.

Stebbing wrote a series of articles in the 1930s criticizing the "nebulous philosophy" of Jeans and Eddington and a book in 1937 that summarized her arguments. She sought to reveal the grounds of their philosophical views to free the reading public and philosophical community that had been trapped by their speculations. They "are not always reliable guides. Their influence has been considerable upon the reading public, upon theologians, and upon preachers; they have even misled philosophers who should have known better." Eddington's fundamental problem was described as "his strong philosophical bent [which] makes him anxious to connect his philosophy of science with his philosophy of life at all costs."[104] This led to his tendency to omit critical information and provide a misleading emphasis.

Jeans and Eddington were both censured for their desire to be entertaining in their writings. More precisely, their problem was that they tried to arouse a reader's emotions, which was an intolerable abuse of the common person's interest in science. While acknowledging that Eddington was an "original thinker," his and Jeans's appeal to emotion reduced them to "the level of revivalist preachers." This, Stebbing announced, was a violation of the scientific spirit.[105] It leaves the reader in a state of mental confusion, unable to distinguish between metaphor, inexactitude, and science. Even attempts at humor are counterproductive, because readers think they understand something when they have really only been entertained.[106]

Stebbing acknowledged that Jeans and Eddington had different arguments, but both were trying to claim that it "is within the competence of physics to establish that there is a God." Jeans was dismissed quickly as "almost pathetic" and an out-of-date pretender to philosophy.[107] Most of the book was dedicated to refuting Eddington or, more precisely, to showing that his expositional strategies made him unreliable as an authority. She attacked his tendency to use commonsense language to make important points, without making it clear when metaphor or analogy was being used. Eddington's folksy descriptions could not be defended as mere illustration, because they were fundamental to the argument being made. This imprecision of language was a sign of his confused thinking.[108] Her criticisms with respect to content revolved around a denial that Eddington's schism between the mental and material worlds was correct or even meaningful. Like Russell, she saw this split as a rejection of empiricism and a misunderstanding between the symbol and the symbolized. Again like Russell, she was piqued by the tendency for Eddington and Jeans to be given positions of great cultural authority: scientists "have long aspired to the mantle of the prophets; now we thrust the mantle upon them."[109] Stebbing saw herself as a defender of knowledge and was frustrated by the elevation of scientists who celebrated that science knew less than it once did.

The only substantial defense Eddington offered against his philosophical critics was in his 1934 Messenger Lectures, delivered that spring at Cornell, which were published shortly after as *New Pathways in Science*. One chapter was devoted to addressing his critics and his "over-enthusiastic friends,"

and he lamented not having had the opportunity to respond in all circumstances.[110] He first defended himself against those reviewers who in some
sense objected to any nontechnical presentation of technical material. His
defense of the project of popularization provides insight into his motivation
and methodology:

> The aim of such books must be to convey exact thought in inexact lan-
> guage. . . . [The author] will not always succeed. He can never succeed without
> the cooperation of the reader. . . . It is not a question of stepping down from the
> austere altitude of scientific contemplation to a plane of greater laxity. To free
> our results from pedantries of expression, and to obtain an insight in which
> the less essential complications do not obtrude, is as necessary in research as
> in public exposition. We strive to reduce what we have ascertained to an exact
> formulation, but we do not leave it buried in its formal expression. We are
> continually drawing it out from its retreat to turn it over in our minds and
> make use of it for further progress; and it is in this handling of the truth that
> the rigor of scientific thought especially displays itself.[111]

So the deep thought and concentration needed to make science comprehensible without technical language was actually useful for scientific research, in
that it forced the constant reevaluation necessary for progress. The values that
supported popularization, then, were the same that supported science as a
whole.

Eddington saw his philosopher critics as all starting from a basic disbelief
in the primacy of mind. His attempts to directly refute the philosophical
arguments of attackers such as the philosopher of religion W. T. Stace were
generally weak, and it seems that Eddington was aware of this weakness. He
instead tried to defend his enterprise as something slightly new, a "scientific
philosophy" that should be treated differently. Philosophy, he said, often
attacked only those problems accessible to the full tools of logic: that is, ones
that could be brought to a rigorous conclusion. Eddington, on the other
hand, said his goal was never the elaboration of a philosophically complete
and final position. Rather, his job was merely to show that "the new scientific
philosophy is not quite the defenceless victim" that philosophers assume.[112]
To him, this disarmed the earlier criticisms of the philosopher C. E. M. Joad,
which were merely about inconsistencies of expression. Eddington felt that
because his task was scientific, it should be judged by scientific standards, not
philosophical ones:

> I do not think that such discrepancies will appear so heinous to a scientist as
> they do to a philosopher. In science we do not expect finality. The theories
> described in the scientific part of this book do not form a complete and

flawless system; there are incoherencies which we cannot remedy until further research gives us new light. It may well be that the scientific theory will be substantially modified in its future progress toward completion; nevertheless we feel justified in claiming that our present imperfect results embody a large measure of truth. I naturally look on scientific philosophy as subject to the same progressive advance.[113]

It did not disturb him if there were loose ends that did not yet fit into a system. There was no surprise when that happened in science; why should scientific philosophy be any different? Formal consistency was not as important in physics as in philosophy, because physics did not have to rely on formal consistency alone. Eddington did not want to "suppress the many-sidedness of the truth" in physics, with the result that he saw himself as easy prey for those seeking inconsistency.[114]

As for the occasional criticism that he had been too dogmatic in his assertions, Eddington blamed the mathematicians: "In summarising conclusions for the general reader, mathematical and physical considerations become fused together, and it is impossible to show without elaboration of technical detail where the dogmatic mathematical deduction ends and the plausible physical inference begins. You may therefore find that a book which on the whole reflects the liberal undogmatic attitude of science is chequered with pronouncements which suggest omniscience and intolerance."[115] One of the philosophers most concerned with Eddington's apparent dogmatism was Russell. Eddington seemed somewhat offended that Russell had treated him so harshly, especially since he considered their philosophies as kindred projects: "I think that he more than any other writer has influenced the development of my philosophical views." He was frustrated that Russell, like many other readers, had suggested that he was proving religion via physics. "I have not suggested that either religion or free will can be deduced from modern physics; I have limited myself to showing that certain difficulties in reconciling them with physics have been removed."[116]

Engagement with the Materialists

The science popularizers who argued for the materialism that Eddington saw as so dangerous did not generally become active until *after* his early popular writings. There were some earlier publications, but for the most part Eddington was responding directly to one or both of two manifestations of Marxist thought. The first possibility was that he learned of Marxist approaches to physics through his colleagues at Cambridge and their enthusiasm for the philosophy. This could have been through word of mouth, casual conversation, and personal communications that have left no documentary record. If Eddington was reacting to this, he would have been reacting to the earliest emergence of these ideas in Cambridge, because Marxism did not gain serious momentum among scientists there until a few years after *Nature of the Physical World*.[117] A more likely target for him was the ill-defined notions of socialist materialism that had been causing tremendous anxiety in Britain in the wake of the Bolshevik rise to power in Russia. This anxiety was particularly present in the religious community, as they saw their very way of life endangered by the idea of a materialist universe. Materialism as a threat to religion was certainly not a new issue in Britain—intellectuals had been grappling with that issue for most of the nineteenth century. The consequence of this was that Christians of all flavors, including Eddington, were already on guard against anyone arguing for materialism. Eddington's arguments about the dangers of mechanism were similar to other liberal Christian thinkers and can probably be traced back indirectly to the Hans Driesch Gifford Lectures of 1907–8.[118] These lectures put forward the basic twentieth-century version of the concern that a materialist outlook would destroy moral responsibility, and their basic framework appeared all over the English-speaking world up to World War Two. Their arguments had been designed to refute X-Club–style materialism but were smoothly adapted to battle Marxism after the Russian revolution.[119]

Churches across Great Britain panicked in the late 1920s over what they saw as encroaching materialist atheism spurred on by the Soviet Union.[120] The General Strike of 1926 was seen by many as a dramatic sign that Christian civilization was in mortal danger from materialist philosophy. Many blamed materialist views for the disaster of the Great War itself.[121] The Quakers were no exception, although their interaction with socialism was somewhat complicated and evolved over time. Eddington's Jesus Lane Meeting had provided meeting space for socialists before and during the Great War as part of a program to raise funds (he and his sister were on the committee that decided to rent out the space). During the war, the Quakers and the socialists found themselves allies as persecuted pacifists—the Cambridge Socialists had even been dragged out of the Jesus Lane Meeting House by the police.[122] This brief alliance soured after the war, when the aggressively atheist Bolshevik regime and its British supporters seemed to threaten all forms of religious belief. The Quakers as a body, and Cambridge Quakers in particular, split internally on how to think about socialism and its more threatening cousin, Marxism. The Cambridge Friends Meeting grappled with these issues in 1927, just as Eddington was revising the manuscript of his Gifford Lectures. The meeting examined the question "What is the function of the Society of Friends as a Christian Group with regard to Industry and organised society?" Their answer to themselves was:

> We believe that the true function of the Christian Church is to lead men individually to Jesus Christ and His way of Life; and by this way alone can human society be redeemed from the disharmony in which it now lies. It is desirable that some, perhaps many, of our members as individuals or groups should be concerned in the details of schemes for the improvement of the economic and social conditions of men, but the true function of our Society... is to lay down general principles of Christian Conduct... [rather] than to enter into the details of economic and industrial reconstruction.[123]

Religion, not Marxism, was agreed to be the route to social improvement. Many Friends felt that their commitment to social justice included issues of economic justice and tried to persuade their meetings to support socialist goals. Very few meetings ever achieved consensus on this issue, leaving the conservative opposition to socialism and Marxism as the official stance of most Friends meetings.

Discussion tied to the basic issue of the proper relationship of religious people to socialism and Marxism was a constant feature of Quaker communities in the late 1920s and 1930s. Eddington could not have avoided thinking about this; in some sense, it was one of the dominating concerns of the day. He was similarly confronted with Marxism at the university. Cambridge was a hotbed of Marxist approaches to science, and the popular writings of J. D. Bernal and others were having an increasing public impact. There is no indication that Eddington read deeply in any materialist writings. His vision of a "moral materialism" appears to have been formed through secondhand conversation and late nineteenth-century mechanical philosophy. The version of materialism he presents in his writings shows no links to actual contemporary materialists, but it is strikingly similar to the version spoken of in the British Quaker community at large.

British materialists themselves were quick to react to Eddington's idealist philosophy of science. He represented all that was wrong with bourgeois science: he was an idealist, a religious believer, and ignored the need for applying scientific ideas to society.[124] For example, Lancelot Hogben's *The Nature of Living Matter* was explicitly, and by request, written to refute the dangerous idealism of Eddington and his fellow idealists.[125] Hogben called Eddington's arguments "profoundly misleading" and "solipsistic." J. D. Bernal, the crystallographer and passionate Marxist, warned that, through Eddington, "a new scientific mythical religion is being built up."[126]

Writers like Hogben, Bernal, Hyman Levy, and V. A. Ambartsumian would eventually go on to be the mainstays of the social relations of science movement, and Eddington remained a popular target for them throughout the 1930s. There are many possible examples of this worth examining, but here I will pursue Christopher Caudwell's *The Crisis in Physics*.[127] Caudwell's Marxist credentials were unimpeachable (he devoted his life to spreading the gospel of Marx and was killed on the Republican side of the Spanish Civil War), and his argument was so purely ideological that it provides a crystal-clear distillation of the standard Marxist objections to Eddington's ideas.

Crisis in Physics was aptly named, as its goal was to articulate a worldwide, interrelated crisis of economics, politics, and "bourgeois physics." Caudwell thought that the writings of contemporary physicists "reveal a general feeling of collapse of the old order, together with a complete helplessness and lack of understanding as to its cause, which is characteristic of certain elements of society in a revolutionary crisis." This crisis destroyed all true synthetic views, and conservatives could react only in ways such as Eddington's "mystical positivistic attitude to all spheres of ideology outside one's little garden."[128] Physics was inherently bourgeois because it placed humans over nature just as capitalism put the owners over the workers; or, in different language, nature was a machine, and the machine was a slave to the bourgeois. The goal of bourgeois physics, as demonstrated by Eddington, was to create a "closed world:" this had the mind outside the world and was designed to dominate the environment.[129] Eddington's attempts to do so "indicate the extraordinary confusion and helplessness of the scientists of to-day" when faced with the breakup of the bourgeois worldview. They were driven to outrageous kinds of reactionism, and Eddington wallowed in "the double decadence of positivism and mysticism."[130] Bertrand Russell and Eddington were both singled out for relying on mathematical manipulation instead of experiment, and they were said to be symptomatic of the drift of theory away from practice. Caudwell closed with the common Marxist accusation that what Eddington *really* wanted in the end was determinism—just determinism run by the brain.[131]

Generally, Eddington proved reluctant to engage his critics in public debate. Consider that most of his rebuttals to nearly a dozen different critics were packed into one chapter of one book. At one point he claimed that his arguments simply needed no continued support: "I have not hitherto replied to any unfavourable criticisms of my book. . . . If my contentions are of value they will ultimately find their proper level without continual parental intervention to save them from determined opponents—and, perhaps it should be added, from over enthusiastic friends."[132] This seems not completely likely at first glance. Certainly he did not feel his scientific contentions should be left to stand without defense—his famous controversies with Jeans and Chandrasekhar make it clear that he was more than willing to fight for his ideas when challenged. And as discussed above, his Messenger Lectures contained some counterattacks against his critics. There is the possibility that Eddington thought of his popular works as falling into a different category from his physics and therefore needing a less aggressive approach.

There is some indication that Eddington was particularly sensitive about *Nature of the Physical World* coming under attack from trained philosophers.[133] As he was formally untrained in philosophy, it may be that he did not feel entirely comfortable dealing with their criticisms. In any case he was right that his book would become a target (e.g., Russell's and Stebbing's attacks), and his only real public response in his Messenger Lectures was hardly a substantial defense. There may have been a personal psychological issue at work here. Many colleagues, students, and friends noted that he was essentially quite shy, and his reluctance to engage in public debate outside his area of direct expertise may have reflected this. This leaves us with the puzzling circumstance of his public exchange with Chapman Cohen (see fig. 6.1), the head of the National Secular Society, editor of the aggressively atheist newspaper the *Freethinker*, and highly visible spokesman for materialism. Cohen devoted over two months of weekly articles to attacking Eddington's views as presented in *Nature of the Physical World* and *Science and the Unseen World*, and Eddington responded at length in the *Freethinker*.

Why was Eddington willing and eager to reply directly and at length to Cohen when he explicitly said he had no interest in intervening to defend his philosophical ideas? Why did he spend more time in debate with Cohen than with any other person? At the beginning of his response, he said this was because Cohen was "a downright opponent; at the same time he is a fair-minded opponent, anxious to avoid misrepresenting my meaning, and too sincere to strive after merely verbal triumphs. In such a case there is an inducement to try to elucidate the position."[134] While one could reasonably call Cohen fair-minded in the sense that he was careful to clearly state his opponents' positions, he was famously gifted at scoring "merely verbal triumphs." There was little in Cohen's extensive and sharp-toothed debating history that justified Eddington's warm assessment. Certainly, he had no moral or intellectual high ground relative to Eddington's serious philosopher critics such as Russell. So why, then, was Eddington so interested in contesting Cohen's criticisms? It was because Cohen, as an atheist materialist, was the exemplar of the social danger *Nature of the Physical World* and *Science and the Unseen World* were written to combat. Eddington's disagreements with Russell, while probably more interesting to Eddington and more dangerous to his academic reputation, turned

FIGURE 6.1
Chapman Cohen.

on matters of far less volatility and social importance. As with his response to Herbert Samuel, Eddington saw his engagement with Cohen as making a substantial contribution to the burning question of interwar Britain: Would the country reclaim its endangered heritage as a moral, religious nation, or would it spiral into the totalitarianism and materialist moral bankruptcy of the Russian revolution?

Chapman Cohen grew up in England in a nonreligious Jewish family. As he described it, he was brought into the world a decade after *The Origin of Species* and grew up alongside the materialism the book inspired. He had no dramatic story of his conversion to atheism and was scornful of atheists who felt the need for such stories. Cohen felt that anticonversion stories only gave credence to religious belief and harmed atheism by putting it on an equal footing with religion. Religion belonged to "the childhood of the race" and should not be compared to the triumph of science.[135]

His approach to religion was simple. He argued that the history of free thought in Britain had been too much concerned with putting atheism and agnosticism on the same level as religion, whereas he sought to put "the Christian army on the defensive from the very first."[136] This was necessary because religion was so deeply ingrained in British society that it needed to be confronted directly. Belief in religion was always being perpetuated under other names because it underlay the "thinly disguised aristocratic form of government" that continued to rule the country.[137] Christianity had a low moral value and was just "a form of camouflaging an unintelligent selfishness." It was a religion that functioned on the capitalist profit model—be good so you will be rewarded later.[138] The religious theory of life was hopelessly wrong. Religious experience was just a kind of abnormal psychological state brought on by social suggestion and physical practice and had no more validity than the visions of an opium addict.[139]

Cohen took over the *Freethinker* from its founder, G. F. Foote, in 1915 and described its mission as being to "employ the resources of Science, Scholarship and Ethics against the claims of the Bible as Divine revelation."[140] He wanted to defend atheism, the highest state of evolution of a society, and this would be achieved by the bound engines of science and materialism. The newspaper received a great deal of negative attention from the religious community: Dean Inge described it as "a newspaper called the *Freethinker*, which exists partly to deny with vehemence the possibility of free thinking."[141] Foote thought the essence of science was the concept of natural law, and this was the base of his belief in materialism. Atheists were supposed to hold that the cause of life and mind was to be found in matter as an emergent property. Any attempt to find an underlying reality beneath matter was simply a remnant of half-understood ideas about God.[142] Cohen explicitly and definitively rejected any sort of separate-spheres argument. Science and religion came from differing interpretations of the same phenomena, and conflict was inevitable. Indeed, conflict was essential to ensure the victory of "civilised" over "uncivilised" thought.[143]

Cohen argued for a thorough and unbreachable determinism. This was somewhat more sophisticated than a Laplacian man-as-machine viewpoint but still allowed no room whatsoever for free will or moral responsibility in the Christian sense. He wrote a manifesto for this position in 1919 in which determinism was described explicitly as the application of the principle of causality to human nature.[144] There was no boundary between mind and matter, so there was no a priori reason to restrict causality. If we knew all the forces acting on a person, "the forecasting of a conduct would become a mere problem in moral mathematics. . . . The Determinist claims, therefore, that his view of human nature is thoroughly scientific."[145] The advanced sciences had already replaced a kind of volitional, animistic interpretation of nature with a mechanistic view insisting on deterministic laws. This had not yet happened in the human sciences, but there could "be no reasonable doubt" that it would.[146] The only resistance to this was from theology and its continuing ability to manipulate society. "Volitionists" had no evidence other than that of consciousness, but what could consciousness ever really tell us? It could do no more than testify to its own states, but not what those states *meant*. The will had no concrete existence that was meaningful. Any sense in which humans were able to make choices was illusory, because any choice would be determined by external forces and social laws.[147] Determinism, thus, was not restricted to mechanical models (though it was closely associated with them). It was, instead, a declaration that all phenomena in the universe, including mind and morality, were subject to causation that could be divined by science.

In addition to his work on the *Freethinker*, Cohen constantly sought out opponents to engage in public debate. He began his career as a professional atheist by standing atop boxes in Victoria Park and addressing hostile crowds. (He even claimed to have been beaten by a Christian mob on more than one occasion.) Cohen still felt that high-profile confrontations were critical to the growth of free thought. As I discussed earlier, Eddington's and Jeans's books triggered an upsurge in claims that materialism and atheism had been wrecked on the shoals of modern physics. This encouraged both defenders of idealism or religion to promulgate their views more widely and defenders of materialism to challenge them.

This led to a 1928 public debate between Cohen and the philosopher C. E. M. Joad on whether materialism had been "exploded" by the new physics.[148] Their confrontation turned largely on ideas put forth by Eddington in *Nature of the Physical World*, and both debaters positioned themselves with respect to that book. Cohen wasted no time in making his position clear: "When I talk of Materialism I mean the conception that the whole of the phenomena of Nature—physical, chemical, moral, mental, and social— are ultimately explicable in terms of the composition of forces. Materialism means that, and I say that science means that or nothing." He said this view was in no way tied to a particular understanding of matter; it was simply a question of naturalism versus supernaturalism.[149] Joad retorted that he was no believer in supernaturalism and that such questions were beside the point. The issue was that physics had "dissolved" matter into a space-time continuum that had no resemblance to the billiard-ball materialist universe. He

denied that Cohen could say materialism did not depend on matter and particularly that mind could be so easily subsumed under natural laws. Joad
virtually recapitulated Eddington's arguments that our positivist knowledge
of matter and tautological physical laws demonstrated the primacy of mind.
He explicitly followed Eddington further in holding that there was a part of
experience below the structuralist metricalism of physics, making room for
aesthetics and personality.[150] Thus, a simple account of the forces at work
was insufficient for a complete description. Cohen was unimpressed with
Joad's invocation of Eddington: "Throwing down Professor Eddington does
not matter a hang. Mr. Joad says I have been knocking down God Almighty
for thirty years. You cannot expect, after knocking down God Almighty, that I
am going to jib at Professor Eddington." He flat out denied that science could
deal only with certain things. If science had trouble explaining something
like aesthetics, it was only because our knowledge was incomplete.[151] Joad
brought in Eddington again, this time intended to finish the debate. "I merely
quoted him in order to show that I, who am not a scientist, can claim support
from persons who are eminent scientists." He read directly from Eddington's
contribution to *Science, Religion, and Reality* as though it was a point of fact:
"at issue is whether Professor Eddington excludes mind, and, what is even
more shocking than mind, spiritual values, or whether he does not. I think it is
perfectly clear that he does not; in fact, he says so. This, then, is not a question
of argument: Mr. Chapman Cohen is wrong."[152] It is remarkable how quickly
Eddington became a definitive source about the existence of spiritual values,
to the point where professional philosophers and theologians were willing to
immediately give him authority over their own lifetimes of experience and
training. In one sense Cohen was completely right, in that many guardians
of traditional values did seize on anything that supported their case (even if
they often did not completely understand what was being said).

At some point in early 1929, Cohen decided that Eddington himself was
the real danger, and there was no point in skirmishing with proxies. He
thought that if he could strike down Eddington, he could show that the
defenders of religion were continuing to grasp at straws. There was also some
pressure to respond from within the atheist community, and Cohen took
this opportunity to demonstrate to many of his allies the danger of relying
exclusively on a billiard-ball, physics-dominated view of nature.[153] Doing
so played right into the hands of misguided scientists like Eddington, and
he needed to show that the physicists could not be allowed to dominate
the discussion. For nearly three months, Cohen devoted his front pages and
headlines to crushing Eddington's idealist religious philosophy, which would

eventually draw Eddington himself to participate directly in the columns of the *Freethinker*.

Cohen placed the highest possible consequences on the country's embrace of Eddington; to the leaders of the church and the nation "it is the truth of Christianity that is at stake." He argued that Eddington was fundamentally wrong in his understanding of materialism, and thus talk about materialism being dead or not being believed by science was just "pulpit jargon."[154] Eddington's materialism was appropriate for the eighteenth century, and could be attacked easily as a straw man. It was essentially the religious view of materialism that had been designed and propagated solely for theological purposes. Cohen reproved his own colleagues for supporting those antiquated views and called for a more general understanding based on deterministic forces: "Determinism is an absolute condition of sane and ordered thinking. It is not merely that in science and sound philosophy, it is a case of Materialism or nothing, it is implied in the structure of our mental life."[155] He was very interested in the social reasons for the wide interest in Eddington's books and was particularly harsh on the very liberal kind of religion allowed for in *Nature of the Physical World* and *Science and the Unseen World. Religion* used in such a broad sense ("from beer-drinking to the much more subtle form of intoxication, theosophic meditation"), he said, was essentially meaningless. Cohen claimed that using such a broad definition was a ploy of the religious superstructure of British society and served two social functions. It gave the social respectability of religion to everyone who wished to claim it, and it allowed "the professional religionist" to claim large numbers of believers in God. Eddington was therefore complicit in the continuing domination of free-thinking, poor, and uneducated people by the theocracy.[156]

The *Freethinker* was extremely harsh on Eddington as a scientist: he had been "defiled" by his contact with religion. *Unseen* as it appeared in *Science and the Unseen World* was a synonym for ignorance, and "religion is the deification of ignorance." Scientific men (with the examples of Isaac Newton and Michael Faraday given) trying to talk about religion became particularly absurd, because their intelligence got in the way. A simple man could state his religion simply, but an intelligent one could not: "In this respect nature has not been kind to Professor Eddington. A man with the brain of a scientific thinker trying to establish a religion, commences his task with a handicap that is fatal to his chances of success."[157]

Eddington's extension of physics to all science was the source of his most profound errors. The example of the Armistice Day observance gave no trouble at all to *science*, but it was outside *physics*. The need to invoke psychology, sociology, and economics to explain a phenomenon was surely not a capitulation of the scientific ability to explain the world. Restriction of science to measurable quantities was an a priori consideration of no merit and belonged more in the pulpit than in the mouth of a scientist. The claim that physics was an exact science had no metaphysical significance; it just meant that we were more aware of the various factors involved. Psychology would one day be just as exact.[158] Cohen argued that the distinction between the physical and psychical processes was critical for Eddington, but "it is plainly and hopelessly wrong." The mind responded just as well to stimulus as bodies did (hit someone or take away their rights, and you will see a reaction). Further, Eddington clearly thought so too, or he would not make arguments that he thought could persuade people. Cohen argued that this was again "the argument from ignorance. . . . It is the helplessness of science which is stressed, not the possession of knowledge."[159]

Cohen acknowledged that he was not technically qualified to contest Eddington's challenges to causation and claimed that he was really concerned with "the way in which they have been welcomed by religious leaders." He attacked these leaders as hypocritical, because they had until recently said that a world governed by laws was proof of God and had done a quick about-face with Eddington. Cohen attacked Eddington himself for accepting quantum theory as a final truth, especially when leading physicists like Einstein expected a return to determinism. The final blow was the accusation that Eddington's science popularizations were nothing but new bottles for old, weak wine: "The truth is, I fancy, that Professor Eddington's opinion as to the current bearing of science is dictated largely by pre-existing beliefs. . . . The confession of Professor Eddington that he can't hope to convert an Atheist is the implication that religious belief has nothing of the nature of reasoned conviction behind it."[160]

The next issue of the newspaper brought with it a rather startling feature, a lengthy column written by A. S. Eddington. Cohen expressed both surprise and pleasure that Eddington had responded to his criticism and gladly made several pages available to the astronomer. Eddington emphasized his purpose in *Nature of the Physical World* and the insufficiency of materialism. He reiterated vigorously that his goal was not to prove religion, which is somewhat surprising since Cohen had explicitly acknowledged this. He described his goal in the Gifford Lectures to have been to address one specific issue: defending religion against the charge that it was incompatible with physical science. Eddington thought this clarification ameliorated several of Cohen's objections. This explained why he restricted his discussion to physical science (explicit in the stated problem), why he was willing to accept scientific results as final (he was responding to critics using a particular group of scientific ideas), and why he took the essential truth of religion for granted ("the soldier whose task is to defend one side of the fort must assume that the defenders of the other side have not been overwhelmed").[161]

Eddington expressed confusion about Cohen's use of the term *materialism*. First, he questioned whether the term was not merely tautological: "We must assume that the Materialist, in asserting the all-sufficiency of physical or mechanistic conceptions, intends to rule out some conceptions as non-physical and non-mechanistic; otherwise he is merely asserting a truism; and in drawing the line the only guide is the boundary of physical science accepted at the present day."[162] He also distinguished here (which he generally did not do in his books) between billiard-ball materialism and a wider sense of the term: "Crude Materialism, which asserts that matter is the sole reality, has been replaced by a modern Materialism which asserts that the world built out of the concepts of physics is the sole reality—that the whole of experience is the interplay of these physical entities fulfilling the laws of physics, and that's all there is to it. That is the position I attempt to refute in my book."[163] Eddington's arguments against materialism in *Nature of the Physical World* were clearly aimed at what he called "Crude Materialism," and he took pains here to extend the validity of his arguments to cover Cohen's more general materialism. In some sense, it appears that Eddington was unwilling (or possibly unable) to distinguish between those two positions. To him, the differences were unimportant insofar as they both supported "moral materialism." In his conclusion he lapsed back to criticizing Crude Materialism and, particularly, the sufficiency of that idea to explain the fundamental qualities of humans: "A particular belief may correspond with a particular configuration of atoms in a brain-cell, but the mechanistic conception of the atoms cannot be transferred into a mechanistic conception of belief. The configuration of the atoms is an indifferent phenomenon; *the belief matters.*"[164]

Unsurprisingly, Cohen penned a lengthy reply. He described himself as an admirer of Eddington, saying that if the professor fell into error, it was only because of his attempts to mix science and religion. He reserved the greatest scorn for the clergymen and journalists using the "weakest and least scientific part" of Eddington's book to defend positions that Eddington himself rejected.[165] He described Eddington as not so much proving his claims about the division of the world into material and spiritual as simply asserting them. This was simply a consequence of his religiously driven search for "an ultimate reality which is beyond experience." Essentially, he distorted science in his misguided quest.[166]

Cohen reprinted his and Eddington's *Freethinker* essays in a book two years later, in which he also took aim at Jeans, Julian Huxley, and Einstein. His aim was to show that, first, these scientists were not representative of their profession's attitude toward religion, and, second, that the old idea of "two truths" was completely untenable. Science and religion belonged to two culture stages, one developed and one primitive. It was, therefore, absurd to link them in the modern world: "It is one thing to say that certain scientific men—as an outcome of their early religious prepossessions—are making overtures to religion, but it is quite another thing to say that science is becoming more religious." The majority of scientists were mechanists, they simply did not receive the same publicity for their ideas from the religion-dominated press.[167] The basic problem with science popularizers was their possession of knowledge without understanding.[168] Further, any defense of two truths was based on a hoped-for ignorance of science and a willful ignorance of the biological and social roots of religion. "Once again, and for the thousandth unanswered time, will anyone show a substantial difference between the visions of a dipsomaniac and those of a Christian saint? The fact that one is produced by overindulgence in alcohol, and the other by over absorption in religion is surely not enough to establish a scientific difference."[169]

Eddington surely played a dangerous game by engaging in debate on his opponents' terms and territory. He clearly felt the stakes were high and that he needed particularly to dissuade the application of materialism to questions of human consciousness and social behavior. Cohen was correct to say that Eddington's understanding of materialism was old-fashioned and probably of religious origin. As discussed at the beginning of this section, Eddington was chasing the ghosts of a long-dead materialism. It is perhaps because of this that he never registered the details of Chapman Cohen's position. In some sense Cohen was an odd choice for Eddington to single out as an opponent. Cohen was unquestionably an atheist and a declared determinist, but he was explicitly *not* a materialist in the sense that Eddington was refuting. Cohen had no particular stake in the billiard-ball view of the universe, but Eddington assumed he did. Eddington simply lumped anyone who called themselves a determinist, mechanist, or materialist into one category associated with the Marxist threat to religious belief. Similarly, Cohen's choice of Eddington as a primary opponent was somewhat odd. There were plenty of scientists and theologians using classical arguments from design and seeking the proof of religion that Cohen found so reprehensible. In contrast, Eddington's position was quite moderate and did not in any way seek to support the authority of the established Church or its hierarchy. What seems to have infuriated Cohen was Eddington's personal stance as a believing scientist and, more importantly, that Eddington's writings were being read and used as though they *were* proofs of religion. Eddington was simply labeled an ally of antideterminism and that was enough.

Thus, both sides of the debate were misunderstanding the other in important and profound ways. Both Eddington and Cohen set out to dispel the myths that they had designated as most hostile to their worldview (materialism and religion, respectively) without making serious effort to appreciate that those myths actually described a range of positions instead of a singularity. They both acknowledged that they agreed on several issues, but there was no attempt to moderate their arguments accordingly. Instead, they both relied on their perceived knock-out arguments. They had significantly different understandings of what *religion, materialism,* and *determinism* meant, and thus it was inevitable that their arguments would pass at right angles.

Conclusion

Eddington's supporters were staggeringly numerous and startlingly varied. His writings were so widely disseminated that parodies of popular hymns were written to celebrate his expertise.[170] Many schoolchildren were inspired to pursue careers in science after reading his books—historian Gerald Holton places Eddington's works among the "tribal books" that taught a generation of scientists what it meant to do science.[171] From the pulpit to Parliament, Eddington's views were proclaimed to be the apex of scientific thought, although most readers did not appreciate how idiosyncratic his ideas were.[172] Despite the varied reasons for this enthusiasm, there were strong common themes. The first was excitement about Eddington's defense of traditional values. In a time when it seemed science had either given up on values (and thus caused the devastation of the Great War) or was actively fighting them (through encroaching socialism), the calm reassurance of a major scientific figure caused a cultural sensation. Even beyond defending the values themselves, Eddington defended the traditional ways of thinking about and addressing values. That is, he said it was perfectly acceptable for citizens of the modern state to base their desires and actions on their innermost religious, spiritual, and aesthetic beliefs, just as they had for centuries. Science, far from imperiling values, was a route to a deeper realization of them: "I believe that science, like art, enables mankind to approach nearer to the realization of the absolute values that alone give an aim and meaning to life.... A life spent in complete devotion to an absolute value is a good life."[173]

The second major theme was that Eddington's arguments supported a distinctly modern understanding of religion. The supporters of liberal theology were enthused that a scientist famous for his work on the most modern aspect of science (relativity) would ally himself with the most modern religionists. He was a living example that the ideals of liberal theology were alive and valid: religion and science could be reinforcing partners in the quest for progress. The death of liberal theology with the coming of the Second World War crushed these hopes, a transition well documented by Peter Bowler.

Eddington's critics were similarly united in certain ways. Three basic categories were salient. First, critics censured him for speaking outside his area of expertise. Scientists were welcome to talk about their research, but there were firm (if sometimes invisible) boundaries past which they had violated their disciplinary obligations. Many attacks claimed he was not talking "scientifically" and therefore should be accorded no authority. Interestingly, nonscientists (mostly philosophers) were much more willing to do this than scientists were. This was complicated by Eddington's own ambivalence about what he was doing in his popular writings. He sometimes said he was speaking as a scientist, sometimes not. Sometimes he described writings like *Nature of the Physical World* as philosophy and sometimes as "scientific philosophy." As when he compared mathematicians and physicists in *Internal Constitution of the Stars*, Eddington's presentation of disciplinary boundaries was epiphenomenal to the values that could move underneath them.

The second widespread criticism was that Eddington, as a scientist, should have nothing to do with religion. He was seen to have violated the boundary between religion and science that had been established with such difficulty in the late Victorian period. His willingness to move between those categories threatened the independence that made science viable in a world still dominated by religion.

Finally, there was the argument that he should not be as influential as he was. The results of science and philosophy were complex and could only be understood by trained experts. Popularization would have no result but to confuse and mislead the public. Attempts to translate specific technical concepts into common language would inevitably be disastrous, no matter how skilled the expositor.

Eddington's various detractors and enthusiasts were key players in the social debates in interwar Britain. These revolved around a series of very broad concepts, such as religion, morality, socialism, and Englishness.[174] All parties involved had their own native categories in which these concepts made sense, but there were large numbers of subgroups in Britain that interpreted the concepts differently. Religion could be the Church of England, Quakers, Catholics, Jews, or practitioners of folk traditions. Consider Eddington's and Jeans's differing beliefs about the nature of religion and the role science should play in it. The poles of the debates around science and religion had been so deeply established in the culture as Tyndallian materialism versus natural theology that Eddington's and Jeans's books were read by many as virtually identical. If they were religious scientists, they must be providing proof of religion; that was what religious scientists did. Indeed, the two popularizers became so closely associated that it was quite common to see their names reflexively paired as "Eddington and Jeans" or "Jeans-Eddington." Few readers were thoughtful enough about the issues to appreciate the serious gulf that separated their ideas. Eddington's liberal approach confused many readers who expected him to fit the mold of natural theology. Many critics of *Nature of the Physical World* (including Bertrand Russell) had either never read it and were simply relying on secondhand assumptions or misinterpreted it as yet another proof of God.

Eddington's defense of religious values rather than religious ontology was typical of the liberal theology of the time. The confusion over exactly what this meant helps point out the weaknesses of liberal religion as a movement. The groups that had always focused on questions of values (mostly Dissenters) found liberal thinking to be straightforward and Eddington to be an excellent spokesman. Most British Christians, however, came from Anglican traditions that were having trouble adapting to a values-oriented approach to religion. In some sense the conversion to a liberal approach was never successful, and the continued reading of all religious scientists as natural theologians indicates that Anglicans still had a powerful desire for orthodox religious belief. Perhaps, then, part of Eddington's success as a popular writer was an accident. There is no question that he had a great gift at explaining physics to the public, but if his readers had understood his ideas better, he would not have registered so strongly with a body of religious practitioners who longed for the reassurances of the past.

Thinking about Values and Science

Eddington spent almost the entire last decade of his life developing what would become his *Fundamental Theory*, an attempt to unify relativity and quantum mechanics through epistemology.[1] This was effectively a theory of everything, an ultimate explanation of all science from one principle. Prima facie, one might think this would be a classic opportunity for linking his religion with his physics. Similarly, Eddington conducted significant cosmological investigations that required deep consideration of the beginning of the universe.[2] Why does my study on religion and science in Eddington's work have nothing to say on these matters? Surely matters of cosmology and ultimate explanation must involve his religion?

It is intriguing that, in the end, they do not. Questions of ontology and metaphysics had no special significance in the Quakerism of Eddington's time, and they were simply irrelevant (or at most, ancillary) to his beliefs. There appear to be no significant traces of his religion in his *Fundamental Theory*. Unlike the seeking approach of his stellar models, his *Fundamental Theory* and its preceding investigations were relentlessly deductive. He required completeness from it in a way he never did from his astrophysical theories. In his own terminology, he was acting like an inflexible mathematician, not an exploratory physicist. One could claim that his *Fundamental Theory* was related to Eddington's religion in that it and his experience-oriented philosophy both required a conscious observer (building on the arguments of chapter 5). But there was no active link; his values did not drive or support the theory's comprehensive explanatory scheme.

Our conventional expectations of how science and religion interact fail us here. Unlike Michael Faraday's waste-hating deity who provided for conservation of energy, Eddington's God did not act in the material world. Similarly, although Johannes Kepler's divine being guaranteed mathematical harmony in nature, Eddington could rely on no such assumptions. The existence of God for Eddington was related to personal religious experience and its implications only. All the values discussed in this book came, eventually, from the Quaker experience of the Inner Light. Scripture did not matter. The metaphysical questions of Creation, divine omnipotence, and the special creation of man had no impact. The authority of a church hierarchy was irrelevant. There were no religious pressures brought to bear by a society invested in particular outcomes to scientific work. Eddington had no religious commitments to a particular type of material universe. Given the approach of much of the historiography on science and religion, these would mean that there was no link between Eddington's religion and his science. But by asking our questions through the vocabulary of values, rather than theological presuppositions, we can see that Eddington's science was thoroughly steeped in religion. This approach has allowed us to investigate an entire field of interaction between science and religion that would have otherwise been missed.

It is necessary to look carefully at where religion and science *do* interact rather than where we would *expect* them to interact. The evidence of Eddington's life shows the importance of this; sometimes religious values were relevant to his science, sometimes they were not. My argument is based on the existence of valence values that can move between science and religion, but I have no stake in whether or not this *has* to happen or whether it does happen all the time. Some values in scientific disciplines have no points of intersection with common values in religion. Valence values, however, give us a new way to look for those points of intersection in situations where the sociocultural role in science is not obvious. Further, valence values themselves are historical entities that emerge from specific contexts and concerns and thus provide a further opportunity to ground science and religion in history. No set of values is transcendental; Eddington the Quaker scientist would have looked very different if he had been born fifty years earlier or later.[3] The valence values that were so important to him were a product of the Quaker Renaissance and a particular location in space and time.

A values-based approach to the history of science has the significant benefit of allowing exploration of the relationship between science and society without needing to decide a priori whether the physical world or the social world dominates. In this book religion was not reduced to science, nor was science just a construct of religion. Each category had a substantial existence unto itself and no attempt was made to make one merely epiphenomenal. Rather, the central methodological claim was that values fashioned the interaction of scientists with both worlds. Because values clearly exist in both categories, looking for values that move between them gives us a way to speak of specific historical interactions between science and culture without reducing either one. This permits us to take scientific claims seriously while still looking for serious and significant relationships with society.

The examples from Eddington's life discussed in this book suggest points where valence values may manifest themselves in science. Methodology within a particular scientific discipline is probably the area of greatest interest to historians and philosophers of science. Eddington's mystical outlook was a valence value that moved easily into the structure of astrophysics and dramatically shaped the methodology of stellar physics. This sort of methodological values often defines disciplinary identities: consider Eddington's discussion in *Internal Constitution of the Stars* about mathematicians versus physicists. I suggest that historians interested in disciplinary formation and change would do well to look for the values that historical actors hold, as they are good candidates for the efficient cause of change. Questions of value-driven methodology in science also allow real and substantial connections between very technical aspects of science and very rich aspects of culture. The internal-external divide can be, in a sense, outflanked.

The effect of values on disciplinary boundaries was also seen here in the discussion of Eddington's popularization. His value of experience shaped his interpretations of relativity and quantum mechanics, and his readers' values shaped their reaction to his delineation of the physical and spiritual worlds. The natural theological values of much of the British clerical community meant they understood Eddington's ideas as a modern version of William Paley's, which was very different from what Eddington intended. On the opposite extreme was Bertrand Russell, whose values of secularity gave him boundaries between science, philosophy, and religion that painted Eddington as a traitor to science and a capitulator to theological domination. Like Eddington's arguments with Jeans, many of his conflicts with Russell were, at root, an incompatibility of values.

Eddington also needed to confront the boundary of his discipline on a different axis, that of the relationship between science and the state. His struggle with the British conscription apparatus was an excellent illustration of different values at work. For him, science intersected politics through the values of internationalism; for his colleagues, science intersected politics through the values of patriotism. This makes it difficult to speak of scientists as a body capitulating to political forces, because different scientists will hold values that expect mediation between science and politics through different issues. In any case, valence values will help bind scientists to the state, but in ways that demand attention to specific details of their values.

Clifford Geertz tells us that values help determine choice of action, and Eddington certainly demonstrates this. Eddington's internationalist values demanded that he act to protest the chauvinism of wartime, and the fact that they were *valence* values helped him shape a scientific project as his protest. The 1919 eclipse expedition was a manifestation of values, some of which were religious. It was an example of multiple values interacting. Here I focused on internationalism (for obvious reasons), but scientific values such as coherence of explanation, mathematical simplicity, and harmony between observation and theory were all crucial. A role for social values does not deny a role for technical ones, or vice versa. Instead, they work together to make a particular goal or project important and viable and sometimes to invest it with a *moral* character—Eddington's defenses of indeterminism and Einstein were not just intellectual positions, but were instead *moral obligations* based on what he felt to be the most essential truths of the relationship between humanity and the divine.

Where does the example of Eddington leave us with respect to the question of values in science? Three major conclusions present themselves. First, values make a difference in science. Practice (in the sense of methodology), choice of mathematical technique, and mode of explanation can be determined by values. Similarly, interpretation is shaped by values. For example, exactly what the Lorentz contraction *meant* was debated from the ramparts of values. And the role of science within the state and society evolved through values. Are scientists supposed to be prophets, sages, weapons designers, or ordinary citizens? The answers during and after the Great War varied according to the values of the different actors involved. The diversity of expectations for scientists was mapped from the diversity of values in Britain. The Admiralty, the Royal Society, the Cambridge Syndics, and the Cambridge Friends all valued different aspects of the scientist. Some overlapped (such as the Royal Society and Cambridge) despite holding to different fundamentals. Considering the overlap brings to light the synchronic development of some of these structures. For example, the Royal Society and the Admiralty disagreed on the role of the scientist at the beginning of the war but agreed by the end. Values change. Historical events alter the fundamentals from which values come and the understanding of the connections that certain values have with the external world. International values in British science were paramount in 1913, then vanished for years, and slowly returned two decades later. Diversity in values exists both temporally and spatially.

Second, values form a bridge between science and society. Cultural studies of science have often suffered from imprecision about the relationship of culture and science. Valence values are a historically precise way of investigating this; they can show in concrete terms what science and culture share. We no longer need to talk about culture influencing science or vice versa and can instead investigate how Geertz's "pronenesses" drive science as well as other cultural activities. As moderators and stimuli of activity, values let us see the invisible common ground between apparently separate spheres. Thus, we can discuss the relationship between science and society without privileging either one. This is particularly useful for clearing away distracting dead wood from discussions of religion and science, where one category is so often reduced to a shadow of the other. It is important to remember that not all values are valence values. Science is based on values, but not all of those necessarily have some link to other areas of culture. Values *can* move back and forth between culture and science, but they do not *have* to do so. Some values do not travel well, and an analysis of values in science must pay close attention to their actual source.

Finally, a biographical approach is useful for seeing and thinking about the development and function of a set of values. It demonstrates how values put forward by various groups (in this case, a religious community) are shaped to fit particular historical circumstances. A case could be made that the values of the Quaker Renaissance were not significantly different from those of the quietistic Quakers but that the implementation of those values was significantly different. Focusing on an individual makes it clear how such an implementation happens, exactly what conditions allow or require the adaptation of a set of values. Here we saw that Eddington's time at Manchester and Cambridge both enriched and modified the values he inherited from his family, J. W. Graham, and Rufus Jones. Looking at individuals' values is also useful because it shows how they function differently in changing circumstances, thus giving the historian several data points of interaction with which to characterize them. Also, sometimes synergies appear among values that require a more holistic approach. Eddington's struggle during the Great War was a case where multiple values worked together. Personal pacifism and internationalism were intertwined in shaping his reaction to events, which required the synoptic approaches of chapters 3 and 4. A study of only internationalism, for instance, could miss the cross-influence of pacifism or conflate Eddington's religious internationalism with the socialists' proletariat version. Viewing an individual's values as a whole helps define how and in which areas valence values function. This is also a methodological opportunity for the historian—when a particular value is seen to be at work, one should try to track it to its source and see what other values are associated with that source. Values generally appear in associated value structures, such as a religious tradition or research group, and can be found operating simultaneously along different facets of an individual.

So biography is helpful for assessing values in science—does the framework of values help with the problems of scientific biography?[4] One constant issue in this genre is the correct level of analysis vis-à-vis the individual and society. How do we mediate between making the individual *sui generis* and allowing the subject to dissolve into a blur of social influences? Values, as a mechanism by which the social works through the individual, can be a useful way to fully embrace context without becoming lost in it. Similarly, biographers of scientists are often unsure how to balance technical content with personal issues and cultural location. Because values appear strongly in both technical and cultural matters, the balance of the two becomes a natural result of examining particular values. Finally, values provide an alternative to a sometimes constraining chronological structure. Organizing a narrative around values can help tease apart stories that might otherwise be confusingly grouped together (such as relativity in chapters 3 and 5) and help pull together stories that might otherwise be separate (such as Eddington's stellar structure and his mysticism). This is, however, a double-edged sword, because important aspects of a person's life might fall outside the values being examined; here, I do not discuss Eddington's unified field theory and mention even his death only in passing. Thus, it is critical that authors be clear about what kind of a work they are writing—a full biography, or a biographical study concerned with some particular issues. There is, of course, plenty of room and need for both.

So in Eddington's story we see a focused study of a historical episode of the interaction of science and religion—how does this help us develop our thinking on this relationship in general? We have now seen how to have a discussion of religion and science that has no particular concern with the "facts" or "truth" of either category. The correctness of, say, Mosaic cosmogony, has little relevance to a values-based study, and as a focus on facts evaporates, so too does the territory of the so-called "military metaphor." Values are not a zero-sum game the way Genesis versus Darwin purportedly is, and it is difficult to sustain an a priori expectation of the conflict between science and religion when the battlefield has room for many parties. By lifting our gaze from the specifics of Creation, values remind us that there is a great deal more to religion than the truth-claims of scripture. Religion is a dynamic entity that relies not just on facts about the world but also integrates practices, attitudes, communities, and other aspects of human life; values remind us that religion has a tremendous impact on human action far beyond the text of a holy book.

If we think that religious belief matters to scientists only insofar as it tells them what conclusions to draw about the dating of rocks, then we have profoundly misunderstood what it means to be religious. The modern debates over the roles of science and religion often founder on exactly these issues, and if we do not pay attention to the aspects of religion that move smoothly in and out of science, we create a stage of siege in which there must be one winner and one loser. And that, quite simply, does not reflect the way science and religion have interacted in the past, and continue to do today.

Consider Francis Collins, one of the heads of the Human Genome Project and a committed Christian. Recall William Provine's claim that began this book, that there is no intellectually honest position for a religious scientist. Is Dr. Collins simply being intellectually dishonest? Is he lying about his faith, or is his science corrupt because he believes in God? Those who consider science and religion only as simple bodies of facts have their possible answers pushed to awkwardly polarized positions. But when we begin to speak in the language of values—What values does Collins feel form the basis of Christianity? Do the values of protecting, cherishing, and healing human life appear in his thinking about both Jesus and the human genome?—we can begin to understand what is actually happening. We are able to ask questions that help make sense of the varied people around us in the world, instead of refighting endless battles. We now have the capability for nuance and specificity. In complex issues like bioethics, we do not have to duel over whether the scientific view or the religious view is correct. We can instead search for the values that create the framework for those views and look for valency among them. Francis Collins does not seem to think the human genome belongs to one camp or the other, and considering values gives us both the obligation and the opportunity to understand how that can be.

And where are we left with Eddington as a historical figure? We now have a bridge between portions of his work that had been separated by biographers and have given precision to the vague sense that Eddington thought of himself as a religious scientist. His technical work had always been separated from his philosophical writings, probably because the authors (usually practicing scientists) were concerned to preserve his legacy as an astronomer by not contaminating it with his controversial, "unscientific" ideas. This resulted in a picture of someone working productively as a reasonable scientist then, on or about January 1927, losing touch with reality. His philosophy and epistemological physics were described as a second phase of his life, bounded by his writings on religion, which bore no resemblance to the first. This sort of schizophrenic view creates strange biography. I am certainly not claiming that there can be no shift in someone's views or outlook, but a basic continuity can be expected. Valence values from Eddington's Quakerism provide this continuity. His Gifford Lectures now become a natural consequence of the same outlook that shaped his stellar models and his observations of the relativistic deflection of light. His structuralist philosophy, apparently deeply flawed and inexplicable, gains internal consistency if we see its roots in the values of religious experience. Eddington, a major figure in the history of twentieth-century physical science, no longer needs to be either a scientific saint or a crazed mystic. His valence values give us a way to take him seriously as both an astronomer and a Quaker without discarding the heart of either.

Without his valence values, Eddington appears as a fractured figure, a scattered collection of ideas, activities, and alliances without any interrelation. Nothing holds him together as a single historical object. It is only with the framework of values that the points of light snap into a constellation that is recognizable as a whole figure. Some of the stars in that constellation are bright and enduring, others have dimmed over time, but it is only in their relationship with each other that they have any role in representation. Values, like the lines between stars, are invisible but essential.

Notes

Introduction

1. William Provine, "Scientists, Face It! Science and Religion Are Incompatible," *Scientist* 2, no. 16 (September 5, 1988): 10.

2. The classic work on historical thinking about science and religion is John Hedley Brooke's *Science and Religion: Some Historical Perspectives* (Cambridge: Cambridge University Press, 1991). Brooke's and Geoffrey Cantor's Gifford Lectures, published as *Reconstructing Nature* (Edinburgh: T & T Clark, 1998), also argue strongly for a historical approach. A concise overview of the historical issues in science and religion can be found in John Hedley Brooke, "Science and Religion," in *Companion to the History of Modern Science*, ed. R. C. Olby et al., 763–82 (London: Routledge, 1990).

3. The subcultures of science are addressed in Peter Galison and David Stump, *The Disunity of Science* (Stanford, CA: Stanford University Press, 1996), and Peter Galison, *Image and Logic* (Chicago: University of Chicago Press, 1997).

4. See Brooke and Cantor, *Reconstructing Nature*, 25, for a discussion of the importance of local interpretations of natural theology, and John Hedley Brooke, "Natural Theology," in *Science and Religion: A Historical Introduction*, ed. Gary Ferngren, 163–75 (Baltimore: Johns Hopkins University Press, 2002), for an introduction to natural theology in general.

5. For examples, see any of the recent statements by physicist Steven Weinberg on religion (e.g., his comments at the 2006 forum on science and religion reported in the *New York Times*, November 21, 2006; or the interview in *Freethought Today*, April 2000, 4–6); Paul Davies's *God and the New Physics* (London: Dent, 1983); and Stephen J. Gould's *Rocks of Ages* (New York: Ballantine, 1999). The common failings here are portraits of religion that bear little resemblance to actual beliefs and practices.

6. Clifford Geertz, *The Interpretation of Cultures* (New York: Basic Books, 1973), 95.

7. A classic study on the effect of religion on actions and life choices is Gerhard Lenski, *The Religious Factor: A Sociological Study of Religion's Impact on Politics, Economics, and Family Life* (Garden City, NY: Doubleday, 1961). As exemplars of how values function in science, see Charles Rosenberg, *No Other Gods: On Science and American Social Thought* (Baltimore: Johns Hopkins University Press, 1976), particularly chap. 8; Robert Kohler, *Lords of the Fly: Drosophila Genetics and the Experimental Life* (Chicago: University of Chicago Press, 1994); and Simon Schaffer, "Metrology, Metrication, and Victorian Values," in *Victorian Science in Context*, ed. Bernard

Lightman, 438–74 (Chicago: University of Chicago Press, 1997). An interesting philosophical essay on the role of values in science can be found in Joseph Grünfeld, *Science and Values* (Amsterdam: B. R. Grüner, 1973), chap. 11.

8. A classic investigation of the impact of religious values on science is Robert K. Merton, *Science, Technology and Society in Seventeenth Century England* (Bruges, Belgium: Saint Catherine Press, 1938).

9. For example, see Richard Westfall's excellent essay "Newton and Christianity," in *Newton*, ed. Westfall and Cohen, 356–70 (New York: Norton, 1995).

10. Note that this is not to say that twentieth-century physical sciences have no *significance* for metaphysics or religious commitments. One need only look at the religious responses to Big Bang cosmologies to see an intersection of this type. I would like to suggest, however, that these interactions have increasingly little impact on the day-to-day work of a practicing scientist. See Helge Kragh, *Matter and Spirit in the Universe: Scientific and Religious Preludes to Modern Cosmology* (London: Imperial College Press, 2004).

11. Alan Gilbert, *The Making of Post-Christian Britain: A History of the Secularization of Modern Society* (London: Longman, 1980); and Adrian Hastings, *A History of English Christianity 1920–2000* (London: SCM Press, 2001), are classic examples of scholarship that takes the decline of British religion as a real and culturally defining process. They rely on church attendance records, institutional practices, and similar sources. In contrast, S. C. Williams, *Religious Belief and Popular Culture in Southwark c. 1880–1939* (Oxford: Oxford University Press, 1999), argues against the decline model by exploring more personal, experiential aspects of religion. José Harris, *Private Lives, Public Spirit: A Social History of Britain, 1870–1914* (Oxford: Oxford University Press, 1993), makes important contributions to understanding the early history of apparent secularization. Harris provides material that suggests a values-oriented investigation would be useful in explaining the apparent disconnect between decline and persistence of working-class religion in the late Victorian and Edwardian periods. Callum Brown, *The Death of Christian Britain: Understanding Secularisation 1800–2000* (London: Routledge, 2001), rejects the gradual secularization narrative of Gilbert and Hastings in favor of a sharp plummet of religion in the 1960s. His reassessment, like Harris's, relies on reconceptualizing what is meant by religion.

12. A productive demonstration of these sorts of links can be found in Boyd Hilton, *The Age of Atonement: The Influence of Evangelicalism on Social and Economic Thought, 1795–1865* (Oxford: Oxford University Press, 1988).

13. In *Reconstructing Nature*, Geoffrey Cantor and John Hedley Brooke appeal to the importance of biography in considerations of religion and science, similar to the studies of James Moore in *The Post-Darwinian Controversies* (Cambridge: Cambridge University Press, 1979); and Cantor's own *Michael Faraday: Sandemanian and Scientist* (London: Macmillan, 1991).

14. A. S. Eddington, *The Expanding Universe* (Cambridge: Cambridge University Press, 1933), 17, and *New Pathways in Science*, 2nd ed. (Cambridge: University Press, 1947), 211. Emphases in original.

15. A. S. Eddington, *The Nature of the Physical World* (Cambridge: Cambridge University Press, 1928), 333.

16. In this sense, my study of Eddington will share much with Crosbie Smith and M. Norton Wise's *Energy and Empire: A Biographical Study of Lord Kelvin* (Cambridge: Cambridge University Press, 1989). I see their biography of William Thomson to be predicated on the idea that a particular set of values was salient throughout his life and serves to reveal important historical connections. For example, his latitudinarianism formed a pattern of behavior and

thought that was manifest not only in debates over Glasgow intellectual life, but also in debates over valid physical explanation: "Common sense" philosophy applied to electromagnetism as well as theology. Further, Thomson saw the British Empire as the guardian of these values, and this led to his lifelong commitment to the defense of its role in British society. What is relevant here is that the authors show how his values of religious latitudinarianism were not simply a set of ideas or beliefs, but instead deeply affected his entire life.

17. For example, see W. H. McCrea, "Arthur Stanley Eddington," *Scientific American* 264, no. 6 (1991): 66–71; C. W. Kilmister, *Sir Arthur Eddington* (Oxford: Pergamon, 1966); G. J. Whitrow, "Sir Arthur Eddington, OM (1882–1944)," *Quarterly Journal of the Royal Astronomical Society* 24 (1983): 258–66.

18. A. Vibert Douglas, *The Life of Arthur Stanley Eddington* (London: Thomas Nelson, 1956). There is also David Evans's *The Eddington Enigma* (self-published, 1998), which is a virtual paraphrase of the Douglas biography, with the addition of some interesting personal recollections.

19. Loren Graham, *Between Science and Values* (New York: Columbia University Press, 1981); David Wilson, "On the Importance of Eliminating 'Science' and 'Religion' from the History of Science and Religion: The Cases of Oliver Lodge, J. H. Jeans, and A. S. Eddington," in *Facets of Faith and Science*, ed. Jitse M. van der Meer (Lanham, MD: Pascal Centre for Advanced Studies in Faith and Science, University Press of America, 1996), 1:27–48. See also A. H. Batten, "A Most Rare Vision: Eddington's Thinking on the Relation between Science and Religion," *Quarterly Journal of the Royal Astronomical Society* 35 (1994), 249–70. Kragh's *Matter and Spirit* discusses Eddington and religion on 103–11. Philosophers have been particularly egregious in this neglect of the whole of Eddington's life and work. Examples of this are C. F. Ritchie's work, in which a casual reader would never know that Eddington had been a practicing scientist, and John Yolton's essay that quoted Eddington's definition of philosophy through practice but made no effort to see how his philosophy might have come from his scientific practice. C. F. Ritchie, *Reflections on the Philosophy of Sir Arthur Eddington* (Cambridge: Cambridge University Press, 1948); John Yolton, *The Philosophy of Science of A. S. Eddington* (The Hague: M. Nijhoff, 1960). Other philosophical perspectives include Herbert Dingle, *The Sources of Eddington's Philosophy* (Cambridge: Cambridge University Press, 1954), which completely misunderstands Eddington's cultural and religious context, and Johannes Witt-Hansen, *Exposition and Critique of the Conceptions of Eddington Concerning the Philosophy of Science* (Copenhagen: G.E.C. Gads Forlag, 1958), which makes an effort to show some of the religious roots of Eddington's philosophy but makes no connection to his science. Pleasant exceptions to philosophical misunderstandings of Eddington include Steven French's "Scribbling on the Blank Sheet: Eddington's Structuralist Conception of Objects," *Studies in History and Philosophy of Modern Physics* 34 (2003), 227–59; and Thomas Ryckman, *The Reign of Relativity: Philosophy in Physics, 1915–1925* (Oxford: Oxford University Press, 2005).

20. This is similar to the approaches taken by Michael Gordin, *A Well-Ordered Thing* (New York: Basic Books, 2004); Adrian Desmond and James Moore, *Darwin* (London: Michael Joseph, 1991); and Alan Rocke, *Nationalizing Science: Adolphe Wurtz and the Battle for French Chemistry* (Cambridge, MA: MIT Press, 2001). On the value, dangers, and difficulties of biographical approaches in the history of science, see Michael Shortland and Richard Yeo, eds., *Telling Lives in Science: Essays on Scientific Biography* (Cambridge: Cambridge University Press, 1996).

21. Christopher Lawrence and Anna-K. Mayer, eds., *Regenerating England: Science, Medicine and Culture in Inter-War Britain* (Atlanta: Editions Rodopi, 2000), is a useful starting point for understanding these issues, particularly the introduction and Anna-K. Mayer's contribution.

Chapter One

1. *Report of the Proceedings of the Conference of Members of the Society of Friends, Held. . .in Manchester from Eleventh to Fifteenth of Eleventh Month, 1895 (London: Headley Bros., 1896)*, 219.

2. Good overviews of Quaker history include Hugh Barbour and J. William Frost, *The Quakers* (New York: Greenwood Press, 1988); Howard Brinton, *Friends for 300 Years* (New York: Harper, 1952); Thomas C. Kennedy, *British Quakerism, 1860–1920: The Transformation of a Religious Community* (Oxford: Oxford University Press, 2001); and Elizabeth Isichei, *Victorian Quakers* (Oxford: Oxford University Press, 1970).

3. A useful overview of British evangelicalism is David W. Bebbington, *Evangelicalism in Modern Britain: A History from the 1730s to the 1980s* (London: Unwin Hyman, 1989). On evangelicalism and intellectual life (particularly science), see Hilton, *Age of Atonement*; David N. Livingstone, D. G. Hart, and Mark A. Noll, eds., *Evangelicals and Science in Historical Perspective* (Oxford: Oxford University Press, 1999); and Aileen Fyfe, *Science and Salvation: Evangelical Popular Science Publishing in Victorian Britain* (Chicago: University of Chicago Press, 2004).

4. Barbour and Frost, *Quakers*, 180.

5. The Richmond Declaration came from a conference of American evangelical Quakers in 1887.

6. Kennedy, *British Quakerism*, 124.

7. Ibid., 134.

8. W. C. Braithwaite, "Some Present-Day Aims of the Society of Friends," *Friends' Quarterly Examiner* (1895): 321–41, as quoted in Kennedy, *British Quakerism*, 146.

9. Kennedy, *British Quakerism*, 146–47.

10. Barbour and Frost, *Quakers*, 222, and Geoffrey Cantor, *Quakers, Jews, and Science: Religious Responses to Modernity and the Sciences in Britain, 1650–1900* (Oxford: Oxford University Press, 2005), 257–62.

11. Kennedy, *British Quakerism*, 149; and Cantor, *Quakers, Jews, and Science*, 267–69.

12. Roger Wilson, "Road to Manchester," in *Seeking the Light: Essays in Quaker History*, ed. J. William Frost and John M. Moore (Haverford, PA: Pendle Hill Publications, 1986), 157.

13. See Kennedy, *British Quakerism*, 150–51, for an overview of speakers at this session.

14. *Conference of Members of the Society of Friends*, 221.

15. Ibid., 224.

16. Thompson was a major figure in the Quaker science of his generation. As a physicist he was unusual—most previous Quakers working in science were naturalists or observational astronomers. See Cantor, *Quakers, Jews, and Science*, for a discussion of Quakers and science before 1895. Thompson is discussed on 240–42, 262–63, and 268–69; and chap. 6 (225–47) analyzes the distinctiveness of Quaker scientific methods.

17. *Conference of Members of the Society of Friends*, 230.

18. Ibid., 234.

19. Cantor, *Quakers, Jews, and Science*, 225–47.

20. *Conference of Members of the Society of Friends*, 235.

21. Ibid., 236–37.

22. Ibid., 237.

23. Ibid., 239.

24. Ibid., 241.

25. Ibid., 242.

26. Quoted in Kennedy, *British Quakerism*, 124.

27. Quoted in ibid., 148. Newman was the editor of the *Friend*, a major Quaker newspaper.

28. Kennedy, *British Quakerism*, 164–66.

29. Douglas, *Life of Arthur Stanley Eddington*, 33.

30. Ibid., 1–4. See also Eddington's entry in the *Dictionary of Quaker Biography* in Friends House Library, Euston Road, London; in the *Dictionary of Scientific Biography* (essentially a compressed version of Douglas's biography); and in the *Dictionary of National Biography*. There is no good critical biography of Eddington (the Douglas biography is inadequate in almost every way but is useful for gathering information in one place).

31. *Friend*, March 20, 1896, 188.

32. Douglas, *Life of Arthur Stanley Eddington*, 8.

33. Ibid., 5.

34. Manchester had been a draw for Dissenting students for many years; it never discriminated religiously, as the Oxbridge universities did. The establishment of Dalton Hall made it especially appealing for Quakers. Cantor, *Quakers, Jews, and Science*, 94. See chap. 3 of Cantor, *Quakers, Jews, and Science*, for a detailed discussion of how religious discrimination affected Quaker education and careers in science.

35. J. W. Graham to Sarah Ann Eddington, July 2, 1899, Trinity College, Wren Library, Uncatalogued Eddington Material (hereafter UEM) 11(2). Permission to quote the UEM provided by the Master and Fellows of Trinity College, Cambridge, and the Arthur Stanley Eddington Memorial Trust. I am grateful to Geoffrey Cantor for pointing out that Graham was probably especially concerned about these issues because he had been forced to withdraw from Cambridge due to ill health.

36. See his entry in the *Dictionary of Quaker Biography* for more information, though there is not much more than presented here. J. W. Graham's papers are at Manchester University but are uncatalogued.

37. John William Graham, "The Meaning of Quakerism," [1900?], J. W. Graham Papers, Manchester University. Reproduced by courtesy of the University Librarian and Director, the John Rylands University Library, the University of Manchester. The laboratory imagery here may be related to George Fox's famous description of his religious experiences: "And this I knew experimentally." Originally from Fox's journal of 1647, quoted in *Quaker Faith and Practice* (London: Yearly Meeting of the Society of Friends, 2004), 19:02.

38. John William Graham, lectures delivered 1900–1905, J. W. Graham Papers.

39. Review of "The Meaning of Quakerism," unknown source, J. W. Graham Papers.

40. John William Graham, *Methodist Recorder*, June 21, 1900.

41. Kennedy, *British Quakerism*, 250.

42. Ibid., 260.

43. Ibid., 304–5.

44. Ibid., 308–9.

45. Robert Kargon, *Science in Victorian Manchester* (Baltimore: Johns Hopkins University Press, 1977), 219.

46. Ibid., 223, 225–26.

47. Ibid., 234.

48. Arthur Schuster, "The Influence of Mathematics on the Progress of Physics," Arthur Schuster Papers, University of Manchester, D7 Inaugural Lecture 1881–82.

49. Arthur Schuster, "International Science," *University Review*, June 1906. I preserve the original British spelling when quoting historical actors.

50. Silvanus Phillips Thompson, *The Methods of Physical Science* (London: Longman, 1877), 28.

51. Ibid., 13.

52. Ibid., 29.

53. Eddington's courses and grades can be found in the archives of Manchester University in the class registers (RA/1/6–8).

54. Douglas, *Life of Arthur Stanley Eddington*, 5. Apparently A. N. Whitehead was the decisive voice on the committee that gave Eddington the scholarship. See Douglas, *Life of Arthur Stanley Eddington*, 10–11.

55. J. W. Graham to S. A. Eddington, August 1902, UEM.

56. *Wrangler* was a term used to denote the students who received top scores on the Cambridge Tripos mathematics examination. The Tripos was a grueling multiday examination, and students who placed highly often went on to distinguished careers in mathematics and natural philosophy. Herman taught Eddington differential geometry, among other important tools, which would become crucial in his later work in relativity. For more on the Cambridge Tripos tradition in general, R. A. Herman in particular, and Eddington's debt to his Cambridge education, see Andrew Warwick's *Masters of Theory: Cambridge and the Rise of Mathematical Physics* (Chicago: University of Chicago Press, 2003), chaps 5 and 9.

57. A sense of what a Tripos education at Cambridge consisted of can be found in J. E. Littlewood, *A Mathematician's Miscellany* (London: Methuen, 1953), 66–83. Littlewood came to Cambridge the same year as Eddington and also studied with Herman.

58. Douglas, *Life of Arthur Stanley Eddington*, 11.

59. *Friend*, June 24 1904, 432–33.

60. Douglas, *Life of Arthur Stanley Eddington*, 7. It has been speculated that Eddington was a homosexual and that he and Trimble were lovers. This is certainly plausible (Eddington displayed no interest in romantic relationships with women), but no evidence has survived to confirm or deny it. Arthur I. Miller, in his *Empire of the Stars: Obsession, Friendship, and Betrayal in the Quest for Black Holes* (Boston: Houghton Mifflin, 2005), attempts to use Eddington's homosexuality to explain a variety of issues, including alleged psychological instability, but this certainly overreaches the thin circumstantial evidence available.

61. Douglas, *Life of Arthur Stanley Eddington*, 32–33.

62. Ibid., 121.

63. Journal of A. S. Eddington, UEM Add. MS 48, 29.

64. Ibid., 32, 35.

65. Ibid., 32.

66. Graham's presidential address, "The Calling of the Teacher," is preserved at the Friends House Library, Euston Road, London, Friends' Guild of Teachers Archive. Thanks to the Library of the Religious Society of Friends for permission to quote these materials.

67. "Religious Instruction in Our Schools: A Letter," [1902?], 1, Friends' Guild of Teachers Archive.

68. S. P. Thompson's presidential address, "Illustrative Teaching," 1902, 1, Friends' Guild of Teachers Archive.

69. "Religious Instruction in Our Schools," 5, Friends' Guild of Teachers Archive.

70. Eddington's presidential address to the guild was titled "The Purpose of Science" and appeared in *Friends' Quarterly Examiner* 246 (April 1928): 89–110.

71. W. M. H. Christie to Secretary of the Admiralty, January 8, 1906, W. M. H. Christie Papers, RGO 7, University Archives, Cambridge. Used by permission of the Particle Physics and Astronomy Research Council and the Syndics of Cambridge University Library.

72. W. M. H. Christie to Secretary of the Admiralty, January 23, 1906, Christie Papers.

73. A. S. Eddington to W. M. H. Christie, February 6, 1906, Christie Papers.

74. David Dewhirst, "The Greenwich-Cambridge Axis," *Vistas in Astronomy* 20 (1976): 109–11.

75. Journal of A. S. Eddington, 41.

76. Ibid., 50–56.

77. Cadbury was one of the Quaker chocolate magnates.

78. Kennedy, *British Quakerism*, 171–73.

79. *Westmoreland Gazette*, August 8, 1908.

80. *Kendal Times*, August 7, 1908.

81. Journal of A. S. Eddington, 55–56.

82. *Westmoreland Gazette*, August 8, 1908.

83. *Friend*, May 15, 1908, 317–18.

84. Rufus M. Jones, *Spiritual Energies in Everyday Life* (New York: Macmillan, 1922), 135–44. *Primitive Church* is a term used to refer to the religious movement that thrived during Jesus' life.

85. Rufus M. Jones, "Beginnings of Quakerism" (1912), in *Quakerism: A Spiritual Movement*, 119–38 (Philadelphia: Philadelphia Yearly Meeting of Friends, 1963).

86. Rufus M. Jones, *Studies in Mystical Religion* (London: Macmillan, 1909), xv.

87. Rufus M. Jones, *The Inner Life* (New York: MacMillan, 1917), 113.

88. Ibid., 115.

89. Jones, *Studies in Mystical Religion*, 152.

90. Ibid., xv.

91. *Rufus Jones: Essential Writings*, ed. Kerry Walters (New York: Orbis Books, 2001), 22.

92. Jones, *Inner Life*, vi.

93. Ibid., v.

94. Ibid., 10–11.

95. Ibid., 17.

96. Quoted in L. S. Kenworthy, "Rufus M. Jones, Luminous Friend," in *Living in the Light: Some Quaker Pioneers of the 20th Century*, vol. 1, ed. L. S. Kenworthy (Kennett Square, PA: Friends General Conference and Quaker Publications, 1984), 124.

97. Rufus M. Jones, *The New Quest* (New York: Macmillan, 1928), 80–81.

98. Ibid., 97.

99. The AFSC made such an impact with its activities that it shared the 1947 Nobel Peace Prize with the Friends Service Council, its British counterpart. This was the first time the prize had been given to an organization.

100. *Westmoreland Gazette*, August 8, 1908.

101. Journal of A. S. Eddington, 55.

102. Eddington carried out an expedition to Malta to make a longitude determination of that island and also went to Brazil in 1912 to make observations of an eclipse.

103. His work on statistical cosmology peaked with the publication of his *Stellar Movements and the Structure of the Universe* (London: Macmillan, 1914). This phase of astronomical thought is covered in Erich Robert Paul, *The Milky Way Galaxy and Statistical Cosmology, 1890–1924* (Cambridge: Cambridge University Press, 1993); and Robert J. Smith, *The Expanding Universe* (Cambridge: Cambridge University Press, 1982).

104. The Smith's Prize was originally an extra post-Tripos examination intended to give the strongest Wranglers (see n. 56 above) an opportunity to further distinguish themselves. After 1883 the prestigious prize was awarded for the best original essay on any subject in mathematics or natural philosophy. Winners often went on to become fellows at Trinity. See June Barrow-Green, "'A Corrective to the Spirit of Too Exclusively Pure Mathematics': Robert Smith (1689–1786) and His Prizes at Cambridge University," *Annals of Science* 56 (1999): 271–316.

105. Eddington's journal has an impressive list of his reading, cross-referenced by author and year read.

106. A. Schuster to A. S. Eddington, November 9, 1909 (UEM 1/5); and A. S. Eddington to A. Schuster, November 15, 1909 (UEM 1/6).

107. See, for example, A. S. Eddington to Sarah Ann Eddington, August 5, 1913 (UEM 1/25). See chap. 3 for more on Dyson and internationalism.

108. Douglas, *Life of Arthur Stanley Eddington,* 35.

109. See Cambridge University Library, Royal Greenwich Observatory Archives, F. W. Dyson Papers MS.RGO.8 144, for various correspondence between Dyson and the Syndics of the observatory with regard to the decision to appoint Eddington as director. The difficulty was chiefly due to an individual who felt he was better qualified to be director. The Dyson papers are quoted with the permission of the Particle Physics and Astronomy Research Council and the Syndics of Cambridge University Library. Also see Douglas, *Life of Arthur Stanley Eddington,* 30, 90.

110. Douglas, *Life of Arthur Stanley Eddington,* 89.

111. Cambridge, Huntingdon, and Lynn Monthly Meeting, Cambridgeshire County Record Office R59/26/5/4, March 18, 1914, and October 15, 1914. These minutes and those of the Jesus Lane Preparative Meeting (R69/44) record the activities of the Eddingtons within the Meeting over the next three decades. Thanks to the Cambridgeshire Archives Service for their assistance with the material in their record office.

112. John S. Rowntree, *The Society of Friends: Its Faith and Practice* (London: Headley Brothers, 1908), 6, 25, 30.

113. Minutes of the Jesus Lane Preparative Meeting, Cambridgeshire County Record Office, R69/44, January 11, 1914, and January 15, 1914.

114. Silvanus Phillips Thompson, *The Quest for Truth* (Bishopsgate: Headley Brothers, 1915), 48–49.

115. Ibid., 92.

Chapter Two

1. John Kavanaugh, ed., *The Quaker Approach* (New York: Putnam's Sons, 1953), 166. Also Rowntree, *Society of Friends.*

2. Ibid., 178.

3. Silvanus Phillips Thompson, "The Mystery of Nature," *Friends' Quarterly Examiner* 9 (1875): 405–22.

4. Thompson, *Quest for Truth,* 49–50.

5. Ibid., 92.

6. A. S. Eddington, *Science and the Unseen World* (New York: Macmillan, 1929), 22–23. The "clouds of uncertainty" reference is probably an allusion to Lord Kelvin's Baltimore Lectures. My thanks to Geoffrey Cantor for pointing this out.

7. Ibid., 39.

8. Ibid., 41.

9. Ibid., 77–78.

10. Ibid., 45–47.

11. A. S. Eddington, *Philosophy of Physical Science* (Cambridge: Cambridge University Press, 1939), 223.

12. Ibid., 222.

13. Eddington, *Science and the Unseen World*, 42–43.

14. A. S. Eddington, *Expanding Universe* (Cambridge: Cambridge University Press, 1933), 21.

15. Eddington, *Science and the Unseen World*, 89.

16. Eddington, *Philosophy of Physical Science*, 5.

17. Margery Abbott, *A Certain Kind of Perfection* (Wallingford, PA: Pendle Hill Publications, 1997), 8.

18. Eddington, *Expanding Universe*, 39.

19. This approach to science can be found in Eddington's work on stellar movements, but the investigations into stellar structure show the relationship to his religious values much more clearly. See Paul, *Milky Way Galaxy and Statistical Cosmology*, for details of Eddington's investigations on stellar movements and statistical cosmology. Robert W. Smith also has forthcoming work on the same subject.

20. Auguste Comte, *Cours de Philosophie Positive*, in Gertrud Lenzer, *Auguste Comte and Positivism* (London: Transaction Publishers, 1998), 130–31.

21. A. S. Eddington, "Star," in *Encyclopaedia Britannica*, 11th ed., 788. I would like to thank an anonymous referee for alerting me to Eddington's authorship of this article.

22. Karl Hufbauer, *Exploring the Sun: Solar Science since Galileo* (Baltimore: Johns Hopkins University Press, 1991), 97–108; and Owen Gingerich, ed., The *General History of Astronomy*, vol. 4, part A, *Astrophysics and Twentieth-Century Astronomy to 1950* (Cambridge, Cambridge University Press, 1984). T. J. Cowling, "Development of the Theory of Stellar Structure," *Quarterly Journal of the Royal Astronomical Society* 7 (1966): 121–37, provides an overview of the major players and events in the development of these theories, but is written without any historical sensibility. Leon Mestel, "Arthur Stanley Eddington: Pioneer of Stellar Structure Theory," *Journal of Astronomical History and Heritage* 7, no. 2 (2004): 65–73, is also a useful survey of early stellar structure theory. Note that Hufbauer and David DeVorkin, *Henry Norris Russell: Dean of American Astronomers* (Princeton, NJ: Princeton University Press, 2000), both use manuscript sources that I explore here.

23. Warwick, *Masters of Theory*, 260. Paul Dirac, the pioneering quantum physicist, also attended Merchant Taylors'.

24. E. A. Milne, *Sir James Jeans* (Cambridge: Cambridge University Press, 1952), 1–7. Scholarship on Jeans has generally focused on his contributions to radiation theory: Rob Hudson, "James Jeans and Radiation Theory," *Studies in the History and Philosophy of Science* 20 (1989): 57–76; and G. Gorham, "Planck's Principle and Jeans's Conversion," *Studies in the History and Philosophy of Science* 22 (1991): 471–97. Warwick discusses Jeans's textbooks and his work on non-Euclidean geometry. Warwick, *Masters of Theory*, 381–88, 464–65. A short but useful biographical treatment is Robert W. Smith, "Sir James Hopwood Jeans 1877–1946," *Journal of the British Astronomical Association* 88 (1977): 8–17. E. A. Milne's biography, *Sir James Jeans* (Cambridge:

Cambridge University Press, 1952), is more comprehensive but is somewhat polemical; Milne had vigorous disagreements with both Eddington and Jeans on various technical matters, and much of the book is Milne justifying his positions.

25. James H. Jeans, "Review of R. Emden's *Gaskugeln*," *Astrophysical Journal* 30 (1909): 72–74.

26. James H. Jeans, *Problems of Cosmogony and Stellar Dynamics* (Cambridge: Cambridge University Press, 1919), vi.

27. Ralph Kenat, "Physical Interpretation: Eddington, Idealization, and the Origin of Stellar Structure Theory" (PhD diss., University of Maryland College Park, 1987, AAT 8725406).

28. Ibid., 353.

29. A. S. Eddington, "The Radiative Equilibrium of the Sun and Stars," *Monthly Notices of the Royal Astronomical Society* (hereafter *MNRAS*) 77 (1917): 16–35.

30. "Meeting of the Royal Astronomical Society, 1916 December 8," *Observatory* 509 (January 1917): 38. Many of the items in *The Observatory* do not have titles or attributed authors. When such are present, they will be cited.

31. Eddington, "Radiative," 17.

32. Ibid., 17. Kappa is the opacity of a given substance (the likelihood that a photon will be absorbed while passing through a unit amount of material). Epsilon is the energy production (usually in the form of photons) at any given point in the star. The product of the two can be thought of as the fraction of energy produced in the star that immediately interacts with the surrounding stellar material, either thermally or as radiation pressure.

33. Ibid., 22–23.

34. Ibid., 34–35.

35. "Meeting of the Royal Astronomical Society, 1916 December 8," 40–43.

36. A. S. Eddington, "Further Notes on the Radiative Equilibrium of the Stars," *MNRAS* 77 (1917): 596–97.

37. Ibid., 599.

38. Ibid., 611. An overview of astrophysicists' attempts to solve the solar energy problem can be found in Karl Hufbauer, "Astronomers Take Up the Stellar-Energy Problem, 1917–1920," *Historical Studies in the Physical Sciences* 11 (1981): 277–303; and Jeanne Bishop, "Golden Era of Theoretical Physics: The Black Box of Stellar Energy," *Griffith Observer* 42 (1978): 3–17.

39. Eddington, "Further Notes," 600–601.

40. Ibid., 604–5.

41. A. S. Eddington, "The Pulsation Theory of Cepheid Variables," *Observatory* 516 (August 1917): 290–93.

42. "Meeting of the Royal Astronomical Society, Friday, 1917 November 9," *Observatory* 520 (December 1917): 432.

43. J. H. Jeans, "The Equations of Radiative Transfer of Energy," *MNRAS* 78 (1917): 28.

44. Ibid., 37.

45. Douglas, *Life of Arthur Stanley Eddington*, 110. Douglas also comments that this anecdote shows Eddington's amazingly fast computational skills.

46. Milne, *Sir James Jeans*, 31.

47. Eddington to Russell, August 12, 1917, H. N. Russell Papers, Princeton University Library.

48. James Jeans, "Radiation of the Stars," *Nature* 99 (July 5, 1917): 365.

49. A. S. Eddington, "Radiation of the Stars," *Nature* 99 (August 2, 1917): 445.

50. "Meeting of the Royal Astronomical Society, Friday, 1918 January 11," *Observatory* 523 (February 1918): 78–79.

51. A. S. Eddington, "The Radiative Equilibrium of the Stars: A Reply to Mr. Jeans' Criticism," *MNRAS* 78 (December 1917): 113–15, 113.

52. "Meeting of the Royal Astronomical Society, Friday, 1918 March 8," *Observatory* 525 (April 1918): 159.

53. A. S. Eddington, "On the Pulsations of a Gaseous Star and the Problems of the Cepheid Variables. Part I," *MNRAS* 79 (1919): 2.

54. Ibid., 8, 12.

55. Ibid., 16.

56. Ibid., 20.

57. Eddington to Russell, August 12, 1917, H. N. Russell Papers, Princeton University Library.

58. A. S. Eddington, "On the Pulsations of a Gaseous Star and the Problems of the Cepheid Variables. Part II," *MNRAS* 79 (1919): 177.

59. Eddington, "Pulsations II," 179.

60. "Meeting of the Royal Astronomical Society, Friday, 1919 January 10," *Observatory* 536 (February 1919): 72.

61. J. H. Jeans, "The Problem of the Cepheid Variables," *Observatory* 536 (February 1919): 88–89.

62. J. H. Jeans, "The Internal Constitution and Radiation of Gaseous Stars," *MNRAS* 79 (1919): 319, 320, 321.

63. Ibid., 322.

64. "Meeting of the Royal Astronomical Society, Friday, 1919 March 14," *Observatory* 538 (April 1919): 149–50.

65. Hufbauer, *Exploring the Sun*, 99.

66. A. S. Eddington, "On the Relation between the Masses and Luminosities of the Stars," *MNRAS* 84 (1924): 308. Joann Eisberg's PhD thesis, "Eddington's Stellar Models and Early Twentieth-Century Astrophysics," explored some of the observational and institutional background to the mass-luminosity relation (History of Science Dept., Harvard University, 1991, AAT 9123018).

67. Eddington, "Relation between the Masses and Luminosities," 325.

68. Ibid., 328–29, 332.

69. "Meeting of the Royal Astronomical Society, Friday, 1925 January 9," *Observatory* 609 (February 1925): 30.

70. J. H. Jeans, "On the Masses, Luminosities, and Surface-Temperatures of the Stars," *MNRAS* 85 (1925): 196.

71. Ibid., 196–97.

72. Ibid., 208–9. Note that Jeans is agnostic about whether there is even an empirical relation between mass and luminosity.

73. "Meeting of the Royal Astronomical Society, Friday, 1925 January 9," 30.

74. Ibid., 31.

75. Ibid., 32.

76. A. S. Eddington, *The Internal Constitution of the Stars* (Cambridge: Cambridge University Press, 1926), 3–4.

77. *Observatory* 611 (April 1925), 93–94.

78. Eddington, *Internal Constitution of the Stars*, 21.

79. Ibid., 121–22.

80. Ibid., 25.

81. Ibid., 101–3.

82. Russell to Eddington, March 22, 1932, Henry Norris Russell Papers, Princeton University Library.

83. Eddington, *Internal Constitution of the Stars*, 164–70.

84. Ibid., 296–97. Eddington's flexibility on whether he is defending the "astronomer's" or "physicist's" approach reflects the delicate disciplinarity of astrophysics in this period. Despite his willingness to use either title, he is consistent in the value structure he is describing.

85. J. H. Jeans, "A Suggested Explanation of Radio-activity," *Nature* 70 (1904): 101.

86. H. N. Russell, "The Sources of Stellar Energy," *Publications of the Astronomical Society of the Pacific* 31 (July 1919): 209, 209–10.

87. Eddington, *Internal Constitution of the Stars*, 300.

88. A. S. Eddington, "The Source of Stellar Energy," *Supplement to Nature*, no. 2948 (May 1, 1926): 29.

89. Eddington, *Internal Constitution of the Stars*, 301.

90. Ibid., 315.

91. Ibid., 306.

92. Ibid., 393. See also n. 6 of this chapter.

93. The definitive work on Russell is DeVorkin, *Henry Norris Russell* (Princeton, NJ: Princeton University Press, 2000). Other contributors to the American branch of theoretical astrophysics are discussed in Donald Osterbrock, "Herman Zanstra, Donald H. Menzel, and the Zanstra Method of Nebular Astrophysics," *Journal of the History of Astronomy* 32 (2001): 93–108.

94. The first major historical study of this controversy has appeared recently in Miller's *Empire of the Stars*. Miller is quite correct to conclude that Eddington rejected Chandrasekhar's methods because they were incompatible with his own work on unifying quantum mechanics and relativity. However, his strong implication that Eddington must have also had racist motivations is unsupported and is extremely unlikely. Indeed, as a Quaker he was a committed internationalist and a vocal critic of national stereotypes (see chap. 3) and even spent several years as president of the National Peace Council, whose primary goal was to secure better treatment for Indians within the British Commonwealth. There is no need or reason to invoke racism as an explanation in this case, although it is clear that Chandrasekhar experienced professional obstacles from racism several other times in his career. There may have been personal tensions between the two men, but there is little evidence that Eddington treated Chandra any differently from the way he treated other rivals such as Jeans or E. A. Milne. Further, they appeared to remain on genial terms for the rest of their lives, and the senior astronomer supported Chandra for numerous posts and awards—unlikely acts for someone driven by racist hatred. Miller describes well the lifelong impact of this dispute on Chandrasekhar, but this can surely be accounted for simply by the emotional trauma of a sensitive young man having his ideas rejected by an intellectual hero. Chandra also seemed to have a lifelong tendency to imagine slights from his colleagues, even manifesting in his failure to feel that the Nobel Prize was adequate recognition for his work. Chandrasekhar's personal account of the dispute is presented in (among other places) his *Eddington, the Most Distinguished Astrophysicist of His Time* (Cambridge: Cambridge University Press, 1983). His version of the story has entered scientific folklore and is passed regularly from professors to graduate students in physics and astronomy (and appears again in Miller). Another detailed account can be found in Kameshwar Wali's biography *Chandra* (Chicago: University of Chicago Press, 1991), but Wali's defense of Chandra and assumptions of Eddington's maliciousness and incomprehensibility border on the ridiculous. See also Sakura Gooneratne's PhD dissertation, "The White Dwarf Affair: Chandrasekhar, Eddington and the

Limiting Mass" (University College London, University of London, 2005). Alex Ipe's PhD diss. analyzes this controversy from a philosophical-sociological viewpoint. "Plausibility and the Theoreticians' Regress: Constructing the Evolutionary Fate of Stars" (Dept. of Sociology and Anthropology, Carleton University, 2002; AAT NQ67048).

95. A. S. Eddington, "On 'Relativistic Degeneracy,'" *MNRAS* 95 (1935): 195.

96. Interestingly, Eddington's later work on unifying relativity and quantum mechanics was similar to the deductivism he criticized in Jeans and Chandrasekhar. See chap. 5 and the conclusion to this book on how these different approaches can be reconciled.

97. Philosophical treatments of Eddington's work include S. French, "Scribbling on the Blank Sheet: Eddington's Structuralist Conception of Objects," *Studies in History and Philosophy of Modern Physics* 34 (2003): 227–59; Ritchie, "Reflections on the Philosophy of Sir Arthur Eddington"; and Yolton, *Philosophy of Science of A. S. Eddington*. Other perspectives are Dingle's *Sources of Eddington's Philosophy*, which either ignores or misunderstands the role of experience in Eddington's religious and cultural context, and Witt-Hansen, *Exposition and Critique of the Conceptions of Eddington*, which makes an effort to show some of the religious roots of Eddington's philosophy but makes no connection to his science.

98. Eddington, *Philosophy of Physical Science*, vii.

99. Eddington, *New Pathways in Science*, 24.

100. MS of "The Stars and Their Movements," Perth, August 1, 1914, UEM, 1 29.

101. Eddington, *Science and the Unseen World*, 24.

102. Eddington, *New Pathways in Science*, 290.

103. Eddington, *Nature of the Physical World*, 352–53.

104. Eddington, *Expanding Universe*, 180.

105. Eddington, *Science and the Unseen World*, 77.

106. Eddington, *New Pathways in Science*, 279–80.

107. Ibid., 291.

108. A. S. Eddington, *Stars and Atoms* (London: Oxford University Press, 1927), 53.

109. A. S. Eddington, "The Internal Constitution of the Stars," *Observatory* 557 (October 1920): 357. This was Eddington's presidential address to Section A of the British Association for the Advancement of Science.

110. Eddington, *Philosophy of Physical Science*, 222.

111. Eddington, *Science and the Unseen World*, 87, 88.

112. Eddington, *Stars and Atoms*, 40–41.

113. Eddington, "Purpose of Science," 92.

114. "The Teaching of the Life Sciences," 22, Friends' Guild of Teachers Archive.

115. Relativity was a passion for Eddington for a variety of value-driven reasons, including its importance for internationalism, personal interest, experience, and scientific exploration. The theory appears multiple times in this book in these different contexts. See chaps. 3, 5, and 6.

116. A. S. Eddington, *Report on the Relativity Theory of Gravitation* (London: Fleetway Press, 1919), v.

117. A. S. Eddington, *Space, Time and Gravitation: An Outline of the General Relativity Theory* (Cambridge: Cambridge University Press, 1920), 183.

118. Eddington, *Nature of the Physical World*, 35.

119. Ibid., 60 (emphasis original), 58.

120. Ibid., 61–62.

121. J. W. N. Sullivan, *Contemporary Mind: Some Modern Answers* (London: Humphrey Toulmin, 1934), 126.

122. I do not intend to suggest that Eddington was unique among contemporary scientists in advocating pragmatic, fallibilist approaches. Rather, I have structured this chapter to emphasize the novelty of these approaches in the astronomy of the time, which was still dominated by those trained in classical mechanics. As I have shown, Eddington repeatedly emphasized that his methods would be accepted as quite typical by laboratory physicists.

123. For example, see S. S. Schweber, *QED and the Men Who Made It* (Princeton, NJ: Princeton University Press, 1994). The importance of pragmatism in American physics is also addressed by Schweber in "The Empiricist Temper Regnant: Theoretical Physics in the United States, 1920–1950," *Historical Studies in the Physical Sciences* 17 (1986): 55–98.

Chapter Three

1. S. Chandrasekhar, *Truth and Beauty* (Chicago: University of Chicago Press, 1987), 115.

2. W. H. McCrea, "Einstein: Relations with the RAS," *Quarterly Journal of the Royal Astronomical Society* 20, no. 3 (1979): 251–60. Similar attitudes can be found in Sir Gavin de Beer, *The Sciences Were Never at War* (London: Nelson, 1960). The belief that British scientists remained friendly with their German counterparts during the war is widespread. For example, see the otherwise excellent Stuart Wallace, *War and the Image of Germany: British Academics 1914–1918* (Edinburgh: John Donald, 1988), 195; and Robert Reid, *Tongues of Conscience: War and the Scientist's Dilemma* (London: Walker, 1969), 47–48. David Edgerton, "British Scientific Intellectuals and the Relations of Science, Technology, and War," in *National Military Establishments and the Advancement of Science and Technology*, ed. Paul Forman and José Sánchez-Ron, 1–35 (London: Kluwer, 1996), is a useful overview of British scientists' failure to adhere to their internationalist rhetoric and also of their ability to creatively remember events such that they fit with that rhetoric (e.g., McCrea).

3. Douglas, *Life of Arthur Stanley Eddington*, 90, 91.

4. *Observatory* 479 (October 1914): 397.

5. *Observatory* 480 (November 1914): 430.

6. Douglas, *Life of Arthur Stanley Eddington*, 92.

7. *Observatory* 485 (March 1915): 155.

8. *Observatory* 481 (December 1914): 466.

9. Pickering to Dyson, October 8, 1914, F. W. Dyson Papers, MS.RGO.8/104, Cambridge University Library, Royal Greenwich Observatory Archives. Hereafter, Dyson Papers.

10. Dyson to Pickering, October 20, 1914, Dyson Papers, RGO.8/104.

11. Strömgren to Dyson, November 6, 1914; Dyson to Pickering, November 9, 1914, Dyson Papers, RGO.8/104.

12. Dyson to Postmaster General, November 18, 1914; Post office to Dyson, November 27, 1914, Dyson Papers, RGO.8/104.

13. For example, on March 6, 1916, the Astronomer Royal received a curt telegraph from the Chief Censor's Office: "With reference to a telegram in code, addressed by you to Stromgren, Copenhagen, will you kindly let me know the exact meaning of the telegram, and whether you have authority to use private code to Copenhagen. The telegram was transmitted." Dyson Papers, RGO.8/104.

14. Plummer to Dyson, November 26, 1914; Plummer to Dyson, December 3, 1914, Dyson Papers, RGO.8/104.

15. R. T. A. Innes at Johannesburg to Dyson, December 10, 1914, Dyson Papers, RGO.8/104.

16. *Times* (London), August 4, 1914; *Daily News* (London), August 5, 1914.

17. Arthur Marwick, *The Deluge: British Society and the First World War* (London: Bodley Head, 1965), 49.

18. Royal Society Council Minutes, vol. 10, 1908–1914, November 1, 1917, 258. They replied that "it is contrary to the practice of the Society to express an opinion in its Corporate capacity upon questions involving political considerations." Thanks to the Royal Society for permission to quote from material in their archives.

19. *Times* (London), November 1, 1914. See also Albert Marrin, *The Last Crusade: The Church of England in the First World War* (Durham, NC: Duke University Press, 1974), 92–97, for more on the demonization of individual Germans.

20. A. Ruth Fry, *A Quaker Adventure* (London: Nisbet, 1926), xvii. For a more historical perspective on the origins and changing interpretation of the peace testimony, see Hugh Barbour, "The 'Lamb's War' and the Origins of the Quaker Peace Testimony," in *The Pacifist Impulse in Historical Perspective*, ed. Harvey Dyck, 145–58 (Toronto: University of Toronto Press, 1996).

21. Rufus M. Jones, *A Service of Love in Wartime* (New York: Macmillan, 1920), 3–4.

22. Jones, *Service*, 65–66.

23. British scientists' embrace of anti-German attitudes is also examined in Lawrence Badash, "British and American Views of the German Menace in World War I," *Notes and Records of the Royal Society of London* 34 (1979): 91–121. The wartime views of British academics in general is explored in great detail in Wallace, *War and the Image of Germany*.

24. The firsthand reports of the destruction of the library can be found in Martin Gilbert, *A History of the 20th Century*, vol. 1, *1900–1933* (New York: William Morrow, 1997), 345.

25. G. F. Nicolai, *The Biology of War* (New York: Century, 1919), ix.

26. Ibid., xiv.

27. Ibid., xvii–xix.

28. *Observatory* 500 (May 1916): 241. For the German perspective on the Manifesto, see J. L. Heilbron, *The Dilemmas of an Upright Man: Max Planck as Spokesman for German Science* (Berkeley: University of California Press, 1986).

29. A. J. P. Taylor, *English History 1914–1945* (Oxford: Oxford University Press, 1965), 19.

30. *Observatory* 483 (January 1915): 47.

31. *Observatory* 485 (March 1915): 143–45.

32. *Observatory* 492 (October 1915): 409.

33. *Observatory* 489 (July 1915): 306.

34. [H. H. Turner], "From an Oxford Note-Book," *Observatory* 488 (June 1915): 265.

35. *Observatory* 492 (October 1915): 413.

36. [H. H. Turner], "From an Oxford Note-Book," *Observatory* 500 (May 1916): 240.

37. Eddington to Hills, January 27, 1914, RAS MS Grove Hills 2/1 14, Royal Astronomical Society (RAS), London. Eddington wrote, "I have been considering the problem of the next Foreign Secretary....The conclusion I have come to is that it ought to be Turner. [He] is so intimately in touch with international organisations and foreign astronomers generally that he has an exceptional claim. Moreover he is tremendously loyal to the Society." All letters from Eddington are used with the kind permission of the Eddington Trustees.

38. J. H. Morgan, "German Atrocities: An Official Investigation"; quoted in [H. H. Turner], "From an Oxford Note-Book," *Observatory* 500 (May 1916): 241–42.

39. [H. H. Turner], "From an Oxford Note-Book," *Observatory* 500 (May 1916): 241.

40. A. S Eddington, "The Future of International Science," *Observatory* 501 (June 1916): 271.

41. By this point in the war, it appears that Eddington was the only British member still active in the Astronomische Gesellschaft. He received their notices through Copenhagen, and he

was apparently the source of the reports of the work of German astronomers that occasionally appeared in the *Observatory*.

42. Eddington, "Future of International Science," 271.

43. Ibid., 272.

44. A. S. Eddington, *Observatory* 502 (July 1916): 314.

45. Eddington to Larmor, June 7, 1916, Royal Society Library, Larmor MS 603-9 403. The letter continues: "I wonder if you have really thought what 'viewing our country from the Germans' standpoint' really involves. In the converse case it would mean that Germans should examine the evidence of their atrocities, casting aside any disposition to trust their rulers and their army, and rejecting their natural confidence that German accounts are truthful and ours untruthful—so that they might be expected to arrive at the same horrified judgment which moderate English opinion has arrived at. They must also be prepared to trust the judgment they form, although they may feel that it leads to an attitude harmful to the military success of their country. Surely anything like this on our side would be discouraged, and in the main rightly so. I presume that there are many accusations against us and our allies in the German papers, and believed in Germany, which are not reported over here. They are doubtless lies, nobody would believe them, and nobody is interested in them. We feel there is no need to seek evidence to refute them. But then we are clearly not attempting to view things from the Germans' standpoint. My argument was that similar considerations would prevent the Germans from adopting our standpoint—though of course we think that in doing so they are prevented from seeing the truth.. . .I would rather the Observatory had avoided this topic altogether but Turner forced my hand."

46. [H. H. Turner], "From an Oxford Note-Book," *Observatory* 502 (July 1916): 23–24.

47. A. S. Eddington, "Karl Schwarzschild," *Observatory* 503 (August 1916): 337–39.

48. A. S. Eddington to Annie Jump Cannon, July 3, 1915, Annie Jump Cannon Papers, Harvard University Archives, HUGFP 125.12 Box 2 HA1UPX.

49. Edwin Frost, *Observatory* 505 (October 1916): 435–36.

50. [H. H. Turner], "From an Oxford Note-Book," *Observatory* 506 (November 1916): 476–77.

51. *Report of the War Victims Relief Committee of the Society of Friends* (London: Spottiswoode, 1914–1919), 1:5, 3:44.

52. Jones, *Service*, 45–46.

53. Anna Braithwaite Thomas, *St. Stephen's House: Friends' Emergency Work in England, 1914–1920* (London: Emergency Committee for the Assistance of Germans, Austrians and Hungarians in Distress, 1935), 11, 12, 13.

54. Ibid., 14.

55. Ibid., 15.

56. Minutes of the Cambridge, Huntingdon, and Lynn Monthly Meeting, Cambridgeshire County Record Office, September 1, 1918. Letter from Olive Ludlam. R59/26/5/4.

57. A. Thomas, *St. Stephen's House*, 22.

58. Ibid., 1–17.

59. Ibid., 131–32.

60. Ibid., 43–44.

61. Jones, *Service*, 252.

62. A. Thomas, *St. Stephen's House*, 20.

63. Jones, *Service*, 251–52.

64. Cambridge, Huntingdon, and Lynn Monthly Meeting Minutes, November 1918, Cambridgeshire County Records, R59/26/5/5.

65. A. Marwick, *Deluge*, 48–49.

66. *Times* (London), November 24, 1918.

67. Frank Dilnot, *England after the War* (New York: Doubleday, 1920), 184.

68. John Forbes, *The Quaker Star under 7 Flags, 1917–27* (Philadelphia: University of Pennsylvania Press, 1962), 93. Margaret Macmillan's *Peacemakers: The Paris Conference of 1919 and Its Attempt to End War* (London: J. Murray, 2001), is an excellent resource for understanding the Conference.

69. Forbes, *Quaker Star*, 95–98, 91.

70. Edward Thomas, *Quaker Adventures* (London: Fleming H. Revell, 1935), 22–30, 1–22.

71. Fry, *Quaker Adventure*, 312, 315, 330–31.

72. Ibid., 331–32, 355.

73. De Sitter's motivations for and strategies in communicating relativity to Eddington were complex. For an insightful analysis, see Warwick, *Masters of Theory*, chap. 9.

74. A. S. Eddington, "Some Problems of Astronomy (XIX. Gravitation)," *Observatory* 484 (February 1915): 93–98.

75. Eddington saw religious significance in relativity as well. See chaps. 5 and 6.

76. Eddington to de Sitter, October 13, 1916; quoted in Pierre Kerszberg, *The Invented Universe* (Oxford: Oxford University Press, 1989), 99.

77. Einstein to de Sitter, January 23, 1917, in *The Collected Papers of Albert Einstein*, vol. 8A, ed. R. Schulmann et al. (Princeton, NJ: Princeton University Press, 1997), 383–84, document 290.

78. For an overview of the development and reception of general relativity, see D. Howard and J. Stachel, eds., *Einstein and the History of General Relativity* (Boston: Birkhauser, 1989), and Jürgen Renn, ed., *The Genesis of General Relativity* (Dordrecht: Springer, 2007).

79. [H. H. Turner], "From an Oxford Note-Book," *Observatory* 524 (March 1918): 128.

80. "America and German Science," letter to the editor, *Nature* 102 (February 6, 1919): 446–47.

81. Badash, "British and American Views," 96–98.

82. [H. H. Turner], "From an Oxford Note-Book," *Observatory* 524 (March 1918): 147.

83. Such rejections of German thought are found throughout the British intellectual community during the war. Wallace, *War and the Image of Germany*, documents this across the academic spectrum; Marrin, *Last Crusade*, 97–118, discusses the treatment of German theology in particular.

84. Quoted in [H. H. Turner], "From an Oxford Note-Book," *Observatory* 522 (January 1918): 71.

85. Strömgren to Wesley, July 22, 1917; and Royal Astronomical Society Council Minutes (hereafter RASCM), vol. 11, November 9, 1917.

86. RASCM, December 14, 1917. For example, see Eddington to Wesley, October 13, 1915, RAS Library.

87. RASCM, May 9, 1919.

88. Ibid., April 9, 1920, and May 14, 1920.

89. Ibid., February 11, 1921.

90. Eddington to Crommelin, December 28, 1919, RAS Library. The RAS archives have many such requests from former Central Powers countries.

91. Eddington to Wesley, January 30, 1921, RAS Library.

92. Bureau de Longitudes to Dyson, May 18, 1917. The RAS stopped using the service in 1921. RASCM, January 14, 1921, RGO.8/104.

93. Plummer to Dyson, July 26, 1917, Dyson Papers, RGO.8/104.

94. Unknown to Dyson, November 26, 1917, Dyson Papers, RGO.8/104.

95. Draft of Dyson to Strömgren, November 14, 1917, Dyson Papers, RGO.8/104. Copy with handwritten annotations also sent to H. H. Turner.

96. Royal Society Council Minutes, June 20, 1918.

97. For a useful overview of issues of nationalism in science, see Elisabeth Crawford, *Nationalism and Internationalism in Science, 1880–1939: Four Studies of the Nobel Population* (Cambridge: Cambridge University Press, 1992).

98. F. W. Dyson, "On the Opportunity Afforded by the Eclipse of 1919 May 29 of Verifying Einstein's Theory of Gravitation," *MNRAS* 77 (March 1917): 445. For more details on the expedition and its scientific background, see John Earman and Clark Glymour, "Relativity and Eclipses: The British Expeditions of 1919 and Their Predecessors," *Historical Studies in the Physical Sciences* 11 (1980): 49–85. See the conclusion of this chapter and n. 181 below for commentary on Earman and Glymour's analysis of Eddington's work on the expedition.

99. See Earman and Glymour, "Relativity and Eclipses," for more on observations of the deflection before and after 1919. Freundlich's many observations to test relativity's predictions are covered in Klaus Hentschel, *The Einstein Tower* (Stanford, CA: Stanford University Press, 1997).

100. Eddington to Weyl, August 18, 1920, ETH-Bibliothek Zürich, Hs 91:523. Used by permission of Dr. Michael Weyl and the ETH-Bibliothek Zürich.

101. F. W. Dyson to Annie Jump Cannon, February 20, 1915, Annie Jump Cannon Papers, Harvard University Archives, HUGFP 125.12 Box 2 HA1UPX.

102. "Young Papers," O.11.22, UEM.

103. "Gravitation and Light," *Observatory* 461(May 1913): 231.

104. Einstein's first calculation was based solely on the equivalence principle and yielded a value equal to an explicitly ballistic calculation, as would be expected from Newtonian theory. This is also the origin of the first "half" of the general relativistic deflection; the second "half" comes from the curvature of space near the sun relative to distant space. The possibility of Newtonian gravity affecting light had been investigated a handful of times (notably by Henry Cavendish) over the previous two centuries, but the idea was never taken particularly seriously. For more, see Clifford Will, "General Relativity at 75: How Right Was Einstein?" *Science* 250 (1990): 770–76.

105. Eddington was involved in the efforts to reinstate Russell after his dismissal for authoring a peace pamphlet. For more on Russell and the war, see G. H. Hardy, *Bertrand Russell and Trinity* (Cambridge: Cambridge University Press, 1970).

106. Eddington, *Report on the Relativity Theory of Gravitation*. The *Report* was read surprisingly widely, and *Nature*'s reviewer called it "the most remarkable publication during the war." *Nature* 103 (March 6, 1919): 2.

107. J. Evershed, "The Einstein Effect and the Eclipse of 1919 May 29," *MNRAS* 78 (1917): 269–70.

108. "RAS Meeting," *Observatory* 526 (May 1918): 215.

109. Ibid., 323. Lindemann later became Lord Cherwell and was Churchill's science advisor. See F. W. F. S. Bikenhead, *The Prof in Two Worlds* (London: Collins, 1961); and Robert Fox and Graeme Gooday, eds., *Physics in Oxford* (Oxford: Oxford University Press, 2005).

110. *Observatory* 517 (July 1917): 356.

111. Charles St. John, "The Principle of Generalized Relativity and the Displacement of Fraunhofer Lines toward the Red," *Astrophysical Journal* 96 (1917): 249.

112. For more on the culture of eclipse expeditions, see Alex Pang, "The Social Event of the Season: Solar Eclipse Expeditions and Victorian Culture," *Isis* 84 (1993): 252–77; and Alex Pang, *Empire and the Sun: Victorian Eclipse Expeditions* (Palo Alto, CA: Stanford University Press, 2002).

113. Minutes of the Joint Permanent Eclipse Committee (JPEC) Meetings, RAS Papers 54.1, November 10, 1917.

114. F. W. Dyson, "On the Opportunity Afforded by the Eclipse of 1919 May 29 of Verifying Einstein's Theory of Gravitation," *MNRAS* 77 (March 1917): 445.

115. JPEC minutes, June 14, 1918.

116. F. W. Dyson, A. S. Eddington, and C. Davidson, "A Determination of the Deflection of Light by the Sun's Gravitational Field, from Observations Made at the Total Eclipse of May 29, 1919," *Philosophical Transactions of the Royal Society of London*, ser. A, 220 (1920): 295.

117. JPEC minutes, November 8, 1918.

118. See S. Chandrasekhar, "Verifying the Theory of Relativity," *Notes and Records of the Royal Society of London* 30 (1976): 250. The events surrounding Eddington's conscription hearings are dealt with in detail in chap. 4.

119. JPEC minutes, February 14, 1919.

120. JPEC minutes, February 14, 1919, and January 10, 1919.

121. The origin of the "Newtonian" deflection in this troika of possibilities is unclear. In 1917 Lodge presented a calculation of the deflection assuming Newtonian gravitation held good, which yielded a value half of Einstein's. Earman and Glymour, in "Relativity and Eclipses," suggest that Eddington invented the half deflection to give the expedition the flavor of a crucial experiment; it is not clear whether they were aware of Lodge's calculation.

122. A. S. Eddington, "The Total Eclipse of 1919 May 29 and the Influence of Gravitation on Light," *Observatory* 537 (March 1919): 119–22. The Michelson-Morley experiments in the 1880s attempted to measure the velocity of the earth's movement with respect to the hypothetical ether. The experiments returned null results, causing some puzzlement at the time and spurring various reinterpretations of electromagnetic theory. The results were later interpreted as evidence for relativity. The historiography on the Michelson-Morley experiment is massive. A useful overview of the experiment can be found in L. Swenson, *The Ethereal Ether: A History of the Michelson-Morley Aether-Drift Experiment, 1880–1930* (Austin: University of Texas Press, 1972); and Gerald Holton, "Einstein, Michelson, and the 'Crucial Experiment,'" in *Thematic Origins of Scientific Thought*, 261–352 (Cambridge, MA: Harvard University Press, 1973), is a classic appraisal of the role the experiment actually played in the history of relativity.

123. Although the fighting in Europe ended in November 1918, all the combatants were technically still at war until the deliberations in Paris were finished and the treaties signed.

124. Eddington to Sarah Ann Eddington, two letters, March 11 and March 15, 1919, UEM.

125. Eddington to Sarah Ann Eddington, April 29, 1919, UEM.

126. Eddington to Winifred Eddington, May 5, 1919, UEM.

127. Eddington to Sarah Ann Eddington, June 21, 1919, UEM.

128. *Observatory* 540 (June 1919): 256.

129. Dyson, Eddington, and Davidson, "Determination of the Deflection of Light," 298.

130. One of these tickets survives in the Royal Greenwich Observatory (RGO) archives. Cambridge University Library, RGO Archives, MS.RGO.8.150.

131. *Observatory* 540 (June 1919): 256.

132. Dyson, Eddington, and Davidson, "Determination of the Deflection of Light," 309.

133. Ibid., 320–28.

134. Ibid., 309–12.

135. Ibid., 299–309.

136. Eddington to Dyson, October 3, 1919, Dyson Papers, RGO.8/150.

137. Eddington to Dyson, October 21, 1919, Dyson Papers, RGO.8/150. Letters such as this demonstrate that Eddington carefully analyzed his data and sought to avoid even a suggestion of favoritism. Thus, the letters are important evidence refuting the assumption that he manipulated the data in favor of Einstein. In any case, several other people were involved in the expedition and the data analysis, so a wide-scale conspiracy would be necessary (in addition to counterfeiting the photographic plates distributed to other astronomers). There are also criticisms that Eddington's error analysis was poor; most of these claims are confused in assuming that he averaged all three sets of results, which he explicitly did not do.

138. Such a joint meeting was the standard way to report results from expeditions performed under the auspices of the Joint Committee.

139. A. N. Whitehead, *Science and the Modern World* (New York: Macmillan, 1947), 15.

140. "Joint Eclipse Meeting of the Royal Society and the Royal Astronomical Society," *Observatory* 545 (November 1919): 389–98.

141. "Meeting of the Royal Astronomical Society, Friday, 1919 December 12," *Observatory* 547 (January 1920): 37–38.

142. Frank Dyson, "Relativity and the Eclipse Observations of May, 1919," *Nature* 106 (February 17, 1921): 786.

143. "Meeting of the Royal Astronomical Society, Friday, 1919 December 12," *Observatory* 547 (January 1920): 41.

144. JPEC minutes, November 14, 1919. The original plates have since been lost, but many of the copies still exist.

145. H. N. Russell, "Note on the Sobral Eclipse Photographs," *MNRAS* 81 (1920): 154–64.

146. For an interesting electromagnetic explanation, see H. A. Wilson, "An Electromagnetic Theory of Gravitation," *Physical Review* 17 (1921): 54.

147. "RAS Meeting," *Observatory* 547 (January 1920): 33–44, 56. See also folder 123 of the Dyson papers, which contains correspondence regarding the interpretation of the eclipse results. For more of the competing explanations for the deflection, see Donald Moyer, "Revolution in Science: The 1919 Eclipse Test of General Relativity," in *On the Path of Albert Einstein*, ed. A. Perlmutter and L. F. Scott, 55–101 (London: Plenum Press, 1979).

148. See chaps. 5 and 6 for more on these issues.

149. Charles St. John, "Displacement of Solar Lines and the Einstein Effect," *Observatory* 547 (January 1920): 158.

150. J. Larmor, letter to the editor, *Nature* 104 (December 25, 1919): 412.

151. Eddington had been worried about the lack of spectroscopic evidence for some time (see Eddington to W. S. Adams, January 28, 1918, in the Hale Papers, box 154, Huntington Library), and this may have helped shape his "law not theory" strategy.

152. F. Schlesinger to Dyson, February 16, 1920, Dyson Papers, RGO.8/123.

153. For an interesting analysis of the relative importance of the eclipse results and Mercury's motion. Stephen G. Brush, "Prediction and Theory Evaluation: The Case of Light Bending," *Science* 246 (1989): 1124–29. Philosophers interested in the importance of novelty in theory evaluation continue to use the 1919 expedition as a case study. See, for example, Deborah

G. Mayo, *Error and the Growth of Experimental Knowledge* (Chicago: University of Chicago Press, 1996), 133–37, 278–93; and Robert Hudson, "Novelty and the 1919 Eclipse Experiments," *Studies in the History and Philosophy of Modern Physics* 34 (2003): 107–29. In addition to issues of novelty, Mayo and Hudson disagree about Eddington and Dyson's interpretation of the eclipse results. See n. 181 below.

154. Dyson to unknown, April 19, 1923, Dyson Papers, RGO.8/123.

155. [H. H. Turner], "From an Oxford Note-Book," *Observatory* 560 (January 1921): 234–35. Later expeditions by the Lick and other observatories returned values for the deflection very close to Einstein's prediction, and the quality of the data was significantly better than the 1919 expedition. This was enough to convince most of the remaining skeptics of the reality of deflection. For more, see Jeffrey Crelinsten, "William Wallace Campbell and the 'Einstein Problem': An Observational Astronomer Confronts the Theory of Relativity," *Historical Studies in the Physical Sciences* 14 (1983): 1–91.

156. *Times* (London), November 7, 1919.

157. *Times* (London), November 8, 1919. See also *Nature* 104 (January 22, 1920): 541, which points out that Einstein was "called" to Berlin from his more pleasant posts in Zurich and Prague.

158. *Times* (London), November 28, 1919.

159. Eddington's strategy for presenting relativity to the media and the public was carefully planned and began well before the November Joint Meeting. This is dealt with in detail in Alistair Sponsel, "Constructing a 'Revolution in Science': The Campaign to Promote a Favorable Reception for the 1919 Solar Eclipse Experiments," *British Journal for the History of Science* 4, no. 35 (December 2002): 439–67. The author wishes to thank Mr. Sponsel for many valuable conversations and helpful criticism.

160. A. S. Eddington, "Einstein's Theory of Space and Time," *Contemporary Review* 116 (1919): 639.

161. "The Teaching of International Relations in Schools," Friends' Guild of Teachers Archive. For example, a statement of the Friends' Guild of Teachers (of which Eddington was a member and officer) reads: "The study of International Relationships should not be confined to any one subject; possibly even more marked in Science or Mathematics, which deal with universal truths, than in other subjects." 6.

162. Eddington to Einstein, December 1, 1919, in *The Collected Papers of Albert Einstein*, vol. 9, ed. Diana Kormos Buchwald et al. (Princeton, NJ: Princeton University Press, 2004), 262–63, document 186.

163. Ibid.

164. Eddington to Strömgren, November 1919. Quoted in Henrietta Hertzsprung-Kapteyn, "J. C. Kapteyn," *Space Science Reviews* 64 (1993): 81.

165. A. S. Eddington, "Das Strahlungsgleichgewicht der Sterne," *Zeitschrift fur Physik* 7 (1921): 531. It is not clear who translated the article.

166. MS of "Radiative Equilibrium of the Stars," August 1921, UEM.

167. RASCM, November 14, 1919.

168. RASCM, December 12, 1919.

169. RASCM, January 9, 1920.

170. R. J. Tayler, ed., *History of the RAS*, vol. 2 (London: Blackwell Scientific, 1987), 20.

171. Eddington to Einstein, January 21, 1920, in *Collected Papers of Albert Einstein*, vol. 9, 369–70, document 271.

172. Ludlam to Einstein, January 23, 1920, in *Collected Papers of Albert Einstein*, vol. 9, 378, document 279.

173. Eddington to Einstein, January 22, 1926, Albert Einstein Archives, ALS 9-287. Thanks to the Jewish National & University Library, The Hebrew University of Jerusalem, Israel.

174. Ronald Clark, *Einstein: The Life and Times* (New York: World Publishing, 1965), 238.

175. W. Benn, "Alien Influence in England," *Contemporary Review* 116 (1919): 637.

176. Clark, *Einstein*, 272–77.

177. Royal Society Council minutes, January 24, April 25, and October 3, 1918.

178. For more on the International Research Council, as well as the American attitude toward German science during and after the war, see Daniel Kevles, "Into Hostile Political Camps," *Isis* 62 (1971): 47–60; Badash, "British and American Views"; and B. Schroeder-Gudehus, "Challenge to Transnational Loyalties: International Organisations after the First World War," *Science Studies* 3 (1973): 98–118. For more on the International Astronomical Union (IAU) in general and particularly in this period, see Adriaan Blaauw, *History of the IAU: The Birth and First Half-Century of the International Astronomical Union* (Boston: Kluwer Academic, 1994). There is a great deal of correspondence in the RGO archives regarding the exclusion and admission of the Central Powers into the IAU, but a more in-depth analysis is outside the scope of this book.

179. *Times* (London), January 28, 1939, 12. This appeal was broadcast in German by the BBC before it was made available in English. The other signatories were Lord Willingdon, Lord Derby, Lord Dawson of Penn, Lord Horder, Lord Macmillan, Lord Stamp, Montagu Norman, H. A. L. Fisher, G. M. Trevelyan, Lord Eustace Percy, Michael Sadler, Dr. Vaughn Williams, Sir William Bragg, Sir Edwin Lutyens, Kenneth Clark, Lord Burghley, and John Masefield.

180. The National Peace Council was particularly interested in a harmonious relationship between the United States and the United Kingdom and in the liberation of India. Eddington contributed an introductory essay to the National Peace Council pamphlet *The British Commonwealth and the US in the Post-War World*, Peace Aims Pamphlet no. 10 (London: National Peace Council, 1942). The minutes from National Peace Council meetings while Eddington was an officer can be found in "National Peace Council Minutes 28 September 1938–2 May 1944" and "Executive Minutes 8 July 1940–8 June 1944." The records of the National Peace Council are held in the archives of the London School of Economics, LSE/NPC/1/5 and 2/6, respectively.

181. See Harry Collins and Trevor Pinch, *The Golem: What Everyone Should Know about Science* (Cambridge: Cambridge University Press, 1993), 27–56; and Earman and Glymour's conclusion to "Relativity and Eclipses" (their paper is the source for much of Collins and Pinch's treatment of the expedition). Both pairs argue that Eddington had no justification for assigning little weight to the Sobral astrographic, though their views of science are quite different. Both criticisms fail to deal with the observers' stated reasons for treating the data as they did and do not acknowledge that Eddington and his colleagues, as professional, trained astronomers, had extensive experience in determining the accuracy and self-consistency of a measurement. Further, the astronomical community, with similar levels of experience and skill, had ample opportunity to check and evaluate their work. (Mayo makes related arguments in *Error and the Growth of Experimental Knowledge*, 285–87.) Hudson, "Novelty and the 1919 Eclipse Experiments," accepts Earman and Glymour's accusation and levels two main criticisms: first, that the astronomical community seemed insufficiently critical of the results and, second, that there was insufficient diligence on the part of Eddington and Dyson in analyzing the Sobral astrographic results. The former claim ignores the ample evidence that many leading astronomers did reserve judgment until seeing copies of the photographic plates themselves (e.g., Schlesinger, quoted above). The latter is strange because Eddington and Dyson *did* perform a full error analysis on the plates, including a full correction for the aberration that resulted in Einstein-favorable numbers. The mythology of a conspiracy to support Einstein is widespread and influential. For examples of

such criticisms outside science studies, see C. W. F. Everitt, "Experimental Tests of General Relativity: Past, Present and Future" in *Physics and Contemporary Needs*, vol. 4, ed. Riazuddin, 529–55 (New York: Plenum, 1980); Ian McCausland, "Anomalies in the History of Relativity," *Journal of Scientific Exploration* 13 (1999): 271; and P. Marmet, *Einstein's Theory of Relativity versus Classical Mechanics* (Gloucester, Ont.: Newton Physics Books, 1997), 189–96.

Chapter Four

1. Taylor, *English History*, 4.

2. Ibid., 1.

3. The military disasters of the Boer War provided much stimulus for these discussions. For a representative contemporary work, see George F. Shee, *The Briton's First Duty: The Case for Conscription* (London: Grant Richards, 1901).

4. See, for instance, the anonymous pamphlet *Some Arguments for the Maintenance of Voluntary Service* (London: St. Clements Press, [1914?]).

5. John Stevenson, *British Society 1914–45* (London: A. Lane, 1984), 47.

6. A. Marwick, *Deluge*, 35.

7. A copy of this game is on display at the Scottish War Museum in Edinburgh Castle, Edinburgh.

8. Alan Wilkinson, *The Church of England and the First World War* (London: SPCK, 1978), 32–33; and Marrin, *Last Crusade*, 74–75, 180–86.

9. A. J. Hoover, *God, Germany, and Britain in the Great War: A Study in Clerical Nationalism* (London: Praeger, 1989), 21–25.

10. Stevenson, *British Society*, 52–53.

11. A. Marwick, *Deluge*, 35.

12. J. W. Graham to Richard Graham, October 19, 1914. Even before registration began, J. W. Graham warned his son that he heard that "a large firm in Reading are printing for the Government placards ordering the registration of all males between fifteen and thirty-five years of age." He said that did not necessarily mean conscription would be starting immediately, but it was time to start thinking about what to do in regard to it. J. W. Graham to Richard Graham, October 16, 1914, J. W. Graham Papers, Manchester University Library. Some people, when registering, gave a preemptory declaration that, if conscripted, they would not serve. John W. Graham, *Conscription and Conscience: A History, 1916–1919* (London: Allen & Unwin, 1922), 52.

13. A. Marwick, *Deluge*, 77.

14. Mass mailed letter from the War Office, October 1915, War Tracts vol. 12 (5). Friends House Library.

15. Vickers, the arms manufacturer, requested that a badge system for munitions workers be instituted so their workers would not come under pressure to enlist. For examples of this sort of social pressure, see Denis Hayes, *Conscription Conflict* (London: Sheppard Press, 1949), 165–66, 195.

16. Ibid., 194.

17. Central Tribunal Minutes, April 6, 1916, United Kingdom Public Record Office (hereafter PRO) MH 47 1.

18. See "Ministry of Health, Schedule," November 8, 1922, PRO MH 47 3, for details on the decision to set up the tribunals. The official report read that "machinery should be set up for determining questions as to the retention in civil life of men who had attested under the scheme."

19. A. Marwick, *Deluge*, 78.

20. Stevenson, *British Society*, 64.

21. J. W. Graham, *Conscription*, 53.

22. Friends Peace Committee, *Lest We Forget* (Leominster, UK: Orphans' Printing Press, [1917?]), 9.

23. Hayes, *Conscription Conflict*, 205.

24. A. Marwick, *Deluge*, 79.

25. No-Conscription Fellowship (NCF), *Repeal the Act*, pamphlet (London: NCF, n.d.). The NCF's newsletter, *The Tribunal*, provides excellent and comprehensive documentation of the experience of COs from the tribunals to the jailhouse.

26. For an overview of modern pacifism and the Quaker role in it, see Peter Brock, *Twentieth-Century Pacifism* (London: Van Nostrand Reinhold, 1970). There were other religious groups with principled objections to fighting. See Marrin, *Last Crusade*, 143–76.

27. J. W. Graham to Richard Graham, October 19, 1914; J. W. Graham to Richard Graham, June 5, 1915. Both in J. W. Graham Papers. Nearly a thousand Quaker men would end up fighting in the war. Kennedy, *British Quakerism*, 313.

28. J. W. Graham, *Conscription*, 75.

29. Yearly Meeting, quoted in Ibid., 162.

30. Ibid., 50.

31. Yearly Meeting, quoted in Ibid., 163.

32. Royal Society Council Minutes, November 5, 1914, 10:475.

33. Royal Society Council Minutes, January 21, 1915, 11:17.

34. A. Marwick, *Deluge*, 228. For more, see the *Final Report of the Committee on Commercial and Industrial Policy after the War*, Cd 9035 (London: HM Stationery Office, 1918).

35. Royal Society Council Minutes, April 22, 1915, 11:54. For an example, see the correspondence between the War Committee and chemical manufacturers trying to replicate German glues. Royal Society Library, CD 917-919, "Advice to Chemical Manufacturers."

36. Peter Alter, *The Reluctant Patron: Science and the State in Britain, 1850–1920* (Berg: Oxford, 1987), 61–62. Alter questions whether laissez-faire is really an appropriate description for British government at this time, but concedes that it was difficult to convince the state of the necessity of support. See also Ian Varcoe, "Scientists, Government and Organized Research in Great Britain 1914–1916: The Early History of the DSIR," *Minerva* 8 (1970): 198–99; D. S. L. Cardwell, "Science and World War I," *Proceedings of the Royal Society* ser. A, 342 (1975): 447–56; and R. MacLeod, "The Origins of the DSIR: Reflections on Ideas and Men, 1915–1916," *Public Administration* 48 (1970): 23–48.

37. Michael Pattison, "Scientists, Inventors and the Military in Britain, 1915–1919: The Munitions Inventions Department," *Social Studies of Science* 13 (1983): 523.

38. A. Marwick, *Deluge*, 229; and Pattison, "Scientists, Inventors and the Military," 527–28.

39. Philip Gummett, *Scientists in Whitehall* (Manchester: Manchester University Press, 1980), 24.

40. J. L. E. Dreyer, ed., *History of the Royal Astronomical Society*, vol. 1, *1820–1920* (London: Royal Astronomical Society, 1923), 30. These disputes over the role of scientists during wartime are also discussed in Michael Whitworth, *Einstein's Wake: Relativity, Metaphor, and Modernist Literature* (Oxford: Oxford University Press, 2001), 113–21.

41. *Cambridgeshire Times*, June 2, 1916.

42. Alter, *Reluctant Patron*, 96–97.

43. Ibid., 73.

44. A. Marwick, *Deluge*, 230.

45. For instance, see RASCM, vol. 11, January 12, 1917, and June 8, 1917. See Varcoe, "Scientists, Government and Organized Research," for a survey of the sort of projects undertaken by the Department for Scientific and Industrial Research.

46. Varcoe, "Scientists, Government and Organized Research," 202.

47. "Science and the Services," Royal Society Library, CD 816.

48. Ibid., CD 816a and 816b. They praise German geological and French acoustical knowledge.

49. Pattison, "Scientists, Inventors and the Military," 534.

50. Roy MacLeod, "The Chemists Go to War: The Mobilization of Civilian Chemists and the British War Effort, 1914–1918," *Annals of Science* 50 (1993): 455–81. See Alter, *Reluctant Patron*, 220–21, for more on scientists and other intellectuals (including H. G. Wells) complaining of the low status of scientists.

51. Cambridge Observatory Syndicate Minutes, 1896–1971, December 6, 1915, Cambridge University Archives, UA Obsy A1 iii. Quoted with the permission of the Syndics of Cambridge University Library.

52. Cambridge Observatory Syndicate Minutes, December 6, 1915.

53. Royal Society Council Minutes, February 22, 1917, 11:214–15. Eddington, as RAS secretary, helped at least one astronomer obtain an exemption, though it is not clear whether he did this in his official capacity as a representative of the RAS. See Eddington to Wesley, RAS Letters, September 2, 1916. Guy Hartcup comments that by January 1918 there were 312 chemists and biologists on the deferment list, but I have been unable to determine and verify his archival source. Hartcup, *The War of Invention: Scientific Developments 1914–1918* (London: Brassey's Defence Publishers, 1988), 23. It is certainly the case that individual scientists received individual exemptions throughout the war (e.g., Eddington).

54. For more on Moseley, see J. L. Heilbron, *H. G. J. Moseley: The Life and Letters of an English Physicist, 1887–1915* (Berkeley: University of California Press, 1974).

55. Philip Snowden, MP, *The Military Service Act Fully and Clearly Explained*, pamphlet (London: National Labour Press, n.d.), 5.

56. Central Tribunal Minutes, PRO MH 47 1.

57. Central Tribunal Minutes, April 6, 1916, PRO MH 47 1.

58. Central Tribunal Minutes, January 4, 1916; January 13, 1916, PRO MH 47 1.

59. "Report of the Central Tribunal Appointed under the Military Service Act 1916," (1919), 5–6, PRO MH 47.

60. J. W. Graham, *Conscription*, 65.

61. Snowden, *Military Service Act* (emphasis original).

62. No-Conscription Fellowship, *The NCF: A Summary of Its Activities*, pamphlet (London: NCF, 1917), 6.

63. Brock, *Twentieth-Century Pacifism*, 43.

64. W. H. W, *A Guide to the Conscientious Objector and the Tribunals*, pamphlet (n.p.: n.d.), 6 (Cambridge University Library "War Tracts" archive). Note that contemporary writers sometimes also used *CO* to stand for "commanding officer."

65. J. W. Graham, *Conscription*, 71.

66. Hayes, *Conscription Conflict*, 208.

67. Dr. Alfred Salter, "The Religion of a C.O."; reproduced in J. W. Graham, *Conscription*, 47, 49.

68. *Cambridge Daily News*, February 25, 1916.

69. *Cambridge Daily News*, March 1, 1916.

70. "Selected Case Studies," PRO MH 47 3.

71. "Employers Seeking Exemptions for Employees," PRO MH 47 120.

72. Co-ordinating Committee for Research into the Use of the University for War (CCRUUW), *Cambridge University and War*, pamphlet (Cambridge: CCRUUW, [1917?]), 16–17.

73. J. D. Symon, *Universities' Part in the War*, pamphlet (n.p., 1915). Cambridgeshire Public Library C 45.5, 727–28

74. *Cambridge Daily News*, March 6, 1916.

75. Symon, *Universities' Part in the War*, 735.

76. CCRUUW, *Cambridge University and War*, 4.

77. Quoted in Wallace, *War and the Image of Germany*, 74.

78. CCRUUW, *Cambridge University and War*, 5–7.

79. Ibid., 8–12; Symon, *Universities' Part in the War*, 736.

80. CCRUUW, Cambridge University and War, 15.

81. Ibid., 3.

82. Ibid., cover.

83. Ibid., 13.

84. A Cambridge B.A. [pseud.], "Cambridge and Conscription," *University Socialist*, Michaelmas 1913, 104–7.

85. *Cambridge Daily News*, March 4, 1916.

86. Hardy, *Bertrand Russell and Trinity*, 20–22.

87. Ibid., 3n.

88. Ibid., 9.

89. Cambridge, Huntingdon, and Lynn Monthly Meeting Minutes, November 11, 1915, Cambridgeshire County Record Office, R59/26/5/4.

90. Jesus Lane Preparative Meeting Minutes, January 1, 1916, February 14, 1917, Cambridgeshire County Record Office, R69/44.

91. Jesus Lane Preparative Meeting Minutes, March 12, 1916, R69/44. It so happened that the minutes for this meeting were recorded by Eddington. This was the case several times a year and suggests that he was present at most Monthly Meetings.

92. Letter from the editor of *The Christian World* to J. W. Graham, May 19, 1916.

93. Cambridge, Huntingdon, and Lynn Monthly Meeting Minutes, January 13, 1915, R59/26/5/4. Their position was made even more nebulous by their willingness to let unpopular groups (such as the University Socialists) use the Meeting House. See the Jesus Lane Preparative Meeting Minutes for November 5, 1916, R69/44. Soon after the Armistice, a mob broke into the Meeting House and dragged off the socialists meeting inside.

94. *Cambridge Daily News*, February 25, 1916.

95. *Cambridge Daily News*, March 6, 1916.

96. *Cambridge Magazine*, March 4, 1916, 359.

97. *Cambridge Daily News*, March 4, 1916.

98. *Cambridge Daily News*, March 16, 1916.

99. CCRUUW, Cambridge University and War, 19.

100. Though they were not, apparently, too young to fight. Friends Peace Committee, *Lest We Forget*, 13.

101. *Cambridge Daily News*, March 18, 1916.

102. Wormwood Scrubs was the destination of most imprisoned COs, to the point where the largest Quaker meeting in London during the war was held inside the prison. Kennedy, *British Quakerism*, 349.

103. After the war, the government expressed some misgivings (though not publicly) about rearresting men for what was essentially the same offense. See PRO MH 47 "Report of the Central Tribunal Appointed under the Military Service Act 1916." 1919, 20.

104. Minutes of the Cambridge, Huntingdon and Lynn Monthly Meeting. January 9 1918. R59/26/5/4.

105. William H. Marwick, *Ernest Bowman Ludlam* (London: Friends House Service Committee, 1960). Also see his entry in the *Dictionary of Quaker Biography* at Friends House, London. There were a few other similar cases of individuals claiming both conscientious objection and national importance. See PRO MH 47 66, "Case Papers of C.O.s," for the case of James Francis Dunworth, 26, a metallurgical researcher. He stated: "On the grounds of my Religious Conviction I object to any participation whatever in military service whether combatant or non-combatant. Also as a University trained Chemist, I should be more useful to the nation practising my profession than as a soldier." His application was refused. The tribunal decided that if he was doing work having to do with the production of shells, then his conscientious objection could not be genuine.

106. J. W. Graham, *Conscription*, 221.

107. CCRUUW, *Cambridge University and War*, 37–38.

108. Later regulations officially changed this, but the order was only followed in certain camps. It was generally not enforced. Those COs that accepted work of national importance were under civilian control.

109. Hayes, *Conscription Conflict*, 260.

110. No-Conscription Fellowship, *Two Years Hard Labour for Refusing to Disobey the Dictates of Conscience*, pamphlet (London: NCF, [1918?]), 70.

111. Central Tribunal Minutes, April 6, 1916, PRO MH 47 1.

112. For the details of Russell's case, see Hardy, *Bertrand Russell and Trinity*; and Wallace, *War and the Image of Germany*, chaps. 8 and 9.

113. CCRUUW, *Cambridge University and War*, 39.

114. Hardy, *Bertrand Russell and Trinity*, 42. See also 49–51 for a longer appeal.

115. CCRUUW, *Cambridge University and War*, 39.

116. J. W. Graham to Richard Graham, October 28, 1916, and November 1, 1916, J. W. Graham Papers.

117. Brock, *Twentieth-Century Pacifism*, 16. This caused significant worry in the scientific community, as it seemed their few hard-won exemptions would be revoked. See Royal Society Council Minutes, vol. 11, 1915–1920, April 25, 1918, 304. There is a letter to the Royal Society from Prof. William Bone warning that the new act would be bringing scientifically qualified men into the army. He was careful to say he did not want any special privileges, but on the contrary wanted to make sure that scientific men were used to their best capability. The Royal Society Council referred it to discussion.

118. Cambridge Observatory Syndicate Minutes, 1896–1971, March 12, 1918, UA Obsy A1 iii.

119. Eddington to Lodge, July 22, 1918, Lodge Papers, University College London Library Services, Special Collections, MS ADD 89.

120. *Cambridge Daily News*, June 14, 1918.

121. *Cambridge Daily News*, June 28, 1918.

122. S. Chandrasekhar, "The Richtmeyer Memorial Lecture—Some Historical Notes," *American Journal of Physics* 37, no. 6, 579–80. He tells a similar version of the story in "Verifying the Theory of Relativity," *Notes and Records of the Royal Society of London* 30 (1976): 249–60.

123. *Cambridge Daily News*, July 15, 1918.

124. "Report of the Central Tribunal Appointed under the Military Service Act 1916," 1919, 11–12, PRO MH 47. Participants with special or technical skills were supposed to be put to use in their fields. Apparently, this did not ever happen. See PRO MH 47 1, June 28, 1916.

125. *Cambridge Daily News*, June 28, 1918.

126. This action may have been unprecedented. I can find no evidence for it happening elsewhere, but the conscription records are in such poor condition that it is difficult to know for certain.

127. *Cambridge Daily News*, July 12, 1918.

128. CCRUUW, *Cambridge University and War*, 37.

129. Cambridge, Huntingdon and Lynn Monthly Meeting Minutes, March 12, 1919, R59/26/5/5.

130. J. W. Graham, *Conscription*, 311.

131. Ibid., 322.

132. Charles L. Mowat, *Britain between the Wars* (Chicago: University of Chicago Press, 1955), 6.

133. "Confidential: Conscientious Objectors," Central Tribunal Minutes, June 16, 1916, PRO MH 47 1.

134. J. W. Graham, *Conscription*, 326.

135. Middlesex Appeal Tribunal Minutes, document circular no. 293, PRO MH 47 5.

136. Middlesex Appeal Tribunal Minutes, November 21, 1918, memorandum by the chairman.

137. This is particularly relevant today with the growing strength of fundamentalism in America. The values of religious fundamentalists often conflict with the pluralistic values of the state, and it is difficult to resolve this conflict without simply ignoring the values of one party.

Chapter Five

1. Stanley Goldberg's *Understanding Relativity: Origin and Impact of a Scientific Revolution* (Boston: Birkhäuser, 1984) is a good first stop for understanding the reception of relativity, as is Jürgen Renn, ed., *The Genesis of General Relativity* (Dordrecht: Springer, 2007). For reception in different national contexts, see Thomas Glick, *The Comparative Reception of Relativity* (Boston: Reidel, 1987). An overview of interpretations of relativity can be found in Klaus Hentschel, *Interpretationen und Fehlinterpretationen der Speziellen und der Allgemeinen Relativitaetstheorie durch Zeitgenossen Albert Einsteins* (Basel: Birkhäuser, 1990). Loren Graham explores Eddington's and Fock's approaches to relativity in his *Between Science and Values*. A useful collection of essays on relativity is Peter Galison, David Kaiser, and Michael Gordin, eds., *Science and Society: The History of Modern Physical Science in the Twentieth Century*, vol. 1, *Making Special Relativity*, and vol. 2, *Making General Relativity* (New York: Routledge, 2001). Warwick's *Masters of Theory* provides important insight into relativity's dissemination and use. Ryckman, *Reign of Relativity*, looks at some of the early philosophical approaches to relativity.

2. W. de Sitter, "On Einstein's Theory of Relativity and Its Astronomical Consequences. First Paper," *MNRAS* 76:699–728; and "On Einstein's Theory of Relativity and Its Astronomical Consequences. Second Paper," *MNRAS* 77:155–84. For the physics community's reliance on de

Sitter and Eddington, see J. S. Ames, "Einstein's Law of Gravitation," *Science* 51, no. 1315 (March 12, 1920): 253–61, 253.

3. James Jeans, "Einstein's Theory of Gravitation," *Observatory* 509 (1917): 57. This exchange between Eddington and Jeans is also analyzed in Warwick, *Masters of Theory*.

4. Jeans, "Einstein's Theory of Gravitation," 58.

5. A. S. Eddington, "Einstein's Theory of Gravitation," *Observatory* 510 (1917): 93. The term *mystical* used here (by See, the original correspondent to the journal) has no relationship to Eddington's Quaker mysticism addressed in chap. 2. It simply refers to an obscure and mysterious subject.

6. Eddington, *Report on the Relativity Theory of Gravitation*.

7. See Warwick, *Masters of Theory*, chap. 9, esp. 467–69. Eddington's bibliography of his sources for writing the *Report on the Relativity Theory of Gravitation* is a marvelous resource for seeing who was working productively on general relativity at the time.

8. Eddington, *Report on the Relativity Theory of Gravitation*, v.

9. Ibid., 8–9, 17.

10. A. S. Eddington, "Gravitation and the Principle of Relativity I," *Nature*, March 1918, 17.

11. Eddington, *Report on the Relativity Theory of Gravitation*, 28.

12. Ibid., 29.

13. Ibid., 83.

14. For the history of early cosmology, see Kerszberg, *Invented Universe*; R. W. Smith, *Expanding Universe*; and Ole Molvig, "Cosmological Revolutions: Relativity, Astronomy, and the Shaping of a Modern Universe" (PhD diss., Princeton University, Department of History, 2004).

15. Eddington, *Report on the Relativity Theory of Gravitation*, 91. For other contemporary statements of Eddington on whether Einstein explained gravitation, see A. S. Eddington, "Gravitation and the Principle of Relativity II," *Nature*, March 1918, 34–36.

16. Review of *Report on the Relativity Theory of Gravitation*, by A. S. Eddington, *Nature* 103, no. 2575 (March 6, 1919): 2.

17. Ames, "Einstein's Law of Gravitation," 259.

18. For example, see "The Theory of Relativity," *American Mathematical Monthly* 28, no. 4 (April 1921): 175. For more on the importance of having a common understanding of relativity to which to refer, see Andrew Warwick, "Cambridge Mathematics and Cavendish Physics: Cunningham, Campbell and Einstein's Relativity, 1905–1911, Part I: The Uses of Theory," *Studies in History and Philosophy of Science* 23 (1992): 625–56; and "Cambridge Mathematics and Cavendish Physics: Cunningham, Campbell and Einstein's Relativity, 1905–1911, Part II: Comparing Traditions in Cambridge Physics," *Studies in History and Philosophy of Science* 24 (1993): 1–25.

19. J. H. Jeans et al. "Discussion on the Theory of Relativity," *Proceedings of the Royal Society of London*, ser. A, 97, no. 681 (March 1, 1920): 66.

20. Jeans et al., "Discussion," 73–79.

21. A. S. Eddington, "The Meaning of Matter and the Laws of Nature According to the Theory of Relativity," *Mind* 29, no. 114 (April 1920): 145–58.

22. Ibid., 145.

23. Ibid., 145–46, 147.

24. Ibid., 148–49.

25. Note that Eddington uses $G_{\mu\nu}$ for the contracted Riemann-Christoffel tensor, often expressed in modern notation as $R_{\mu\nu}$.

26. Eddington, "Meaning of Matter and the Laws of Nature," 150–51, 152.

27. Ibid., 153.

28. Ibid., 154, 155.

29. Ibid., 156–57, 157–58.

30. Ibid., 158.

31. The papers from the relativity symposium at the congress were published as A. S. Eddington et al., "The Philosophical Aspect of the Theory of Relativity," *Mind* 29, no. 116 (October 1920): 415–45.

32. Ibid., 415, 416.

33. Ibid., 418–20.

34. Ibid., 423–29.

35. Ibid., 430–36.

36. Ibid., 437–44.

37. R. F. Alfred Hoernle, "The Oxford Congress of Philosophy," *Philosophical Review* 30, no. 1 (January 1921): 60. See also W. P. Montague, "The Oxford Congress of Philosophy," *Journal of Philosophy* 18, no. 5 (March 3, 1921): 118–29.

38. Hoernle, "Oxford Congress of Philosophy," 61.

39. For a brief overview of British idealism and the state of British philosophy at the time, see David Bell, "Philosophy," in *The Twentieth Century Mind: History, Ideas, and Literature in Britain*, ed. C. B. Cox and A. E. Dyson, 174–24 (Oxford: Oxford University Press, 1972).

40. A casual bibliographer would be excused in thinking that Dingle's *Sources of Eddington's Philosophy* would be germane to this discussion. However, it makes no attempt at a historical examination of Eddington's influences and has little to offer.

41. Journal of A. S. Eddington, Trinity College, Wren Library, UEM, Add. MS 48, 6–11, 135–49. The lists of books read appear in several iterations, arranged by genre, author, and year read.

42. Henri Poincaré, *Science and Hypothesis* (New York: Dover, 1952), 140–50. This edition is identical to the first English printing. For more on Poincaré, particularly his conventionalism, see Peter Galison, *Einstein's Clocks, Poincaré's Maps* (New York: Norton, 2003). The differences between Einstein and Poincaré on relativity are discussed in Oliver Darrigol, "The Mystery of the Einstein-Poincaré Connection," *Isis* 95 (2004): 614–26.

43. Karl Pearson, *The Grammar of Science*, 2nd ed. (London: Adam & Charles Black, 1900), 9, 13–14.

44. Ibid., 40–44, 61–63, 91.

45. James Ward, *Naturalism and Agnosticism*, 2nd ed. (London: A. C. Black, 1915), 11–14.

46. Ibid., 12, 46–48, 42–44, 328–87.

47. On Eddington attending Whitehead's lectures, see Douglas, *Life of Arthur Stanley Eddington*, 10. A. N. Whitehead, *An Introduction to Mathematics* (London: Williams & Norgate, 1911).

48. Bertrand Russell, *The Problems of Philosophy* (London: Williams & Norgate, 1910), particularly chaps. 2 and 3, 26–57.

49. Bertrand Russell, *Scientific Method in Philosophy* (Oxford: Clarendon Press, 1914)

50. Eddington, *New Pathways in Science*, 305.

51. For general information on Russell, see Ray Monk, *Bertrand Russell: The Spirit of Solitude* (New York: Free Press, 1996); and *Bertrand Russell: The Ghost of Madness* (New York: Free Press, 2001).

52. Bertrand Russell, *Introduction to Mathematical Philosophy* (London: Allen & Unwin, 1919).

53. Ibid., 55. Relations and their structure are discussed in detail on 42–62.

54. Ibid., 59, 60.

55. William Kingdon Clifford, *The Common Sense of the Exact Sciences* (New York: Appleton, 1894). The relevant passages for geometry and force can be found on 216–26.

56. Ludwik Silberstein, *The Theory of Relativity* (London: Macmillan, 1914), and *The Theory of General Relativity and Gravitation* (New York: Van Nostrand, 1922). Eddington and Silberstein seemed to have a personal antipathy as well: Silberstein was the butt of an oft-mistold quip of Eddington's involving the "three people in the world who understand relativity." Silberstein's 1922 book makes the impressive achievement of making almost no reference to Eddington's work on relativity.

57. Warwick, "Cambridge Mathematics," part 1.

58. E. Cunningham, *Relativity and the Electron Theory* (London: Longmans, 1915), 49.

59. W. de Sitter, "On Einstein's Theory of Relativity and Its Astronomical Consequences. First Paper," *MNRAS* 76 (1916): 700.

60. For an overview of the influence of positivism, and particularly Mach, on Einstein's early work, see Gerald Holton, "Mach, Einstein, and the Search for Reality," in *Thematic Origins of Scientific Thought* (Cambridge, MA: Harvard University Press, 1973). Peter Galison investigates the technological culture surrounding Einstein's instrumentalism in *Einstein's Clocks, Poincaré's Maps*.

61. Eddington was marked on the course records at Manchester as having studied a term of German, but by the time he was corresponding with Einstein he apologized for needing to write in English. It is not clear how well he could read German; he apparently made his way through many papers in German in the process of writing his *Report on the Relativity Theory of Gravitation* (see its bibliography), but he may have had assistance. He clearly had language assistance at some point, as there were also papers in Italian in the bibliography, of which he had no knowledge. His biographer, A. V. Douglas, claimed he had significant skills in several languages, but I have found little evidence of this. Eddington explicitly says in a letter to Einstein in 1919 that he cannot correspond in German, suggesting that his skills in that language were rudimentary at best.

62. Moritz Schlick, *Space and Time in Contemporary Physics*, trans. Henry Brose (New York: Oxford University Press, 1920). It is again unclear what level of proficiency Eddington had in reading German, making it difficult to assess at what point he encountered certain texts.

63. Ibid., 1, 22–27, 82–84.

64. Erwin Freundlich, *The Foundations of Einstein's Theory of Gravitation*, trans. Henry Brose (Cambridge: Cambridge University Press, 1920), viii, xvi. Turner wrote the preface and mentions that he tried to get Eddington to write it. Apparently Eddington insisted on Turner authoring the preface, and it seems likely that this was a strategy to get Turner (a vocal anti-German during the war) publicly collaborating with a German. For more on Freundlich, see Hentschel, *Einstein Tower*. Eddington's support for the translation of the book is also discussed in Whitworth, *Einstein's Wake*, 46.

65. Freundlich, *Foundations of Einstein's Theory*, 8–9, viii.

66. Eddington to Weyl, July 10, 1921, Hs 91:525 ETH Library, Zurich.

67. A. S. Eddington, "A Generalisation of Weyl's Theory of the Electromagnetic and Gravitational Fields," *Proceedings of the Royal Society of London*, ser. A, 99, no. 697 (May 2, 1921): 104–22.

68. Reviewers often commented on the eager audience. For example, see *Times Literary Supplement*, July 29, 1920, 481.

69. Eddington, *Space, Time and Gravitation*, 1–13.

70. Ibid., 1–3.

71. Ibid., 10–11, 9–10.

72. Ibid., 24.

73. Ibid., 29.

74. Ibid., 43.

75. Ibid., 87, 89–92.

76. Ibid., 183.

77. Ibid., 184.

78. Ibid., 197.

79. Ibid., 198n.

80. Ibid., 200–201.

81. Review of *Space, Time and Gravitation*, by A. S. Eddington, *Times Literary Supplement*, July 29, 1920, 481.

82. Review of *Space, Time and Gravitation*, by A. S. Eddington, *Nature* 106, no. 2678 (February 24, 1921): 822–23.

83. H. R. Smart, review of *Space, Time and Gravitation*, by A. S. Eddington, *Philosophical Review* 31, no. 4 (July 1922): 414, 415.

84. William McDougall, "Purposive Striving as a Fundamental Category of Psychology," *Scientific Monthly* 19, no. 3 (September 1924): 305–12, quotes from 311.

85. S. Alexander, *Space, Time, and Deity*, 2 vols., 2nd ed. (London: Macmillan, 1927), xli, 58–87.

86. Viscount Haldane, *The Reign of Relativity* (London: John Murray, 1921), vii, 3–7.

87. Ibid., 97, 59–66, 117, 108–9, 124.

88. Ibid., 236, 389–90.

89. For example, see Viscount Haldane, *The Philosophy of Humanism* (New Haven, CT: Yale University Press, 1922), and his 1902–1904 Gifford Lectures, *The Pathway to Reality* (New York: Dutton, 1905). J. E. Creighton applauds Haldane's service to religion in his review of *The Reign of Relativity*, by Viscount Haldane, *Philosophical Review* 31, no. 3 (May 1922): 288–93.

90. Oliver Lodge, "Geometrisation of Physics, and Its Supposed Basis on the Michelson-Morley Experiment," *Nature* 106, no. 2677 (February 17, 1921): 796. The last sentence is clearly a reference to Eddington's theory of matter, which suggests that Lodge was accepting Eddington's interpretation.

91. H. Wildon Carr, *The General Principle of Relativity in Its Philosophical and Historical Aspect* (London: Macmillan, 1920). For a critique see Edward Kasner's review in *Journal of Philosophy* 19, no. 8 (April 1922): 220–22.

92. J. E. Turner, "Dr. Wildon Carr and Lord Haldane on Scientific Relativity," *Mind* 31, no. 121 (January 1922): 40–52.

93. J. E. Turner, "Some Philosophic Aspects of Scientific Relativity," *Journal of Philosophy* 18, no. 8 (April 14, 1921): 213.

94. J. E. Turner, "Dr. Wildon Carr and Lord Haldane," 42.

95. H. Wildon Carr, "Einstein's Theory and Philosophy," *Mind* 31, no. 122 (April 1922): 170. Turner and Carr continued their exchange for some time.

96. "Idealistic Interpretation of Einstein's Theory," *Proceedings of the Aristotelian Society* 22 (1922): 123–38. An overview of the meeting can be found in Thomas Greenwood, "Einstein and Idealism," *Mind* 31, no. 122 (April 1922): 205–7.

97. "Idealistic Interpretation of Einstein's Theory," 123.

98. Ibid., 128–30.

99. Ibid., 134–38.

100. Ibid., 133.

101. A. N. Whitehead, "The Philosophical Aspects of the Theory of Relativity," *Proceedings of the Aristotelian Society* 22 (1922): 215–23.

102. Whitehead's philosophy is far too difficult and expansive to cover here, even cursorily. For an overview of Whitehead's work, see Victor Lowe, *Alfred North Whitehead: The Man and His Work* (Baltimore: Johns Hopkins University Press, 1985), particularly chaps. 5–8. Good introductions to process philosophy and its influences are Nicholas Rescher, *Process Philosophy: A Survey of Basic Issues* (Pittsburgh: University of Pittsburgh Press, 2000); and Thomas E. Hosinski, *Stubborn Fact and Creative Advance* (Lanham, MD: Rowman & Littlefield, 1993). Thanks to Dr. Donna Bowman for her help in sorting through the massive Whitehead literature.

103. For instance, see Oliver Reiser, "A Monism of Creative Behavior," *Journal of Philosophy* 21, no. 18 (August 1924): 477–91.

104. H. Wildon Carr, "Metaphysics and Materialism," *Nature* 108 (October 20, 1921): 247–48.

105. Hugh Elliot, "Relativity and Materialism," *Nature* 108 (December 1, 1921): 432.

106. A. S. Eddington, *Mathematical Theory of Relativity* (Cambridge: Cambridge University Press, 1923), v.

107. Ibid., 1, 2–3, 4.

108. Ibid., 5, 5–6.

109. Ibid., 49.

110. Ibid., 105.

111. Ibid., 119. In a later interview, Eddington spoke of the personal pleasure this exercise brought him: "I am chiefly interested, now, in watching the scientific picture of the world unfold. As science advances it brings a new beauty and harmony into the scientific picture of the world and that, I find now, is the chief reward of scientific work." Sullivan, *Contemporary Mind*, 124.

112. *Mathematical Theory of Relativity*, 196–202, quote on 198.

113. Eddington, "Generalisation of Weyl's Theory."

114. *Mathematical Theory of Relativity*, 41.

115. Ibid., 106.

116. Ibid., 240.

117. Ibid., 38, 147, 238.

118. Philip Franklin, review of *Mathematical Theory of Relativity*, by A. S. Eddington, *American Mathematical Monthly* 31, no. 9 (November 1924): 444–47.

119. Review of *Mathematical Theory of Relativity*, by A. S. Eddington, *Philosophical Magazine* 6, no. :45 (1923): 1189–90.

120. Whitworth, *Einstein's Wake*, 32.

121. A. S. Eddington, "The Domain of Physical Science," in *Science, Religion and Reality*, ed. Joseph Needham, 2nd ed, 193–224 (New York: George Braziller, 1955). There is some speculation that Eddington may have chosen his title to echo or correct E. W. Hobson's *The Domain of Natural Science* (New York: Macmillan, 1923), but there seems to be little connection. It seems perfectly plausible that Eddington developed the title of his own accord. Unfortunately, I have been unable to document how Eddington came to either offer or be asked to write this chapter. It would have been useful to see how he decided to write an explicit statement on religion and science, but the record does not seem to have survived.

122. Eddington, "Domain of Physical Science," 193–94.

123. Ibid., 197, 198–200.

124. Ibid., 201–2. For examples of the sort of materialist, naturalistic explanations for religious belief Eddington wanted to combat, see Henry Maudsley, *Body and Mind* (London: Macmillan, 1870); and *Natural Causes and Supernatural Seemings* (London: K. Paul, Trench, 1886).

125. Ibid., 199, 202.

126. Ibid., 204, 206.

127. Ibid., 208–9, 210–11.

128. Ibid., 215.

129. Ibid., 216.

130. Ibid., 219–20.

131. Ibid., 220.

132. Ibid., 221.

133. Ibid., 222.

134. Sullivan, *Contemporary Mind*, 124.

135. S. W. Sykes, "Theology," in *Twentieth Century Mind*, ed. C. B. Cox and A. E. Dyson, 146–70 (Oxford: Oxford University Press, 1972). A series of studies of the history of liberal religion can be found in Peter Kaufman and Spencer Lavan, *Alone Together: Studies in the History of Liberal Religion* (Boston: Beacon, 1979).

136. For example, there was a widespread (though incorrect) belief that the 1921 Modern Churchmen's Union Conference had denied the divinity of Jesus.

137. Frank C. O. Beaman, "Church and Science," *Nineteenth Century* 89 (1921): 467.

138. Sykes, "Theology," 162. See Peter Bowler, *Reconciling Science and Religion* (Chicago: University of Chicago Press, 2001), for a broad and detailed overview of these issues in early twentieth-century Britain.

139. J. W. N. Sullivan, review of *Science, Religion, and Reality*, ed. Joseph Needham, *Times Literary Supplement*, November 12, 1925, 748.

140. For example, see H. F. Wyatt, "Science and the Moral Law," *Nineteenth Century* 91 (1922): 858. The Gifford Lectures are also a rich source for liberal thinking on science and religion.

141. "The Unity of Science and Religion," *Nature* 106, no. 2653 (September 2, 1920): 1–2.

142. *Times* (London), November 15, 1919; November 25, 1919.

143. *Church Times*, September 24, 1920, 291.

144. W. A. Cobbe, "A More Excellent Way," *Nineteenth Century* 92 (July–Dec 1922): 1015.

145. A Wyatt Tilby, "The Reconstruction of Religion," *Nineteenth Century* 91 (January–June 1922): 462. Hastings, *History of English Christianity*, is a good survey of the general transitions in British religious thought.

146. Robert Graves and Alan Hodge, *The Long Week-end, a Social History of Great Britain, 1918–1939* (London: Faber & Faber, 1940), 94.

147. Graves and Hodge, *Long Week-end*, 209.

148. John Galsworthy, preface to *A Modern Comedy* (New York: Scribner's Sons, 1929). Similarly, Alfred F. Havinghurst, *Britain in Transition: The Twentieth Century*, 4th ed. (Chicago: University of Chicago Press, 1985), points to Einstein and religious thinkers like James Frazer as the beginning of the intellectual unrest of the 1920s, although most of the examples given are post-1926.

149. The late appearance of antiscience feeling in 1920s Britain is also noted in Paul Forman, "The Reception of an Acausal Quantum Mechanics in Germany and Britain," in *The Reception*

of Unconventional Science, ed. Seymour H. Mauskopf, 11–50 (Boulder, CO: AAAS, 1979). Forman particularly discusses the British case on pp. 23–38.

Chapter Six

1. On popular science in general, see Stephen Hilgartner, "The Dominant View of Popularization: Conceptual Problems, Political Uses," *Social Studies of Science* 20 (1990): 519–39. Many scientists today seem to continue to hold to the "diffusion model" of popularization, in which expert knowledge is simply distributed to a passive public, and for unsurprising reasons. As Hilgartner points out, it provides great cultural authority for them and often places them above criticism for their popularizations. Also see Peter Broks, *Understanding Popular Science* (Philadelphia: Open University Press, 2006); Roger Cooter and Stephen Pumfrey, "Separate Spheres and Public Places: Reflections on the History of Science Popularization and Science in Popular Culture," *History of Science* 32 (1994): 237–67; and Steven Shapin, "Science and the Public," in *Companion to the History of Modern Science*, ed. R. C. Olby et al., 990–1007 (London: Routledge, 1990).

2. Information on sales of Eddington's popular books can be found in Michael Whitworth, "The Clothbound Universe: Popular Physics Books, 1919–39," *Publishing History* 40 (1996): 53–82; and Whitworth, *Einstein's Wake*. Eddington recorded sales of his books in his journal in the UEM. The religious context of Eddington and his fellow science popularizers is discussed in great detail in Bowler, *Reconciling Science and Religion*.

3. Dorothy L. Sayers, "Absolutely Elsewhere," in *The Complete Stories* (New York: Harper-Collins, 1972), 392. Note that the title of the story refers to Eddington's discussions of relativistic space-time diagrams.

4. The Gifford Lectures invite high-profile thinkers (theologians, scientists, philosophers, historians, etc.) to discuss the topic of natural theology in a series of addresses. Established in Lord Gifford's will in 1885, the lectures have been an important pulpit for issues regarding science and religion for over a century. See Stanley L. Jaki, *Lord Gifford and His Lectures: A Centenary Retrospect* (Edinburgh: Scottish Academic Press, 1986).

5. "Death of Eddington," *Time*, December 4, 1944.

6. See Marcel LaFollette, *Making Science Our Own: Public Images of Science, 1910–1955* (Chicago: University of Chicago Press, 1990), 66–67, 149–51.

7. Shapin, "Science and the Public," 996–99.

8. The details of this movement are chronicled in Gary Werskey, *The Visible College* (London: A. Lane, 1978); and William McGucken, *Scientists, Society, and the State: The Social Relations of Science Movement in Great Britain, 1931–1947* (Columbus: Ohio State University Press, 1984). For more on social relations of science with respect to popularization, see Doug Russell, "Popularization and the Challenge to Science-Centrism in the 1930s," in *The Literature of Science: Perspectives on Popular Science Writing*, ed. Murdo William McRae, 37–53 (Athens: University of Georgia Press, 1993); and Jane Gregory and Steve Miller, *Science in Public: Communication, Culture, and Credibility* (New York: Plenum, 1998).

9. James Andrews, *Science for the Masses: The Bolshevik State, Public Science, and the Popular Imagination in Soviet Russia, 1917–1934* (College Station: Texas A&M University Press, 2003).

10. Eddington, *Nature of the Physical World*.

11. Douglas, *Life of Arthur Stanley Eddington*, 149; and "Books: Sales and Receipts," UEM. See Whitworth, *Einstein's Wake* and "Clothbound Universe"; Gillian Beer, "Eddington and the Idiom of Modernism," in *Science, Reason, and Rhetoric*, ed. Henry Krips et al., 295–315 (Pittsburgh:

University of Pittsburgh Press, 1995); and Kate Price, "Poetry and Physics: Interchange within the Writings of Arthur Eddington, I. A. Richards, William Empson and Laura Riding" (PhD diss., Cambridge University, 2003), for the influence of Eddington's writings.

12. Eddington, *Nature of the Physical World*, vii, viii.

13. Ibid., vi. Loren Graham has examined the manuscript of the revision and found that Eddington struggled with the shift from spoken to printed word. Loren R. Graham, *Between Science and Values* (New York: Columbia University Press, 1981), 80.

14. Eddington, *Nature of the Physical World*, xvii, 60, 143.

15. Ibid., 206–10, 248, 243.

16. Ibid., 254.

17. Ibid., 275.

18. Ibid., 276, 278. He quotes Bertrand Russell, *The Analysis of Matter* (London: Kegan Paul, 1927), 320.

19. Eddington, *Nature of the Physical World*, 220–23, 225.

20. Ibid., 228, 293–94. Eddington's position on determinism is also discussed in Forman, "Reception of an Acausal Quantum Mechanics." Forman pays insufficient attention to Eddington's religious outlook, though, and the resulting picture is incomplete.

21. Eddington, *Nature of the Physical World*, 295, 310, 313.

22. Ibid., xviii–xix.

23. Ibid., 287.

24. Ibid., 324.

25. Ibid., 325.

26. For general information on the strike and its effect, see Keith Laybourn, *The General Strike of 1926* (Manchester: Manchester University Press, 1993).

27. Eddington, *Nature of the Physical World*, 333.

28. Ibid., 281–82, 353, 336.

29. Ibid., 337.

30. Ibid., 350. This statement has been misquoted widely for seventy years. For whatever reason, only the first sentence is usually quoted, making it seem Eddington is espousing the position he is actually mocking. Alan Batten discusses this in "What Eddington Did Not Say," *Isis* 94, no. 4 (December 2003): 656–59.

31. Dean W. R. Inge, *God and the Astronomers* (London: Longmans, Green, 1933), v–vi.

32. Review of *The Will to Be Free*, by Howard V. Knox, *Hibbert Journal* 27 (October 1928–July 1929): 560–61. A review of a biography of the former Prime Minister Arthur Balfour clearly relied on Eddington's views of physical causality. H. M. Stannard, review of *A. J. Balfour: The Earlier Phase*, by Blanche Dugdale, *Times Literary Supplement*, September 19, 1936, 733–34; referenced in L. Susan Stebbing, *Philosophy and the Physicists* (London: Methuen, 1937), 141–42.

33. For example, see F. S. Marvin, review of *The Sciences and Philosophy*, by Viscount Haldane, *Hibbert Journal* 28 (October 1929–July 1930): 170–74. Unsurprisingly, Eddington himself was pleased to see echoes of his own ideas in others' Gifford Lectures: A. S. Eddington, "The Bishop of Birmingham's Gifford Lectures," *Cambridge Review*, June 8, 1933, 488–89. Eddington's *Nature of the Physical World* was reviewed in a variety of journals and newspapers. Whitworth, *Einstein's Wake*, looks closely at these reviews, particularly those in literary publications.

34. Review of *Nature of the Physical World*, by A. S. Eddington, *Times* (London), November 30, 1928, 22.

35. F. S. Marvin, "Review of *Nature of the Physical World*," *Hibbert Journal* 27 (October 1928–July 1929): 564–67.

36. H. W. B. Joseph, "Prof. Eddington on 'The *Nature of the Physical World*,'" *Hibbert Journal* 27 (October 1928–July 1929): 406–23.

37. "The Charge on an Electron: A New Theory," *Times* (London), January 18, 1929, 12, describes his use of a pair of golf balls to illustrate the quantum properties of electrons.

38. Douglas, *Life of Arthur Stanley Eddington*, 107–9, 113–14.

39. "Sir A. Eddington Honoured: Freedom of Native Town," *Times* (London), September 26, 1930, 10. At the ceremony, Eddington commented that he was pleased that Kendal "recognized scientific work as service of public importance," perhaps a reference to his experiences during the Great War over that very issue. Note also that Eddington was one of the youngest recipients of the Order of Merit. Obituary of A. S. Eddington, *News Chronicle* (London), November 23, 1944, 2.

40. Review of *Expanding Universe*, by A. S. Eddington, *Times* (London), February 24, 1933, 8.

41. Whitworth, *Einstein's Wake*, 53.

42. Oliver Lodge, "Eddington's Philosophy," *Nineteenth Century* 105 (1929), 364. Lodge continued to express reservations about relativity, especially the equivalence of all frames of reference.

43. Lodge to Eddington, January 25, 1929, Lodge Papers, University College London Archives, MS ADD 89.

44. Rev. Oliver C. Quick, "Mind, Matter and Professor Eddington," *Nineteenth Century* 105 (1929), 669. Also see C. E. M. Joad, *Philosophical Aspects of Modern Science* (New York: Macmillan, 1932), a straightforward philosophical critique of some of Eddington's ideas.

45. Quick, "Mind, Matter and Professor Eddington," 673.

46. Milne, *Sir James Jeans*, x.

47. Ibid., xi.

48. James H. Jeans, *The Mysterious Universe* (Cambridge: Cambridge University Press, 1930), 140. See also Bowler, *Reconciling Science and Religion*, 110–11, for Jeans and his reception.

49. Jeans, *Mysterious Universe*, 144.

50. Whitworth, *Einstein's Wake*, 52.

51. Ibid., 46.

52. Bowler, *Reconciling Science and Religion*, part 1. Eddington, Jeans, and Lodge are all discussed in D. Wilson, "On Removing 'Science' and 'Religion.'"

53. *Indeterminism, Formalism, and Value*, Aristotelian Society Supplementary Volume X (London: Harrison & Sons, 1931). See also Eddington's address to the British Institute of Philosophy, published as "Physics and Philosophy," *Philosophy* 8 (January 1933), 30–43, and his address to the Mathematical Association, "The Decline of Determinism," *Mathematical Gazette* 218 (May 1932), 66–80. Forman, " Reception of an Acausal Quantum Mechanics," 36, claims determinism had not been a significant issue for British thought before quantum mechanics, a somewhat odd claim that seems incompatible with British physics from Newton to Maxwell.

54. *Indeterminism, Formalism, and Value*, 161–66.

55. Ibid., 173–74.

56. Ibid., 180–81.

57. Ibid., 178–79.

58. Sullivan, *Contemporary Mind*, 125.

59. Herbert Samuel, "Cause, Effect and Professor Eddington," *Nineteenth Century* 113 (1933): 469.

60. Ibid., 471.

61. Ibid., 472–78.

62. A. S. Eddington, "Physics and Determinism," *Nineteenth Century* 113 (1933): 715–18, 719–20.

63. Ibid., 722–23.

64. Eddington, *Science and the Unseen World*. A shorter version was also published as a pamphlet by Friends House.

65. Ibid., 39.

66. Ibid., 25–26.

67. Ibid., 23, 89.

68. Ibid., 28–29.

69. Ibid., 50–51.

70. Ibid., 63.

71. Ibid., 66–67, 67.

72. Review of *Science and the Unseen World*, by A. S. Eddington, *Hibbert Journal* 28 (October 1929–July 1930): 168–69.

73. James Secord discusses this technique in *Victorian Sensation: The Extraordinary Publication, Reception, and Secret Authorship of Vestiges of the Natural History of Creation* (Cambridge: Cambridge University Press, 2000).

74. Friends of Historic Girard, "Emanuel Haldeman-Julius," http://skyways.lib.ks.us/kansas/towns/Girard/ehj.html. Accessed July 14, 2006.

75. E. Haldeman-Julius, reply, in *Why I Believe in God: Science and Religion: As a Scientist Sees It*, by Arthur Stanley Eddington (Girard, KS: Haldeman-Julius Publications, 1930), 23.

76. Ibid., 38, 28–29, 40.

77. Peter Bowler, *Reconciling Science and Religion* (Chicago: University of Chicago Press, 2001), is an invaluable resource for understanding this wave of literature.

78. Kenneth Wolfe, *The Churches and the British Broadcasting Corporation, 1922–1956* (London: SCM Press, 1984); and Asa Briggs, *The History of Broadcasting in the United Kingdom*, vol. 1, *The Birth of Broadcasting* (London: Oxford University Press, 1961).

79. British Broadcasting Corp., ed., *Science and Religion: A Symposium* (London: Gerald Howe, 1931), v. The text of the contributions can also be found in the BBC publication *The Listener*.

80. A. S. Eddington, in *Science and Religion*, ed. BBC, 126, 127–30.

81. J. S. Haldane, in *Science and Religion*, ed. BBC, 41, 49, 46–51.

82. Samuel Alexander, in *Science and Religion*, ed. BBC, 137; Dean Inge, in *Science and Religion*, ed. BBC, 144.

83. Dean Inge, in *Science and Religion*, ed. BBC, 151.

84. L. P. Jacks, in *Science and Religion*, ed. BBC, 160–61. Rudolph Otto, *The Idea of the Holy* (London: Oxford University Press, 1923).

85. L. P. Jacks, in *Science and Religion*, ed. BBC, 164.

86. H. D. A. Major, "The Case for Modernism," *Nineteenth Century* 104 (1928): 631, 637.

87. Ray Monk, *Bertrand Russell: The Ghost of Madness* (New York: Free Press, 2000), shows Russell's convoluted trajectory through the 1920s and 1930s.

88. B. Russell, *Analysis of Matter*.

89. Ibid., 84–89, 90–93, 106–7.

90. Ibid., 136.

91. Ibid., 395–96. At least in the early 1920s, Russell's understanding of the technical aspects of relativity was not always on particularly firm ground. In his *The ABC of Relativity* (London: K. Paul, Trench, Trubner, 1925), it is not clear that he understood significant parts of special relativity, such as relativity of simultaneity.

92. Bertrand Russell, *The Scientific Outlook* (New York: Norton, 1931), 92.

93. Ibid., 14, 30–33.

94. Ibid., 87, 89.

95. Ibid., 93, 94.

96. Ibid., 96, 99.

97. Ibid., 101–2. It is not clear whether Russell considered himself to be timid, or not a professor.

98. Ibid., 112.

99. Ibid., 101.

100. Ibid., 103–7.

101. Ibid., 108, 110.

102. Ibid., 132–33. Russell seems to have developed a personal antipathy to Eddington as well. He reportedly said that he would "rather be in Hell by himself than in Heaven with Eddington." From A. Wood, *Bertrand Russell: The Passionate Sceptic* (London, Allen & Unwin, 1957), 202; quoted in Ryckman, *Reign of Relativity*, 177.

103. It is possible that Russell did think he was an expert on social engineering, but his writings cover such a staggeringly wide number of topics (from logic to sex) that it is difficult to think he considered himself an expert on all of them.

104. Stebbing, *Philosophy and the Physicists*, ix, x.

105. Ibid., 5–6.

106. Ibid., 18. There is some evidence that Eddington's readers were aware, and were pleased, that they did not entirely understand the arguments they were accepting. One reviewer noted with apparent relief that "Professor Eddington does so much of our thinking for us." "New Books and Reprints" column, *Times Literary Supplement*, August 24, 1922, 547.

107. Stebbing, *Philosophy and the Physicists*, 19–21, 37.

108. Ibid., 48–52, 104, 111.

109. Ibid., 143.

110. Eddington, *New Pathways in Science*, ix.

111. Ibid., 278–80.

112. Ibid., 285, 288.

113. Ibid., 290.

114. Ibid., 291.

115. Ibid., 304.

116. Ibid., 305–6.

117. See Werskey, *Visible College*, and McGucken, *Scientists, Society, and the State*, for the establishment of Marxism in Cambridge, and Havinghurst, *Britain in Transition*, 232, for the growth of support for planned economics among British intellectuals. Neal Wood, *Communism and British Intellectuals* (New York: Columbia University Press, 1959), is outdated but provides an overview of the main players, with the Cambridge Marxists discussed on 85–87.

118. Published as Hans Driesch, *Science and Philosophy of the Organism* (London: Black, 1908).

119. The X-Club was an influential group of Victorian scientists who worked for a professionalized, research-oriented scientific community free of religious connections or restraint. Many of their members (including T. H. Huxley, John Tyndall, and Herbert Spencer) typified the naturalistic, often materialistic, scientist that so worried religious thinkers such as James Ward. See Ruth Barton, "'An Influential Set of Chaps': The X-Club and Royal Society Politics 1864–1885," *British Journal for the History of Science* 23, no. 1 (1990): 53–82.

120. Hastings, *History of English Christianity*, discusses the varied reactions of Christians to socialism in the first half of the twentieth century and details the pressure British churches felt from Marxism.

121. A. J. Hoover, *God, Germany, and Britain in the Great War: A Study in Clerical Nationalism* (London: Praeger, 1989), 9–11.

122. Minutes of the Jesus Lane Preparative Meeting, March 1919, Cambridgeshire County Records Office, R69/44.

123. Minutes of Cambridge Meeting, July 14, 1927, and Cambridge Special Preparative Meeting held October 26, 1927, Cambridgeshire County Records Office, R59/25/5/6.

124. This clash between Eddington-style idealist science and Marxism was clear to many contemporary observers and often commented on (e.g., Stebbing, *Philosophy and the Physicists*, ix). Unsurprisingly, Eddington-style idealism was also attacked directly in the Soviet Union; see L. Graham, *Between Science and Values*, chap. 3. Paul Josephson, *Physics and Politics in Revolutionary Russia* (Berkeley: University of California Press, 1991), also discusses Soviet attacks on idealism, particularly on 204, 227–75.

125. Lancelot Hogben, *The Nature of Living Matter* (New York: Knopf, 1931), vii. Other named targets included Whitehead and Carr.

126. J. D. Bernal, *The Social Function of Science* (London: Routledge, 1939), 4.

127. Christopher Caudwell, *The Crisis in Physics* (London: John Lane, 1939). The book was written in the mid-1930s but remained unpublished until after the author's death.

128. Ibid., 27.

129. Ibid., 35, 38, 47.

130. Ibid., 68, 66.

131. Ibid., 228–32.

132. A. S. Eddington, "Nature of the Physical World," *Freethinker*, October 20, 1929, 658.

133. Eddington, *Nature of the Physical World*, vi.

134. Eddington, "Nature of the Physical World," *Freethinker*, 658.

135. Chapman Cohen, *Almost an Autobiography: The Confessions of a Freethinker* (London: Pioneer Press, 1940), 22–37.

136. Ibid., 113.

137. Ibid., 99–106.

138. Ibid., 127.

139. Ibid., 154–57.

140. Ibid., 116.

141. Dean Inge, in *Science and Religion*, ed. BBC, 147.

142. Cohen, *Almost an Autobiography*, 207–14.

143. Chapman Cohen, *God and Evolution* (London: Pioneer Press, 1925), 25–26.

144. Chapman Cohen, *Determinism or Free-Will?* (London: Pioneer Press, 1919).

145. Ibid., 11–12.

146. Ibid., 17.

147. Ibid., 33–42. William James came under particularly heavy attack for his views on free will.

148. Chapman Cohen, *Materialism: Has It Been Exploded? Verbatim Report of Debate between Chapman Cohen and C. E. M. Joad* (London: Watts, 1928).

149. Ibid., 10, 17.

150. Ibid., 24, 31–32, 43–45.

151. Ibid., 53–55.

152. Ibid., 59, 60.

153. Chapman Cohen, "Views and Opinions," *Freethinker*, August 11, 1929, 497.

154. Ibid., 497.

155. Ibid., 498.

156. Chapman Cohen, "Views and Opinions," *Freethinker*, August 18, 1929, 513.

157. Ibid., 514.

158. Chapman Cohen, *Freethinker*, August 25, 1929, 529–30; and September 8, 1929, 561.

159. Chapman Cohen, "Views and Opinions," *Freethinker*, September 8, 1929, 562.

160. Chapman Cohen, "Views and Opinions," *Freethinker*, September 1, 1929, 545, 546.

161. Quoted in Chapman Cohen, *God and the Universe: Eddington, Jeans, Huxley and Einstein* (London: Pioneer Press, 1931), 49.

162. Ibid., 53.

163. Ibid., 52.

164. Ibid., 56; Eddington, "Nature of the Physical World," *Freethinker*, 660.

165. Chapman Cohen, "Views and Opinions," *Freethinker,* October 27 1929, 673.

166. Chapman Cohen, "Views and Opinions," *Freethinker*, November 3 1929, 689, 690.

167. Cohen, *God and the Universe*, 13–14.

168. Cohen, *Almost an Autobiography*, 54.

169. Cohen, *God and the Universe*, 29.

170. Some of the text of such a parody can be found at Evans, *Eddington Enigma*, 171. Personal recollections such as this are the only reason to consult Evans. Otherwise his book is a virtual paraphrase of the Douglas biography (*Life of Arthur Stanley Eddington*).

171. Gerald Holton, *The Advancement of Science, and Its Burdens* (Cambridge: Cambridge University Press, 1986), 165.

172. Bowler, *Reconciling Science and Religion*, 254–55.

173. Sullivan, *Contemporary Mind*, 125–26.

174. "Englishness" has been the subject of too large a body of scholarship to survey here, but for the themes of this chapter one should look at Christopher Lawrence and Anna-K. Mayer, eds., *Regenerating England: Science, Medicine and Culture in Inter-War Britain* (Atlanta: Editions Rodopi, 2000); Martin J. Wiener, *English Culture and the Decline of the Industrial Spirit, 1850–1980* (Cambridge: Cambridge University Press, 1981); Philip Williamson, *Stanley Baldwin: Conservative Leadership and National Values* (Cambridge: Cambridge University Press, 1999); and Susan Pedersen and Peter Mandler, eds., *After the Victorians: Private Conscience and Public Duty in Modern Britain: Essays in Memory of John Clive* (New York: Routledge, 1994). Graves and Hodge, *Long Week-end*; Havinghurst, *Britain in Transition*; Mowat, *Britain between the Wars*; Taylor, *English History*; and N. Wood, *Communism and British Intellectuals*, will all help in establishing a social and cultural context for these issues.

Chapter Seven

1. Eddington's *Fundamental Theory* was published posthumously (Cambridge: Cambridge University Press, 1948). It is not clear whether he ever considered the book to be finished. Noel Slater compiled some of Eddington's manuscripts and papers relating to this in *The Development and Meaning of Eddington's "Fundamental Theory"* (Cambridge: Cambridge University Press, 1957). The commentary section of this is not particularly useful. C. W. Kilmister's *Eddington's Search for a Fundamental Theory: A Key to the Universe* (Cambridge: Cambridge University Press, 1994), is a useful introduction to Eddington's ideas on epistemological physics. Its purpose is to advocate the method, however, and is not particularly helpful for situating *Fundamental Theory* historically or philosophically. The best place to start is probably Helge Kragh's chapter on heterodox physics in his *Quantum Generations* (Princeton, NJ: Princeton University Press, 1999), 218–29, which places Eddington's late work in the context of Dirac, Milne, and other novel approaches to physics in the 1930s. The Douglas biography (*Life of Arthur Stanley Eddington*) also provides an overview.

2. There is no focused study of Eddington's cosmological work. His contributions are discussed in Helge Kragh, *Cosmology and Controversy* (Princeton, NJ: Princeton University Press, 1996); and Kerszberg, *Invented Universe*. His earlier work on statistical cosmology (as opposed to relativistic cosmology) is described in Paul, *Milky Way Galaxy and Statistical Cosmology*. Kragh's *Matter and Spirit* examines the relationship of cosmology and religion, and his discussion of Eddington (103–12) agrees that his religious views played no significant part in his cosmological work.

3. See Cantor, *Quakers, Jews, and Science*, for a sense of what Quaker science looked like before the Manchester Conference.

4. My thoughts on these issues have been largely shaped by Shortland and Yeo, *Telling Lives in Science*; and the "Focus: Biography in the History of Science" section that appeared in *Isis* 97, no. 2 (June 2006): 302–29, which included contributions from Joan Richards, Mary Terrall, Theodore Porter, and Mary Jo Nye.

Bibliographic and Archival Note

Eddington is a difficult subject for historical investigation. The material record he left behind is fragmentary and sometimes confusing, and its scattered nature has been often responsible for some of the incomplete scholarship over the years. The chief culprit here is the mysterious destruction of Eddington's personal papers in 1944. It is not at all clear what the circumstances were surrounding this event. There are several versions of the story: one in which Eddington himself burned his papers before his death, one in which his sister Winifred destroyed them after his death, and one in which his colleague F. J. M. Stratton (who was called in to decide which papers were important) recommended that they be discarded. While it is not apparent how this circumstance came about, the result is that the documents relating to Eddington's personal life are severely restricted. It may be that a complete biography of Eddington is impossible, as large portions of his life are now virtually inaccessible.

The only repository of Eddington's papers is in the Wren Library of Trinity College, Cambridge. They are as yet uncatalogued. The papers are meager and consist only of materials that avoided destruction for whatever reason. The most useful item found there is Eddington's journal, which he kept (with varying consistency) from his Manchester years up through his appointment as Plumian Professor. It is helpful for seeing the day-to-day experience of the young Eddington. In the journal, he recorded dutifully the books he read, indexed by year read and author. It is clearly incomplete beginning around the Great War, but nonetheless is valuable for charting his intellectual life. The Wren papers also include manuscripts of *Nature of the Physical World* and *Fundamental Theory* (the latter annotated by N. B. Slater), a handful of letters to Eddington's family from when he was on the 1912 and 1919 eclipse expeditions, a few letters to colleagues, some juvenile writings, a few manuscripts of lectures, some lecture notes, and a compiled list of his book sales. His famous cycling journal, chronicling a lifetime of trips and vacations, is also held. There is a set of photocopies of the Eddington-Weyl correspondence at Trinity (the originals are at the ETH Library, Zurich, Ms. 91:522). The Douglas biography relies heavily on the Wren collection, although it references some letters that have not survived elsewhere. A note at the Wren implies that a number of personal letters were given to Eddington's relatives after his death, but I have been unable to verify this or track down likely relatives. Dozens of Eddington's manuscripts on relativity, astrophysics, philosophy, and popular science (as well as lecture notes) were sold by Sotheby's in 1975. They are presumably now in the hands of a private collector of unknown identity.

In addition to the Wren Library's holdings, materials relating to Eddington can be found in a few other repositories. His correspondence with Einstein is the largest collection of his letters that have survived and is available through the Albert Einstein Archives (20-540) at the Hebrew University in Jerusalem. Substantial letters have also survived in the Henry Norris Russell Papers at Princeton University, and the Huntington Library has Eddington correspondence with Walter Adams, Edwin Hubble, and George Ellery Hale. The University of Chicago has letters in the Chandrasekhar archive, and a few letters to F. A. Lindemann are in the Cherwell papers at Oxford. The Royal Greenwich Observatory archives, held at Cambridge University Library, have records from Eddington's time as Chief Assistant (including his 1912 expedition). The Dyson papers in the Royal Greenwich Observatory collection have some correspondence with Eddington. The records of the Cambridge Observatory have various administrative items from Eddington's tenure as director, and the observatory itself has some photographs of him. A large number of letters to and from Eddington can be found in the administrative records of the Royal Astronomical Society. These almost exclusively deal with mundane matters of RAS business and are mostly not indexed or organized except by date. The RAS Library has a list of the handful of Eddington letters found in other papers held there. Some of those letters between him and Guy Burniston Brown address his late philosophical views. The minutes of the Joint Permanent Eclipse Committee Meetings have useful details on the planning of the 1919 expedition. The Larmor papers at the Royal Society (Larmor Mss 603–9) have a few letters from Eddington. Herbert Dingle's papers, held at Imperial College, have correspondence between him and Eddington, as well as many letters about the establishment of the Eddington Memorial Lectures and the writing of the Douglas biography. The Oliver Lodge collection at University College London (MS ADD 89) has a fair number of letters. Eddington's letters to Erwin Schrödinger and Willem de Sitter can be found at the Österreichische Zentralbibliothek für Physik, Universität Wien, and Leiden Observatory, respectively. Harvard University's archives hold the E. C. Pickering, Harlow Shapley, and Annie Jump Cannon papers, which have a few letters from Eddington. A complete set of notes taken by L. H. Thomas in Eddington's 1923 class on relativity is in the Thomas Collection at the North Carolina State University Library. Manchester University and Cambridge University have academic records from Eddington's student years. The London School of Economics holds the papers of the National Peace Council, which has records of Eddington's peace work late in life. The Archives for the History of Quantum Physics indexes a handful of letters from Eddington (notably to Wolfgang Pauli) and has a number of interviews that discuss Eddington. There is also some correspondence between Eddington and Kurt Koffka, one of the founders of gestalt psychology. See Molly Harrower, *Kurt Koffka: An Unwilling Self-Portrait* (Gainesville, FL: University Presses of Florida, 1983).

Eddington can be seen and heard in action at the East Anglia Film Archive, which has a peculiar film (made for temperance propaganda) featuring him working at the Cambridge Observatory, and at the National Sound Archive in the British Library, where one can hear his radio broadcast on "Other Worlds" (BBC reference number 21968-9).

Evidence of Eddington's Quaker life can be found in the records of the Friends Meeting in Cambridge, which are held in the Cambridge County Records Office, and at Friends House, Euston Road, London. Friends House has the *Dictionary of Quaker Biography*, which is a comprehensive resource for conducting history on the Society of Friends, and a large file of relevant newspaper clippings and similar items. Jesus Lane Friends Meeting has a complete collection of the Eddington Memorial Lectures and supervises the trustees of the Eddington estate.

A complete bibliography of Eddington's own writings and secondary literature about him has been compiled by Dr. Katy Price of Anglia Polytechnic University. The following bibliography does not aim to be comprehensive along these lines, but rather aims to indicate literature relevant to the questions explored in this book.

Bibliography

Abbott, Margery. *A Certain Kind of Perfection*. Wallingford, PA: Pendle Hill Publications, 1997.

Alexander, S. *Space, Time, and Deity*. 2 vols. 2nd ed. London: Macmillan, 1927.

Alter, Peter. *The Reluctant Patron: Science and the State in Britain, 1850–1920*. New York: Berg, 1987.

Ames, J. S. "Einstein's Law of Gravitation." *Science* 51, no. 1315 (March 12, 1920): 253–61.

Andrews, James. *Science for the Masses: The Bolshevik State, Public Science, and the Popular Imagination in Soviet Russia, 1917–1934*. College Station, TX: Texas A&M University Press, 2003.

Badash, Lawrence. "British and American Views of the German Menace in World War I." *Notes and Records of the Royal Society of London* 34 (1979): 91–121.

Barbour, Hugh. "The 'Lamb's War' and the Origins of the Quaker Peace Testimony." In *The Pacifist Impulse in Historical Perspective*, edited by Harvey Dyck, 145–58. Toronto: University of Toronto Press, 1996.

Barbour, Hugh, and J. William Frost. *The Quakers*. New York: Greenwood Press, 1988.

Barrow-Green, June. "'A Corrective to the Spirit of Too Exclusively Pure Mathematics': Robert Smith (1689–1786) and His Prizes at Cambridge University." *Annals of Science* 56 (1999): 271–316.

Barton, Ruth. "'An Influential Set of Chaps': The X-Club and Royal Society Politics 1864–1885." *British Journal for the History of Science* 23 (1990): 53–82.

Batten, Alan. "A Most Rare Vision: Eddington's Thinking on the Relation between Science and Religion." *Quarterly Journal of the Royal Astronomical Society* 35 (1994): 249–70.

———. "What Eddington Did Not Say." *Isis* 94, no. 4 (December 2003): 656–59.

Beaman, Frank C. O. "Church and Science." *Nineteenth Century* 89 (1921): 467–70.

Bebbington, David W. *Evangelicalism in Modern Britain: A History from the 1730s to the 1980s*. London: Unwin Hyman, 1989.

Beer, Gillian. "Eddington and the Idiom of Modernism." In *Science, Reason, and Rhetoric*, edited by Henry Krips et al., 295–315. Pittsburgh: University of Pittsburgh Press, 1995.

Bell, David. "Philosophy." In *The Twentieth Century Mind: History, Ideas, and Literature in Britain*, edited by C. B. Cox and A. E. Dyson, 174–224. Oxford: Oxford University Press, 1972.

Benn, W. "Alien Influence in England." *Contemporary Review* 116 (1919): 637–38.

Bernal, J. D. *The Social Function of Science*. London: Routledge, 1939.

Biagioli, Mario. *Galileo: Courtier*. Chicago: University of Chicago Press, 1993.

Bikenhead, F. W. F. S. *The Prof in Two Worlds*. London: Collins, 1961.

Bishop, Jeanne. "Golden Era of Theoretical Physics: The Black Box of Stellar Energy." *Griffith Observer* 42 (1978): 3–17.

Blaauw, Adriaan. *History of the IAU: The Birth and First Half-Century of the International Astronomical Union*. Boston: Kluwer Academic, 1994.

Bowler, Peter. *Reconciling Science and Religion*. Chicago: University of Chicago Press, 2001.

Braithwaite, W. C. "Some Present-Day Aims of the Society of Friends." *Friends' Quarterly Examiner* 29 (1895): 321–41.

Briggs, Asa. *The History of Broadcasting in the United Kingdom*. Vol.1, *The Birth of Broadcasting*. London: Oxford University Press, 1961.

Brinton, Howard Haines. *Friends for 300 Years: The History and Beliefs of the Society of Friends since George Fox Started the Quaker Movement*. New York: Harper, 1952.

British Broadcasting Corporation, ed. *Science and Religion: A Symposium*. London: Gerald Howe, 1931.

Brock, Peter. *Twentieth-Century Pacifism*. London: Van Nostrand Reinhold, 1970.

Broks, Peter. *Understanding Popular Science*. Philadelphia: Open University Press, 2006.

Brooke, John Hedley. "Natural Theology." In *Science and Religion: A Historical Introduction*, edited by Gary Ferngren, 163–75. Baltimore: Johns Hopkins University Press, 2002.

———. "Science and Religion." In *Companion to the History of Modern Science*, edited by R. C. Olby et al., 763–82. London: Routledge, 1990.

———. *Science and Religion: Some Historical Perspectives*. Cambridge: Cambridge University Press, 1991.

Brooke, John Hedley, and Geoffrey Cantor. *Reconstructing Nature*. Edinburgh: T & T Clark, 1998.

Brown, Callum. *The Death of Christian Britain: Understanding Secularisation 1800–2000*. London: Routledge, 2001.

Brush, Stephen G. "Prediction and Theory Evaluation: The Case of Light Bending." *Science* 246 (1989): 1124–29.

A Cambridge B.A. [pseudo.]. "Cambridge and Conscription." *University Socialist*, Michelmas 1913, 104–7.

Cantor, Geoffrey. *Michael Faraday: Sandemanian and Scientist: A Study of Science and Religion in the Nineteenth Century*. London: Macmillan, 1991.

———. *Quakers, Jews, and Science: Religious Responses to Modernity and the Sciences in Britain, 1650–1900*. Oxford: Oxford University Press, 2005.

Cardwell, D. S. L. "Science and World War I." *Proceedings of the Royal Society*, ser. A, 342 (1975): 447–56.

Carr, H. Wildon. "Einstein's Theory and Philosophy." *Mind* 31, no. 122 (April 1922): 169–77.

———. *The General Principle of Relativity in Its Philosophical and Historical Aspect*. London: Macmillan, 1920.

———. "Metaphysics and Materialism." *Nature* 108 (October 20, 1921): 247–48.

Caudwell, Christopher. *The Crisis in Physics*. London: John Lane, 1939.

Chandrasekhar, S. *Eddington, the Most Distinguished Astrophysicist of His Time*. Cambridge: Cambridge University Press, 1983.

————. "The Richtmeyer Memorial Lecture—Some Historical Notes." *American Journal of Physics* 37, no. 6, 577–84.

————. *Truth and Beauty*. Chicago: University of Chicago Press, 1987.

————. "Verifying the Theory of Relativity." *Notes and Records of the Royal Society of London* 30 (1976): 249–60.

Clark, Ronald. *Einstein: The Life and Times*. New York: World Publishing, 1965.

Clifford, William Kingdon. *The Common Sense of the Exact Sciences*. New York: Appleton, 1894.

Cobbe, W. A. "A More Excellent Way." *Nineteenth Century* 92 (July–December 1922): 1015–20.

Cohen, Chapman. *Almost an Autobiography: The Confessions of a Freethinker*. London: Pioneer Press, 1940.

————. *Determinism or Free-Will?* London: Pioneer Press, 1919.

————. *God and Evolution*. London: Pioneer Press, 1925.

————. *God and the Universe: Eddington, Jeans, Huxley and Einstein*. London: Pioneer Press, 1931.

————. *Materialism: Has It Been Exploded? Verbatim Report of Debate between Chapman Cohen and C. E. M. Joad*. London: Watts, 1928.

Cohen, I. Bernard, ed., *Puritanism and the Rise of Modern Science: The Merton Thesis*. New Brunswick, NJ: Rutgers University Press, 1990.

Collins, Harry, and Trevor Pinch. *The Golem: What Everyone Should Know about Science*. Cambridge: Cambridge University Press, 1993.

Comte, Auguste. *Cours de Philosophie Positive*. In *Auguste Comte and Positivism*, edited by Gertrud Lenzer. London: Transaction Publishers, 1998.

Cooter, Roger, and Stephen Pumfrey. "Separate Spheres and Public Places: Reflections on the History of Science Popularization and Science in Popular Culture." *History of Science* 32 (1994): 237–67.

Co-ordinating Committee for Research into the Use of the University for War. *Cambridge University and War*. Pamphlet. Cambridge: CCRUUW, [1917?].

Cowling, T. J. "Development of the Theory of Stellar Structure." *Quarterly Journal of the Royal Astronomical Society* 7 (1966): 121–37.

Crawford, Elisabeth. *Nationalism and Internationalism in Science, 1880–1939: Four Studies of the Nobel Population*. Cambridge: Cambridge University Press, 1992.

Crelinsten, Jeffrey. *Einstein's Jury: The Race to Test Relativity*. Princeton, NJ: Princeton University Press, 2006.

————. "William Wallace Campbell and the 'Einstein Problem': An Observational Astronomer Confronts the Theory of Relativity." *Historical Studies in the Physical Sciences* 14 (1983): 1–91.

Cunningham, E. *Relativity and the Electron Theory*. London: Longmans, 1915.

Darrigol, Olivier. "The Mystery of the Einstein-Poincaré Connection." *Isis* 95 (2004): 614–26.

Daston, Lorraine. "The Moral Economy of Science." *Osiris* 10 (1995): 3–24.

Davies, Paul. *God and the New Physics*. London: Dent, 1983.

de Beer, Sir Gavin. *The Sciences Were Never at War*. London: Nelson, 1960.

de Sitter, Willem. "On Einstein's Theory of Relativity and Its Astronomical Consequences. First Paper." *Monthly Notices of the Royal Astronomical Society* 76 (1916): 699–728.

————. "On Einstein's Theory of Relativity and Its Astronomical Consequences. Second Paper." *Monthly Notices of the Royal Astronomical Society* 77 (1916): 155–84.

Desmond, Adrian, and James Moore. *Darwin*. London: Michael Joseph, 1991.

DeVorkin, David. *Henry Norris Russell: Dean of American Astronomers*. Princeton, NJ: Princeton University Press, 2000.

Dewhirst, David. "The Greenwich-Cambridge Axis." *Vistas in Astronomy* 20 (1976): 109–11.

Dilnot, Frank. *England after the War*. New York: Doubleday, 1920.

Dingle, Herbert. *The Sources of Eddington's Philosophy*. Cambridge: Cambridge University Press, 1954.

Douglas, A. Vibert. *The Life of Arthur Stanley Eddington*. London: Thomas Nelson, 1956.

Drees, Willem B. *Religion, Science, and Naturalism*. Cambridge: Cambridge University Press, 1996.

Dreyer, J. L. E., ed. *History of the Royal Astronomical Society*. Vol. 1, *1820–1920*. London: Royal Astronomical Society, 1923.

Driesch, Hans. *Science and Philosophy of the Organism*. London: Black, 1908.

Dyson, F. W. "On the Opportunity Afforded by the Eclipse of 1919 May 29 of Verifying Einstein's Theory of Gravitation." *Monthly Notices of the Royal Astronomical Society* 77 (March 1917): 445.

Dyson, F. W., A. S Eddington, and C. Davidson. "A Determination of the Deflection of Light by the Sun's Gravitational Field, from Observations Made at the Total Eclipse of May 29, 1919." *Philosophical Transactions of the Royal Society of London*, ser. A, 220 (1920): 291–333.

Earman, John, and Clark Glymour. "The Gravitational Redshift as a Test of General Relativity: History and Analysis." *Studies in the History and Philosophy of Science* 11 (1980): 175–214.

———. "Relativity and Eclipses: The British Eclipse Expeditions of 1919 and Their Predecessors." *Historical Studies in the Physical Sciences* 11 (1980): 49–85.

Eddington, A. S. "The Bishop of Birmingham's Gifford Lectures." *Cambridge Review*, June 8, 1933, 488–89.

———. "The Decline of Determinism." *Smithsonian Institution Annual Report 1932*. Washington, DC: Smithsonian, 1933.

———. "The Domain of Physical Science." In *Science, Religion and Reality*, edited by Joseph Needham, 2nd ed., 193–224. New York: George Braziller, 1955.

———. "Einstein's Theory of Gravitation." *Observatory* 510 (1917): 93–95.

———. "Einstein's Theory of Space and Time." *Contemporary Review* 116 (1919): 639–43.

———. *Expanding Universe*. Cambridge: Cambridge University Press, 1933.

———. *Fundamental Theory*. Cambridge: Cambridge University Press, 1948.

———. "Further Notes on the Radiative Equilibrium of the Stars." *Monthly Notices of the Royal Astronomical Society* 77 (1917): 596–97.

———. "The Future of International Science." *Observatory* 501 (June 1916): 271.

———. "A Generalisation of Weyl's Theory of the Electromagnetic and Gravitational Fields." *Proceedings of the Royal Society of London*, ser. A, 99, no. 697 (May 2, 1921): 104–22.

———. "Gravitation and the Principle of Relativity I." *Nature*, March 1918, 15–17.

———. "Gravitation and the Principle of Relativity II." *Nature*, March 1918, 34–36.

———. "The Internal Constitution of the Stars." *Observatory* 557 (October 1920): 341–58.

———. *The Internal Constitution of the Stars*. Cambridge: Cambridge University Press, 1926.

———. Introduction to *The British Commonwealth and the US in the Post-war World*. Peace Aims Pamphlet no. 10. London: National Peace Council, 1942.

———. "Karl Schwarzschild." *Observatory* 503 (August 1916): 337–39.

———. *Mathematical Theory of Relativity*. Cambridge: Cambridge University Press, 1923.

———. "The Meaning of Matter and the Laws of Nature According to the Theory of Relativity." *Mind* 29, no. 114 (April 1920): 145–58.

———. *The Nature of the Physical World*. Cambridge: Cambridge University Press, 1928.

———. *New Pathways in Science*. Cambridge: Cambridge University Press, 1935.

———. "On Relativistic Degeneracy." *Monthly Notices of the Royal Astronomical Society* 95 (1935): 194–206.

———. "On the Pulsations of a Gaseous Star and the Problems of the Cepheid Variables. Part I." *Monthly Notices of the Royal Astronomical Society* 79 (1919): 2–22.

———. "On the Pulsations of a Gaseous Star and the Problems of the Cepheid Variables. Part II." *Monthly Notices of the Royal Astronomical Society* 79 (1919): 177–89.

———. "On the Relation between the Masses and Luminosities of the Stars." *Monthly Notices of the Royal Astronomical Society* 84 (1924): 308–32.

———. *Philosophy of Physical Science*. Cambridge: Cambridge University Press, 1939.

———. "Physics and Determinism." *Nineteenth Century* 113 (1933): 715–23.

———. "The Pulsation Theory of Cepheid Variables." *Observatory* 516 (August 1917): 290–93.

———. "The Purpose of Science." *Friends' Quarterly Examiner* 246 (April 1928): 89–110.

———. "Radiation of the Stars." *Nature* 99 (August 2, 1917): 445.

———. "The Radiative Equilibrium of the Stars: A Reply to Mr. Jeans' Criticism." *Monthly Notices of the Royal Astronomical Society* 78 (December 1917): 113–15.

———. "The Radiative Equilibrium of the Sun and Stars." *Monthly Notices of the Royal Astronomical Society* 77 (1917): 16–35.

———. "RAS Centennary." *Monthly Notices of the Royal Astronomical Society* 82 (1922): 431–43.

———. *Relativity Theory of Protons and Electrons*. Cambridge: Cambridge University Press, 1936.

———. *Report on the Relativity Theory of Gravitation*. London: Fleetway Press, 1918.

———. "Science and Religion." In *Science and Religion: A Symposium*, edited by British Broadcasting Corp., 117–30. London: Gerald Howe, 1931.

———. *Science and the Unseen World*. New York: Macmillan, 1929.

———. "Some Problems of Astronomy (XIX. Gravitation)." *Observatory* 484 (February 1915): 93–98.

———. "The Source of Stellar Energy." *Supplement to Nature*, no. 2948 (May 1, 1926): 3–32.

———. *Space, Time and Gravitation: An Outline of the General Relativity Theory*. Cambridge: Cambridge University Press, 1920.

———. "Star." In *Encyclopaedia Britannica*, 11th ed., 784–93

———. *Stars and Atoms*. London: Oxford University Press, 1927.

———. *Stellar Movements and the Structure of the Universe*. London: Macmillan, 1914.

———. "Das Strahlungsgleichgewicht der Sterne." *Zeitschrift fur Physik* 7 (1921): 351–97.

———. *The Theory of Relativity and Its Influence on Scientific Thought*. Oxford: Clarendon Press, 1922.

———. "The Total Eclipse of 1919 May 29 and the Influence of Gravitation on Light." *Observatory* 537 (March 1919): 119–22.

———. *Why I Believe in God: Science and Religion; As a Scientist Sees It, with a Reply by E. Haldeman-Julius*. Girard, KS: Haldeman-Julius Publications, 1930.

Eddington, A. S., et al. "The Philosophical Aspect of the Theory of Relativity." *Mind* 29, no. 116 (October 1920): 415–45.

Edgerton, David. "British Scientific Intellectuals and the Relations of Science, Technology, and War." In *National Military Establishments and the Advancement of Science and Technology*, edited by Paul Forman and José Sanchez-Ron, 1–35. London: Kluwer, 1996.

Einstein, Albert. R. *The Collected Papers of Albert Einstein*. Vol. 8A, edited by Schulmann et al. Princeton: Princeton University Press, 1997.

———. *The Collected Papers of Albert Einstein*. Vol. 9, edited by Diana Kormos Buchwald et al. Princeton: Princeton University Press, 2004.

Eisberg, J. "Eddington's Stellar Models and Early Twentieth Century Astrophysics." PhD diss., Harvard University, 1991.

Eisler, Riane Tennehaus. *The Chalice and the Blade*. London: Thorsons, 1998.

Elliot, Hugh. "Relativity and Materialism." *Nature* 108 (December 1, 1921): 432.

Evans, David. *The Eddington Enigma*. Self-published, 1998.

Everitt, C. W. F. "Experimental Tests of General Relativity: Past, Present and Future." In *Physics and Contemporary Needs*, vol. 4, edited by Riazuddin, 529–55. New York: Plenum, 1980.

Evershed, J. "The Einstein Effect and the Eclipse of 1919 May 29." *Monthly Notices of the Royal Astronomical Society* 78 (1917): 269–70.

Fine, Arthur. *The Shaky Game: Einstein, Realism, and the Quantum Theory*. Chicago: University of Chicago Press, 1986.

Fölsing, Albrecht. *Albert Einstein: A Biography*. New York: Viking, 1997.

Forbes, John. *The Quaker Star under 7 Flags, 1917–27*. Philadelphia: University of Pennsylvania Press, 1962.

Forman, Paul. "The Reception of an Acausal Quantum Mechanics in Germany and Britain." In *The Reception of Unconventional Science*, edited by Seymour H. Mauskopf, 11–50. Boulder, CO: American Association for the Advancement of Science, 1979.

———. "Weimar Culture, Causality, and Quantum Theory, 1918–27: Adaptation by German Physicists and Mathematicians to a Hostile Intellectual Environment." *Historical Studies in the Physical Sciences* 3 (1971): 1–116.

Fox, Robert, and Graeme Gooday, eds. *Physics in Oxford*. Oxford: Oxford University Press, 2005.

Frank, Philipp. *Interpretations and Misinterpretations of Modern Physics*. Paris: Hermann, 1938.

French, S. "Scribbling on the Blank Sheet: Eddington's Structuralist Conception of Objects." *Studies in History and Philosophy of Modern Physics* 34 (2003): 227–59.

Freundlich, Erwin. *The Foundations of Einstein's Theory of Gravitation*. Translated by Henry Brose. Cambridge: Cambridge University Press, 1920.

Friends Peace Committee. *Lest We Forget*. Leominster, UK: Orphans' Printing Press, [1917?].

Frost, J. William, and John M. Moore, eds. *Seeking the Light: Essays in Quaker History*. Haverford, PA: Pendle Hill Publications, 1986.

Fry, A. Ruth. *A Quaker Adventure*. London: Nisbet, 1926.

Fyfe, Aileen. *Science and Salvation: Evangelical Popular Science Publishing in Victorian Britain*. Chicago: University of Chicago Press, 2004.

Galison, Peter. *Einstein's Clocks, Poincaré's Maps*. New York: Norton, 2003.

———. *Image and Logic*. Chicago: University of Chicago Press, 1997.

Galison, Peter, David Kaiser, and Michael Gordin, eds. *Science and Society: The History of Modern Physical Science in the Twentieth Century*. Vols. 1 and 2. New York: Routledge, 2001.

Galison, Peter, and David Stump. *The Disunity of Science*. Stanford, CA: Stanford University Press, 1996.

Galsworthy, John. *A Modern Comedy*. New York: Scribner's Sons, 1929.

Gascoigne, John. "From Bentley to the Victorians: The Rise and Fall of British Newtonian Natural Theology." *Science in Context* 2 (1988): 219–56.

Geertz, Clifford. *The Interpretation of Cultures.* New York: Basic Books, 1973.

Giere, Ronald N., and Alan W. Richardson *Origins of Logical Empiricism.* Minneapolis: University of Minnesota Press, 1996.

Gilbert, Alan. *The Making of Post-Christian Britain: A History of the Secularization of Modern Society.* London: Longman, 1980.

Gilbert, Martin. *A History of the 20th Century.* Vol. 1, *1900–1933.* New York: William Morrow, 1997.

Gingerich, Owen. "Development of Astronomical Theory and Practice." *Vistas in Astronomy* 20 (1976): 1–9.

———. "Writing the History of Modern Astronomy." *Journal for the History of Astronomy* 11 (1980): 145–46.

———, ed. *The General History of Astronomy.* Vol. 4, part A, *Astrophysics and Twentieth-Century Astronomy to 1950.* Cambridge: Cambridge University Press, 1984.

Glick, Thomas. *The Comparative Reception of Relativity.* Boston: Reidel, 1987.

Goldberg, Stanley. *Understanding Relativity: Origin and Impact of a Scientific Revolution.* Boston: Birkhäuser, 1984.

Gooneratne, Sakura. "The White Dwarf Affair: Chandrasekhar, Eddington and the Limiting Mass." PhD diss., Department of Science and Technology Studies, University College London, 2005.

Gordin, Michael. *A Well-Ordered Thing.* New York: Basic Books, 2004.

Gorham, G. "Planck's Principle and Jeans's Conversion." *Studies in the History and Philosophy of Science* 22 (1991): 471–97.

Gould, Stephen J. *Rocks of Ages.* New York: Ballantine, 1999.

Graham, John W. *Conscription and Conscience: A History, 1916–1919.* London: Allen & Unwin, 1922.

Graham, Loren R. *Between Science and Values.* New York: Columbia University Press, 1981.

———. *Science, Philosophy, and Human Behavior in the Soviet Union.* New York: Columbia University Press, 1987.

Graves, Robert, and Alan Hodge. *The Long Week-end, a Social History of Great Britain, 1918–1939.* London: Faber & Faber, 1940.

Green, V. H. *Religion at Oxford and Cambridge.* London: SCM Press, 1964.

Greenwood, Thomas. "Einstein and Idealism." *Mind* 31, no. 122 (April 1922): 205–7.

Gregory, Jane, and Steve Miller. *Science in Public: Communication, Culture, and Credibility.* New York: Plenum, 1998.

Grubb, Edward. *Quaker Thought and History.* London: Swarthmore Press, 1925.

Grünfeld, Joseph. *Science and Values.* Amsterdam: B. R. Grüner B.V., 1973.

Gummett, Philip. *Scientists in Whitehall.* Manchester: Manchester University Press, 1980.

Hacking, Ian. *Representing and Intervening.* Cambridge: Cambridge University Press, 1983.

Haldane, Viscount. *The Pathway to Reality.* New York: Dutton, 1905.

———. *The Philosophy of Humanism.* New Haven, CT: Yale University Press, 1922.

———. *The Reign of Relativity.* London: John Murray, 1921.

Hall, C. Margaret. *Identity, Religion and Values.* Washington, DC: Taylor & Francis, 1996.

Hardy, G. H. *Bertrand Russell and Trinity.* Cambridge: Cambridge University Press, 1970.

Harris, José. *Private Lives, Public Spirit: A Social History of Britain, 1870–1914* Oxford: Oxford University Press, 1993.

Harrower, Molly. *Kurt Koffka: An Unwilling Self-Portrait.* Gainesville: University Presses of Florida, 1983.

Hartcup, Guy. *The War of Invention: Scientific Developments 1914–1918.* London: Brassey's Defence Publishers, 1988.

Hastings, Adrian. *A History of English Christianity 1920–2000.* London: SCM Press, 2001.

Havinghurst, Alfred F. *Britain in Transition: The Twentieth Century,* 4th ed. Chicago: University of Chicago Press, 1985.

Hayes, Denis. *Conscription Conflict.* London: Sheppard Press, 1949.

Heilbron, J. L. *The Dilemmas of an Upright Man: Max Planck as Spokesman for German Science.* Berkeley: University of California Press, 1986.

———. *H. G. J. Moseley: The Life and Letters of an English Physicist, 1887–1915.* Berkeley: University of California Press, 1974.

Hentschel, Klaus. "The Conversion of St. John: A Case Study on the Interplay of Theory and Experiment." *Science in Context* 6, no. 1 (1993): 137–94.

———. *The Einstein Tower.* Stanford, CA: Stanford University Press, 1997.

———. *Interpretationen und Fehlinterpretationen der Speziellen und der Allgemeinen Relativitaetstheorie durch Zeitgenossen Albert Einsteins.* Basel: Birkhäuser, 1990.

Hertzsprung-Kapteyn, Henrietta. "J. C. Kapteyn." *Space Science Reviews* 64 (1993): 1–93.

Hilgartner, Stephen. "The Dominant View of Popularization: Conceptual Problems, Political Uses." *Social Studies of Science* 20 (1990): 519–39.

Hilton, Boyd. *The Age of Atonement: The Influence of Evangelicalism of Social and Economic Thought.* Oxford: Oxford University Press, 1991.

Hoernle, R. F. Alfred. "The Oxford Congress of Philosophy." *Philosophical Review* 30, no. 1 (January 1921): 57–72.

Hogben, Lancelot. *The Nature of Living Matter.* London: Knopf, 1931.

Holton, Gerald. *The Advancement of Science, and Its Burdens.* Cambridge: Cambridge University Press, 1986.

———. "Mach, Einstein, and the Search for Reality." In *Thematic Origins of Scientific Thought.* Cambridge, MA: Harvard University Press, 1973.

———. *Thematic Origins of Scientific Thought.* Cambridge, MA: Harvard University Press, 1973.

Holton, Gerald, and Yehuda Elkana, eds. *Albert Einstein, Historical and Cultural Perspectives: The Centennial Symposium in Jerusalem.* Princeton, NJ: Princeton University Press, 1982.

Hoover, A. J. *God, Germany, and Britain in the Great War.* London: Praeger, 1989.

Hosinski, Thomas E. *Stubborn Fact and Creative Advance.* Lanham, MD: Rowman & Littlefield, 1993.

Hoskin, Michael A. *General History of Astronomy.* Vols. 1–4. Cambridge: Cambridge University Press, 1984.

———. *Stellar Astronomy: Historical Studies.* Chalfont St. Giles, UK: Science History, 1982.

Howard, D., and J. Stachel, eds. *Einstein and the History of General Relativity.* Boston: Birkhauser, 1989.

Hudson, Rob. "James Jeans and Radiation Theory." *Studies in the History and Philosophy of Science* 20 (1989): 57–76.

———. "Novelty and the 1919 Eclipse Experiments." *Studies in the History and Philosophy of Modern Physics* 34 (2003): 107–29.

Hufbauer, Karl. "Astronomers Take Up the Stellar-Energy Problem, 1917–1920." *Historical Studies in the Physical Sciences* 11 (1981): 277–303.

———. *Exploring the Sun: Solar Science since Galileo.* Baltimore: Johns Hopkins University Press, 1991.

Inge, Dean W. R. *God and the Astronomers.* London: Longmans, Green, 1933.

Ipe, Alex. "Plausibility and the Theoreticians' Regress: Constructing the Evolutionary Fate of Stars." PhD diss., Department of Sociology and Anthropology, Carleton University, 2002.

Isichei, Elizabeth. *Victorian Quakers.* Oxford: Oxford University Press, 1970.

Jacks, L. P. *Sir Arthur Eddington: Man of Science and Mystic.* Cambridge: Cambridge University Press, 1949.

Jaki, Stanley L. *Lord Gifford and His Lectures: A Centenary Retrospect.* Edinburgh: Scottish Academic Press, 1986.

Jammer, Max. *Einstein and Religion.* Princeton, NJ: Princeton University Press, 1999.

———. *The Philosophy of Quantum Mechanics: The Interpretations of Quantum Mechanics in Historical Perspective.* New York: Wiley, 1974.

Jastrow, Robert. *God and the Astronomers.* New York: Norton, 1978.

Jeans, James H. "Einstein's Theory of Gravitation." *Observatory* 509 (1917): 57–58.

———. "The Equations of Radiative Transfer of Energy." *Monthly Notices of the Royal Astronomical Society* 78 (1917): 28–36.

———. "The Internal Constitution and Radiation of Gaseous Stars." *Monthly Notices of the Royal Astronomical Society* 79 (1919): 319–32.

———. *The Mysterious Universe.* Cambridge: Cambridge University Press, 1930.

———. "On the Masses, Luminosities, and Surface-Temperatures of the Stars." *Monthly Notices of the Royal Astronomical Society* 85 (1925): 196–211.

———. "The Problem of the Cepheid Variables." *Observatory* 536 (February 1919): 88–89.

———. *Problems of Cosmogony and Stellar Dynamics.* Cambridge: Cambridge University Press, 1919.

———. "Radiation of the Stars." *Nature* 99 (July 5, 1917): 365.

———. "Review of R. Emden's *Gaskugeln*." *Astrophysical Journal* 30 (1909): 72–74.

Jeans, James H., et al. "Discussion on the Theory of Relativity." *Proceedings of the Royal Society of London,* ser. A, 97, no. 681 (March 1, 1920): 66–79.

Joad, C. E. M. *Philosophical Aspects of Modern Science.* New York: Macmillan, 1932.

Jones, Rufus M. "Beginnings of Quakerism." 1912. In *Quakerism: A Spiritual Movement,* 119–38. Philadelphia: Philadelphia Yearly Meeting of Friends, 1963.

———. *The Inner Life.* New York: Macmillan, 1917.

———. *The New Quest.* New York: Macmillan, 1928.

———. *Rufus Jones: Essential Writings.* Edited by Kerry Walters. New York: Orbis Books, 2001.

———. *A Service of Love in Wartime.* New York: Macmillan, 1920.

———. *Spiritual Energies in Everyday Life.* New York: Macmillan, 1922.

———. *Studies in Mystical Religion.* London: Macmillan, 1909.

Joseph, H. W. B. "Prof. Eddington on 'The *Nature of the Physical World*.'" *Hibbert Journal* 27 (October 1928–July 1929): 406–23.

Josephson, Paul. *Physics and Politics in Revolutionary Russia.* Berkeley: University of California Press, 1991.

Kargon, Robert. *Science in Victorian Manchester.* Baltimore: Johns Hopkins University Press, 1977.

Kaufman, Peter, and Spencer Lavan. *Alone Together: Studies in the History of Liberal Religion.* Boston: Beacon, 1979.

Kavanaugh, John, ed. *The Quaker Approach.* New York: Putnam's Sons, 1953.

Kenat, Ralph. "Physical Interpretation: Eddington, Idealization, and the Origin of Stellar Structure Theory." PhD diss., University of Maryland College Park, 1987. AAT 8725406.

Kennedy, Thomas C. *British Quakerism, 1860–1920: The Transformation of a Religious Community.* Oxford: Oxford University Press, 2001.

Kenworthy, L. S., ed. *Living in the Light: Some Quaker Pioneers of the 20th Century.* Vol. 1. Kennett Square, PA: Friends General Conference and Quaker Publications, 1984.

Kerszberg, Pierre, *The Invented Universe.* Oxford: Oxford University Press, 1989.

Kershner, Howard Eldred. *Quaker Service in Modern War.* New York: Prentice-Hall, 1950.

Kevles, Daniel. "Hale, WWI and the Advancement of Science." *Isis* 59 (1968): 427–37.

———. "Into Hostile Political Camps." *Isis* 62 (1971): 47–60.

———. *The Physicists: The History of a Scientific Community in Modern America.* Cambridge, MA: Harvard University Press, 1995.

Kilmister, C. W. *Eddington's Search for a Fundamental Theory: A Key to the Universe.* Cambridge: Cambridge University Press, 1994.

———. *Sir Arthur Eddington.* Oxford: Pergamon, 1966.

———. *Eddington's Statistical Theory.* Oxford: Clarendon Press, 1962.

Kohler, Robert. *Lords of the Fly: Drosophila Genetics and the Experimental Life.* Chicago: University of Chicago Press, 1994.

Kragh, Helge. *Cosmology and Controversy.* Princeton, NJ: Princeton University Press, 1996.

———. *Matter and Spirit in the Universe: Scientific and Religious Preludes to Modern Cosmology.* London: Imperial College Press, 2004.

———. *Quantum Generations.* Princeton, NJ: Princeton University Press, 1999.

LaFollette, Marcel. *Making Science Our Own: Public Images of Science, 1910–1955.* Chicago: University of Chicago Press, 1990.

Larson, Edward J. *Summer for the Gods: The Scopes Trial and America's Continuing Debate over Science and Religion.* Cambridge, MA: Harvard University Press, 1998.

Lawrence, Christopher, and Anna-K. Mayer, eds. *Regenerating England: Science, Medicine and Culture in Inter-War Britain.* Atlanta: Editions Rodopi, B.V., 2000.

Laybourn, Keith. *The General Strike of 1926.* Manchester: Manchester University Press, 1993.

Lenski, Gerhard. *The Religious Factor: A Sociological Study of Religion's Impact on Politics, Economics, and Family Life.* Garden City, NY: Doubleday, 1961.

Lindberg, David, and Ronald Numbers, eds. *God and Nature.* Berkeley: University of California Press, 1986.

Littlewood, J. E. *A Mathematician's Miscellany.* London: Methuen, 1953.

Livingstone, David N., D. G. Hart, and Mark A. Noll, eds. *Evangelicals and Science in Historical Perspective.* Oxford: Oxford University Press, 1999.

Lodge, Oliver. "Eddington's Philosophy." *Nineteenth Century* 105 (1929): 360–69.

———. "Geometrisation of Physics, and Its Supposed Basis on the Michelson-Morley Experiment." *Nature* 106, no. 2677 (February 17, 1921): 795–800.

Lowe, Victor. *Alfred North Whitehead: The Man and His Work.* Baltimore: Johns Hopkins University Press, 1985.

Lowrance, William W. *Modern Science and Human Values.* New York: Oxford University Press, 1985.

MacLeod, R. "The Chemists Go to War: The Mobilization of Civilian Chemists and the British War Effort, 1914–1918." *Annals of Science* 50(1993): 455–81.

———. "The Origins of the DSIR: Reflections on Ideas and Men, 1915–1916." *Public Administration* 48 (1970): 23–48.

Macmillan, Margaret. *Peacemakers: The Paris Conference of 1919 and Its Attempt to End War.* London: J. Murray, 2001.

Major, H. D. A. "The Case for Modernism." *Nineteenth Century* 104 (1928): 629–38.

Marrin, Albert. *The Last Crusade: The Church of England in the First World War.* Durham, NC: Duke University Press, 1974.

Marvin, F. S. Review of *Nature of the Physical World,* by A. S. Eddington. *Hibbert Journal* 27 (October 1928–July 1929): 564–67.

Marwick, Arthur. *The Deluge: British Society and the First World War.* Basingstoke, UK: Macmillan Education, 1991.

Marwick, William H. *Ernest Bowman Ludlam.* London: Friends House Service Committee, 1960.

Maudsley, Henry. *Body and Mind.* London: Macmillan, 1870.

———. *Natural Causes and Supernatural Seemings.* London: K. Paul, Trench, 1886.

Mayo, Deborah G. *Error and the Growth of Experimental Knowledge.* Chicago: University of Chicago Press, 1996.

McCausland, Ian. "Anomalies in the History of Relativity." *Journal of Scientific Exploration* 13 (1999): 271–90.

McCrea, W. H. "Arthur Stanley Eddington." *Scientific American* 264, no. 6 (1991): 66–71.

———. "Einstein: Relations with the RAS." *Quarterly Journal of the Royal Astronomical Society* 20, no. 3 (1979): 251–60.

McDougall, William. "Purposive Striving as a Fundamental Category of Psychology." *Scientific Monthly* 19, no. 3 (September 1924): 305–12.

McGucken, William. *Scientists, Society, and the State: The Social Relations of Science Movement in Great Britain, 1931–1947.* Columbus: Ohio State University Press, 1984.

Merleau-Ponty, Jacques. *Philosophie et Theorie Physique chez Eddington.* Paris: Besancon, 1965.

Merton, Robert K. *Science, Technology and Society in Seventeenth Century England.* Bruges, Belgium: Saint Catherine Press, 1938.

Mestel, Leon. "Arthur Stanley Eddington: Pioneer of Stellar Structure Theory." *Journal of Astronomical History and Heritage* 7, no. 2 (2004): 65–73.

Milburn, Robert Gordon. *The Logic of Religious Thought; An Answer to Professor Eddington.* London: Williams & Norgate, 1929.

Miller, Arthur I. *Empire of the Stars: Obsession, Friendship, and Betrayal in the Quest for Black Holes.* Boston: Houghton Mifflin, 2005.

Milne, Edward Arthur. *Modern Cosmology and the Christian Idea of God.* Oxford: Clarendon Press, 1952.

———. *Sir James Jeans.* Cambridge: Cambridge University Press, 1952.

Minnaert, W. "International Cooperation in Astronomy." *Vistas in Astronomy* 1 (1955): 5–16.

Mol, Hans. *Identity and Religion.* Beverly Hills, CA: Sage Publications, 1978.

Molvig, Ole. "Cosmological Revolutions: Relativity, Astronomy, and the Shaping of a Modern Universe." Ph.D. diss., Department of History, Princeton University, 2004.

Monk, Ray. *Bertrand Russell: The Ghost of Madness.* New York: Free Press, 2001.

———. *Bertrand Russell: The Spirit of Solitude.* New York: Free Press, 1996.

Montague, W. P. "The Oxford Congress of Philosophy." *Journal of Philosophy* 18, no. 5 (March 3, 1921): 118–29.

Moore, James R. *The Post-Darwinian Controversies: A Study of the Protestant Struggle to Come to Terms with Darwin in Great Britain and America, 1870–1900*. Cambridge:: Cambridge University Press, 1979.

Moore, Walter. *Schrödinger*. Cambridge: Cambridge University Press, 1989.

Morrison, Roy D. "Albert Einstein: The Methodological Unity Underlying Science and Religion." *Zygon: Journal of Religion and Science* 14 (1979): 255–66.

Mowat, Charles L. *Britain between the Wars*. Chicago: University of Chicago Press, 1955.

Moyer, Donald. "Revolution in Science: The 1919 Eclipse Test of General Relativity." In *On the Path of Albert Einstein*, edited by A. Perlmutter and L. F. Scott, 55–101. London: Plenum Press, 1979.

Needham, Joseph. *The Great Amphibium: Four Lectures on the Position of Religion in a World Dominated by Science*. London: Student Christian Movement Press, 1931.

Nicolai, G. F. *The Biology of War*. New York: Century, 1919.

No-Conscription Fellowship. *The NCF: A Summary of Its Activities*. Pamphlet. London: NCF, 1917.

———. *Repeal the Act*. Pamphlet. London: NCF, n.d.

———. *Two Years Hard Labour for Refusing to Disobey the Dictates of Conscience*. Pamphlet. London: NCF, [1918?].

Nye, Mary Jo. *Before Big Science: The Pursuit of Modern Chemistry and Physics, 1800–1940*. New York: Twayne Publishers, 1996.

Otto, Rudolph. *The Idea of the Holy*. London: Oxford University Press, 1923.

Pais, Abraham. *Subtle Is the Lord: The Science and the Life of Albert Einstein*. New York: Oxford University Press, 1982.

Pang, Alex Soojung-Kim. *Empire and the Sun: Victorian Eclipse Expeditions*. Palo Alto, CA: Stanford University Press, 2002.

———. "The Social Event of the Season: Solar Eclipse Expeditions and Victorian Culture." *Isis* 84 (1993): 252–77.

Pattison, Michael. "Scientists, Inventors and the Military in Britain, 1915–1919: The Munitions Inventions Department." *Social Studies of Science* 13 (1983): 521–67.

Paul, Erich Robert. *The Milky Way Galaxy and Statistical Cosmology, 1890–1924*. Cambridge: Cambridge University Press, 1993.

Pearson, Karl. *The Grammar of Science*. 2nd ed. London: Adam & Charles Black, 1900.

Poincaré, Henri. *Science and Hypothesis*. New York: Dover, 1952.

Price, Kate. "Poetry and Physics: Interchange within the Writings of Arthur Eddington, I. A. Richards, William Empson and Laura Riding." PhD diss., Cambridge University, 2003.

Pringle, Cyrus G. *The Record of a Quaker Conscience*. New York: Macmillan, 1918.

Proctor, Robert. *Value-Free Science?* Cambridge, MA: Harvard University Press, 1991.

Provine, William. "Scientists, Face It! Science and Religion Are Incompatible." *Scientist* 2, no. 16 (September 5, 1988): 10.

Quick, Oliver C., Rev. "Mind, Matter and Professor Eddington." *Nineteenth Century* 105 (1929): 669–79.

Raistrick, Arthur. *Quakers in Science And Industry: Being an Account of the Quaker Contributions to Science and Industry during the 17th and 18th Centuries*. York: Sessions Book Trust, 1993.

Reid, Robert. *Tongues of Conscience: War and the Scientist's Dilemma*. London: Walker, 1969.

Renn, Jürgen, ed. *The Genesis of General Relativity*. Dordrecht: Springer, 2007.

Report of the Proceedings of the Conference of Members of the Society of Friends, Held . . . in Manchester from Eleventh to Fifteenth of Eleventh Month, 1895. London: Headley Brothers, 1896.

Rescher, Nicholas. *Process Philosophy: A Survey of Basic Issues*. Pittsburgh: University of Pittsburgh Press, 2000.

Richards, Joan. *Mathematical Visions: The Pursuit of Geometry in Victorian England*. Boston: Boston University Press, 1988.

Ritchie, C. F. *Reflections on the Philosophy of Sir Arthur Eddington*. Cambridge: Cambridge University Press, 1948.

Rocke, Alan. *Nationalizing Science: Adolphe Wurtz and the Battle for French Chemistry*. Cambridge, MA: MIT Press, 2001.

Rosenberg, Charles. *No Other Gods: On Science and American Social Thought*. Baltimore: Johns Hopkins University Press, 1976.

Rowntree, John S. *The Society of Friends: Its Faith and Practice*. London: Headley Brothers, 1908.

Russell, Bertrand. *The Analysis of Matter*. London: Kegan Paul, 1927.

———. *Introduction to Mathematical Philosophy*. London: Allen & Unwin, 1919.

———. *The Problems of Philosophy*. London: Williams & Norgate, 1910.

———. *Scientific Method in Philosophy*. Oxford: Clarendon Press, 1914.

———. *The Scientific Outlook*. New York: Norton, 1931.

Russell, Doug. "Popularization and the Challenge to Science-Centrism in the 1930s." In *The Literature of Science: Perspectives on Popular Science Writing*, edited by Murdo William McRae, 37–53. Athens: University of Georgia Press, 1993.

Russell, Henry Norris. *Fate and Freedom*. New Haven, CT: Yale University Press, 1927.

———. "Note on the Sobral Eclipse Photographs." *Monthly Notices of the Royal Astronomical Society* 81 (1920): 154–64.

———. "The Sources of Stellar Energy." *Publications of the Astronomical Society of the Pacific* 31 (July 1919): 205–11.

Ryckman, Thomas. *The Reign of Relativity: Philosophy in Physics, 1915–1925*. Oxford: Oxford University Press, 2005.

Samuel, Herbert. "Cause, Effect and Professor Eddington." *Nineteenth Century* 113 (1933): 469–78.

Sayers, Dorothy L. "Absolutely Elsewhere." In *The Complete Stories*, 392–408. New York: Harper-Collins, 1972.

Schaffer, Simon. "Metrology, Metrication, and Victorian Values." In *Victorian Science in Context*, edited by Bernard Lightman, 438–74. Chicago: University of Chicago Press, 1997.

———. "Where Experiments End: Tabletop Trials in Victorian Astronomy." In *Scientific Practice: Theories and Stories of Doing Physics*, edited by Jed Buchwald, 257–99. Chicago: University of Chicago Press, 1995.

Schlick, Moritz. *Space and Time in Contemporary Physics*. Translated by Henry Brose. New York: Oxford University Press, 1920.

Schroeder-Gudehus, B. "Challenge to Transnational Loyalties: International Organisations after the First World War." *Science Studies* 3 (1973): 98–118.

Schwabe, Calvin. *Quakerism and Science*. Pendle Hill, PA: Pendle Hill Publications, 1999.

Schweber, S. S. "The Empiricist Temper Regnant: Theoretical Physics in the United States, 1920–1950." *Historical Studies in the Physical Sciences* 17 (1986): 55–98.

———. *QED and the Men Who Made It*. Princeton, NJ: Princeton University Press, 1994.

Secord, James. *Victorian Sensation: The Extraordinary Publication, Reception, and Secret Authorship of Vestiges of the Natural History of Creation.* Cambridge: Cambridge University Press, 2000.

Shapin, Steven. "Science and the Public." In *Companion to the History of Modern Science*, edited by R. C. Olby. et al., 990–1007. London: Routledge, 1990.

Shapin, Steven, and Simon Schaffer. *Leviathan and the Air Pump: Hobbes, Boyle, and the Experimental Life.* Princeton, NJ: Princeton University Press, 1985.

Shee, George F. *The Briton's First Duty: The Case for Conscription.* London: Grant Richards, 1901.

Shortland, Michael, and Richard Yeo, eds. *Telling Lives in Science: Essays on Scientific Biography.* Cambridge: Cambridge University Press, 1996.

Silberstein, Ludwik. *The Theory of General Relativity and Gravitation.* New York: Van Nostrand, 1922.

———. *The Theory of Relativity.* London: Macmillan, 1914.

Slater, Noel. *The Development and Meaning of Eddington's "Fundamental Theory."* Cambridge: Cambridge University Press, 1957.

Smart, H. R. Review of *Space, Time and Gravitation*, by A. S. Eddington. *Philosophical Review* 31, no. 4 (July 1922): 414–15.

Smith, Crosbie, and M. Norton Wise. *Energy and Empire: A Biographical Study of Lord Kelvin.* Cambridge: Cambridge University Press, 1989.

Smith, Robert W. *The Expanding Universe.* Cambridge: Cambridge University Press, 1982.

Snowden, Philip, MP. *The Military Service Act Fully and Clearly Explained.* Pamphlet. London: National Labour Press, n.d.

Sponsel, Alistair. "Constructing a 'Revolution in Science:' The Campaign to Promote a Favorable Reception for the 1919 Solar Eclipse Experiments." *British Journal for the History of Science* 35, no. 4 (2002): 439–67.

St. John, Charles. "Displacement of Solar Lines and the Einstein Effect." *Observatory* 547 (January 1920): 158–62.

———. "The Principle of Generalized Relativity and the Displacement of Fraunhofer Lines toward the Red." *Astrophysical Journal* 96 (1917): 249–65.

Stace, W. T. "Sir Arthur Eddington and the Physical World." *Philosophy* 9 (1934): 39–50.

Stachel, John. "Eddington and Einstein." In *The Prism of Science, Israel Colloquium*, vol. 2, edited by E. Ullmann-Margalit, 225–50. Dordrecht: Reidel, 1986.

Stanley, Matthew. "An Expedition to Heal the Wounds of War: The 1919 Eclipse Expedition and Eddington as Quaker Adventurer." *Isis* 94 (2003): 57–89.

Stebbing, L. Susan. *Philosophy and the Physicists.* London: Methuen, 1937.

Stevenson, John. *British Society 1914–45.* London: A. Lane, 1984.

Stratton, F. J. M. "The History of the Cambridge Observatories." *Annals of the Solar Physics Laboratory, Cambridge* 1 (1949).

Sullivan, J. W. N. *Contemporary Mind: Some Modern Answers.* London: Humphrey Toulmin, 1934.

Sutherland, George Arthur. *Dalton Hall, a Quaker Venture.* London: Bannisdale Press, 1963.

Swenson, L. *The Ethereal Ether: A History of the Michelson-Morley Aether-Drift Experiment, 1880–1930.* Austin: University of Texas Press, 1972.

Sykes, S. W. "Theology." In *Twentieth Century Mind: History, Ideas, and Literature in Britain*, edited by C. B. Cox and A. E. Dyson, 146–70. Oxford: Oxford University Press, 1972.

Symon, J. D. *Universities' Part in the War.* Pamphlet. N.p., 1915. Cambridgeshire Public Library.

Tayler, R. J., ed. *History of the RAS.* Vol. 2. London: Blackwell Scientific, 1987.

Taylor, A. J. P. *English History 1914–1945.* Oxford: Oxford University Press, 1965.

Thomas, Anna Braithwaite. *St. Stephen's House: Friends' Emergency Work in England, 1914–1920.* London: Emergency Committee for the Assistance of Germans, Austrians and Hungarians in Distress, 1935.

Thomas, Edward. *Quaker Adventures.* London: Fleming H. Revell, 1935.

Thompson, Silvanus Phillips. *The Methods of Physical Science.* London: Longman, 1877.

———. "The Mystery of Nature." *Friends' Quarterly Examiner* 9 (1875): 405–22.

———. *The Quest for Truth.* Bishopsgate: Headley Brothers, 1915.

Tilby, A Wyatt. "The Reconstruction of Religion," *Nineteenth Century* 91 (January–June 1922): 462–67.

Turner, J. E. "Dr. Wildon Carr and Lord Haldane on Scientific Relativity." *Mind* 31, no. 121 (January 1922): 40–52.

———. "Some Philosophic Aspects of Scientific Relativity." *Journal of Philosophy* 18, no. 8 (April 14, 1921): 210–16.

United Kingdom Home Office, *Final Report of the Committee on Commercial and Industrial Policy after the War.* Cd 9035. London: HM Stationery Office, 1918.

Van Der Meer, Jitse M., ed. *Facets of Faith and Science.* Lanham, MD: Pascal Centre for Advanced Studies in Faith and Science, University Press of America, 1996.

Varcoe, Ian. "Scientists, Government and Organized Research in Great Britain, 1914–1916: The Early History of the DSIR." *Minerva* 8 (1970): 192–217.

Von Kluber, H. "Determination of Einstein's Light Deflection." *Vistas in Astronomy* 3 (1960): 47–77.

W., W. H. *A Guide to the Conscientious Objector and the Tribunals.* Pamphlet. N.p., n.d. Cambridge University Library "War Tracts" archive.

Wali, Kameshwar. *Chandra.* Chicago: University of Chicago Press, 1991.

Wallace, Stuart. *War and the Image of Germany: British Academics 1914–1918.* Edinburgh: John Donald, 1988.

Ward, James. *Naturalism and Agnosticism.* 2nd ed. London: A. C. Black, 1915.

Warwick, Andrew. "Cambridge Mathematics and Cavendish Physics: Cunningham, Campbell and Einstein's Relativity, 1905–1911, Part I: The Uses of Theory." *Studies in History and Philosophy of Science* 23 (1992): 625–56.

———. "Cambridge Mathematics and Cavendish Physics: Cunningham, Campbell and Einstein's Relativity, 1905–1911, Part II: Comparing traditions in Cambridge physics." *Studies in History and Philosophy of Science* 24 (1993): 1–25.

———. *Masters of Theory: Cambridge and the Rise of Mathematical Physics.* Chicago: University of Chicago Press, 2003.

Weber, Max. "Science as Vocation." In *On Universities*, 54–62. Chicago: University of Chicago Press, 1973.

Werskey, Gary. *The Visible College: The Collective Biography of British Scientific Socialists of the 1930s.* London: A. Lane, 1978.

Westfall, Richard S. "Newton and Christianity." In *Newton*, edited by Richard S. Westfall and I Bernard Cohen, 356–70. New York: Norton, 1995.

Whitehead, A. N. *An Introduction to Mathematics.* London: Williams & Norgate, 1911.

———. "The Philosophical Aspects of the Theory of Relativity." *Proceedings of the Aristotelian Society* 22 (1922): 215–23.

———. *Science and the Modern World*. New York: Macmillan, 1947.

Whitrow, G. J. "Sir Arthur Eddington, OM (1882–1944)." *Quarterly Journal of the Royal Astronomical Society* 24 (1983): 258–66.

Whittaker, E. T. *From Euclid to Eddington: A Study of Conceptions of the External World*. New York: Dover, 1949.

Whitworth, Michael. "The Clothbound Universe: Popular Physics Books, 1919–39." *Publishing History* 40 (1996): 53–82.

———. *Einstein's Wake: Relativity, Metaphor, and Modernist Literature*. Oxford: Oxford University Press, 2001.

Wiener, Martin J. *English Culture and the Decline of the Industrial Spirit, 1850–1980*. Cambridge: Cambridge University Press, 1981.

Wilkinson, Alan. *The Church of England and the First World War*. London: SPCK, 1978.

Will, Clifford. "General Relativity at 75: How Right Was Einstein?" *Science* 250 (1990): 770–76.

Williams, S. C. *Religious Belief and Popular Culture in Southwark c.1880–1939*. Oxford: Oxford University Press, 1999.

Wilson, David. "On the Importance of Eliminating 'Science' and 'Religion' from the History of Science and Religion: The Cases of Oliver Lodge, J. H. Jeans, and A. S. Eddington." In *Facets of Faith and Science*, edited by Jitse M. van der Meer, 27–48. Lanham, MD: Pascal Centre for Advanced Studies in Faith and Science, University Press of America, 1996.

Wilson, Roger. "Road to Manchester." In *Seeking the Light: Essays in Quaker History*, edited by J. William Frost and John M. Moore, 145–62. Wallingford, PA: Pendle Hill Publications, 1986.

Witt-Hansen, Johannes. *Exposition and Critique of the Conceptions of Eddington Concerning the Philosophy of Science*. Copenhagen: G.E.C. Gads Forlag, 1958.

Wolfe, Kenneth. *The Churches and the British Broadcasting Corporation, 1922–1956*. London: Student Christian Movement Press, 1984.

Wood, A. *Bertrand Russell: The Passionate Sceptic*. London: Allen & Unwin, 1957.

Wood, Neal. *Communism and British Intellectuals*. New York: Columbia University Press, 1959.

Wooley, Richard. "The Stars and the Structure of the Galaxy." *Quarterly Journal of the Royal Astronomical Society* 11 (1970): 403–28.

Wyatt, H. F. "Science and the Moral Law." *Nineteenth Century* 91 (1922): 858.

Yolton, John. *The Philosophy of Science of A. S. Eddington*. The Hague: M. Nijhoff, 1960.

Index